Mathematical and Numerical Analysis of Nonlinear Evolution Equations

Mathematical and Numerical Analysis of Nonlinear Evolution Equations

Advances and Perspectives

Editor

Carlo Bianca

MDPI • Basel • Beijing • Wuhan • Barcelona • Belgrade • Manchester • Tokyo • Cluj • Tianjin

Editor
Carlo Bianca
Laboratoire Quartz,
ECAM-EPMI
France

Editorial Office
MDPI
St. Alban-Anlage 66
4052 Basel, Switzerland

This is a reprint of articles from the Special Issue published online in the open access journal *Mathematics* (ISSN 2227-7390) (available at: https://www.mdpi.com/journal/mathematics/special_issues/Mathematical_Numerical_Analysis_Nonlinear_Evolution_Equations_Advances_Perspectives).

For citation purposes, cite each article independently as indicated on the article page online and as indicated below:

LastName, A.A.; LastName, B.B.; LastName, C.C. Article Title. *Journal Name* **Year**, *Article Number*, Page Range.

ISBN 978-3-03943-272-1 (Hbk)
ISBN 978-3-03943-273-8 (PDF)

© 2020 by the authors. Articles in this book are Open Access and distributed under the Creative Commons Attribution (CC BY) license, which allows users to download, copy and build upon published articles, as long as the author and publisher are properly credited, which ensures maximum dissemination and a wider impact of our publications.

The book as a whole is distributed by MDPI under the terms and conditions of the Creative Commons license CC BY-NC-ND.

Contents

About the Editor vii

Preface to "Mathematical and Numerical Analysis of Nonlinear Evolution Equations" ix

Carlo Bianca and Marco Menale
Mathematical Analysis of a Thermostatted Equation with a Discrete Real Activity Variable
Reprinted from: *Mathematics* 2020, 8, 57, doi:10.3390/math8010057 1

Carlo Bianca and Marco Menale
A Convergence Theorem for the Nonequilibrium States in the Discrete Thermostatted Kinetic Theory
Reprinted from: *Mathematics* 2019, 7, 673, doi:10.3390/math7080673 9

Bruno Carbonaro and Marco Menale
Dependence on the Initial Data for the Continuous Thermostatted Framework
Reprinted from: *Mathematics* 2019, 7, 602, doi:10.3390/math7070602 23

Hasanen A. Hammad and Manuel De la Sen
Fixed-Point Results for a Generalized Almost (s,q)−Jaggi F-Contraction-Type on b−Metric-Like Spaces
Reprinted from: *Mathematics* 2020, 8, 63, doi:10.3390/math8010063 35

Mohammed AL Horani, Angelo Favini and Hiroki Tanabe
Direct and Inverse Fractional Abstract Cauchy Problems
Reprinted from: *Mathematics* 2019, 7, 1016, doi:10.3390/math7111016 57

Mohammed Al Horani, Mauro Fabrizio, Angelo Favini and Hiroki Tanabe
Fractional Cauchy Problems for Infinite Interval Case-II
Reprinted from: *Mathematics* 2019, 7, 1165, doi:10.3390/math7121165 67

Yassine Benia, Marianna Ruggieri and Andrea Scapellato
Exact Solutions for a Modified Schrödinger Equation
Reprinted from: *Mathematics* 2019, 7, 908, doi:10.3390/math7100908 93

Mikhail Kolev
Mathematical Analysis of an Autoimmune Diseases Model: Kinetic Approach
Reprinted from: *Mathematics* 2019, 7, 1024, doi:10.3390/math7111024 103

Pierluigi Colli, Gianni Gilardi and Jürgen Sprekels
A Distributed Control Problem for a Fractional Tumor Growth Model
Reprinted from: *Mathematics* 2019, 7, 792, doi:10.3390/math7090792 117

Youcef Belgaid, Mohamed Helal and Ezio Venturino
Analysis of a Model for Coronavirus Spread
Reprinted from: *Mathematics* 2020, 8, 820, doi:10.3390/math8050820 149

Zizhen Zhang, Soumen Kundu and Ruibin Wei
A Delayed Epidemic Model for Propagation of Malicious Codes in Wireless Sensor Network
Reprinted from: *Mathematics* 2019, 7, 396, doi:10.3390/math7050396 179

About the Editor

Carlo Bianca is Full Professor at the graduate school ECAM-EPMI, Cergy (France). He received his PhD degree in Mathematics for Engineering Science at Polytechnic University of Turin (Italy). His research interests are in the areas of applied mathematics and, in particular, in mathematical physics including the mathematical methods and models for complex systems, mathematical billiards, chaos, anomalous transport in microporous media and numerical methods for kinetic equations. He has published research articles in reputed international journals of mathematical and engineering sciences. He is referee and editor of numerous mathematical journals.

Preface to "Mathematical and Numerical Analysis of Nonlinear Evolution Equations"

Nowadays, research activity in mathematics has acquired a new mission. Mathematical methods have been proposed and employed in an attempt to understand the behavior and evolution of a particle system, especially those of complex systems. The definition of particle has been expanded to also include the entities derived from living matter, e.g., human cells, virus, pedestrians, and swarms. A particle is not a mere entity but is now assumed to be able to perform a strategy, function, or interaction and thereby acquire the denomination of being 'active' or, even less accurate, 'intelligent'. Bearing all the above in mind, an important research activity in mathematics addresses the modeling of complex systems. In particular, an interplay among researchers coming from different fields has emerged, thus allowing the birth of an applied science termed applied mathematics. Applied mathematics is based on the derivation and application of mathematical frameworks for the modeling of a particle system.

The historical frameworks based on ordinary differential equations (ODEs), partial differential equations (PDEs), kinetic theory (or more in general statistical mechanisms), continuum mechanics, and statistics have been the first employed in some problems related to biology, epidemics, and economics. Each historical framework has shown some advantages and some disadvantages in the study of a complex systems, thus requiring the derivation of further generalized frameworks which should be adapted to the system under consideration. Accordingly, various new frameworks have been proposed based on generalized kinetic theories and fractional calculus.

In this context, the term evolution equation can be considered as a general framework whose solution is a function describing the time evolution of a microscopic, mesoscopic, or macroscopic quantity related to the system. On the one hand, the mathematical analysis allows obtaining information on the qualitative behaviors of the system, including the existence of solutions, asymptotic behaviors, and nonlinear dynamics. On the other hand, numerical and computational analysis furnishes methods to obtain quantitative information about the solutions and the possibility to compare the time evolution of the solution of an evolution equation with empirical data (tuning problem).

This book is a Special Issue reprint. Specifically, it comprises the articles of a Special Issue that I have recently organized in the journal Mathematics (MDPI). The Special Issue is been devoted to researchers working in the fields of pure and applied mathematics and physics and, in particular, to researchers who are involved in the mathematical and numerical analysis of nonlinear evolution equations and their applications.

The first part of the book deals with new proposed mathematical frameworks and their mathematical analysis mainly addressed toward the existence of solutions of related initial and initial-boundary value problems, stability, and asymptotic analysis. Among the new mathematical frameworks presented in the book, the recent developments of the discrete thermostatted kinetic theory for far-off-equilibrium complex systems are presented. In the new discrete framework, the existence and uniqueness of the solution of the related Cauchy problem and of the related non-equilibrium stationary state are established, the rigorous proof that the solution of the discrete thermostatted kinetic model catches the stationary solutions as time approaches infinity is also presented and, finally, the continuous dependence on initial data is also established. In this context, some methods of nonlinear analysis, such as the fixed-point technique, have an important role especially in the analysis of solutions of the mathematical frameworks; accordingly, recent fixed-point results in generalized metric spaces are proposed in the book. Recent mathematical frameworks coming from fractional calculus theory are also part of the present book. Specifically, fractional abstract Cauchy frameworks for possibly degenerate equations in Banach spaces are mathematically analyzed, and some related inverse problems are stated and studied. Applications are also discussed.

The second part of the book is concerned with applications to some complex systems in biology, epidemics and, also, engineering. In detail, a new mathematical model based on a generalized kinetic theory is proposed for an autoimmune disease; numerical results are presented and discussed from a medical viewpoint. The modeling of tumor growth is also taken into account in this book by stating an optimal control problem of a system of three evolutionary equations involving fractional powers of three self-adjoint, monotone, unbounded linear operators having compact resolvents; in particular, the first-order necessary conditions for optimality of a cost function of tracking type are derived. In the context of the coronavirus pandemic, a new mathematical model is also presented via a compartmental dynamical system; its equilibria are investigated for local and global stability; numerical simulations show that contact restrictive measures and an intermittent lockdown policy are able to delay the epidemic's outbreak (if taken at a very early stage). Finally, a delayed SEIQRS-V epidemic model for propagation of malicious codes in a wireless sensor network is presented and analyzed, local stability and existence of Hopf bifurcation are performed, and numerical simulations are presented in order to analyze the effects of some parameters on the dynamical behavior.

I am sure that the reader will find the new results, methods, and models to be of great interest. I hope that you, the reader, will benefit from the contents of this book in the development and pursuit of your own research activity.

Carlo Bianca
Editor

Article

Mathematical Analysis of a Thermostatted Equation with a Discrete Real Activity Variable

Carlo Bianca [1,2] and Marco Menale [1,3],*

1   Laboratoire Quartz EA 7393, École Supérieure d'Ingénieurs en Génie Électrique, Productique et Management Industriel, 95092 Cergy Pontoise CEDEX, France; c.bianca@ecam-epmi.com
2   Laboratoire de Recherche en Eco-innovation Industrielle et Energétique, École Supérieure d'Ingénieurs en Génie Électrique, Productique et Management Industriel, 95092 Cergy Pontoise CEDEX, France
3   Dipartimento di Matematica e Fisica, Università degli Studi della Campania "L. Vanvitelli", Viale Lincoln 5, I-81100 Caserta, Italy
*   Correspondence: marco.menale@unicampania.it

Received: 28 November 2019; Accepted: 17 December 2019; Published: 2 January 2020

Abstract: This paper deals with the mathematical analysis of a thermostatted kinetic theory equation. Specifically, the assumption on the domain of the activity variable is relaxed allowing for the discrete activity to attain real values. The existence and uniqueness of the solution of the related Cauchy problem and of the related non-equilibrium stationary state are established, generalizing the existing results.

Keywords: real activity variable; thermostat; nonlinearity; complex systems; Cauchy problem

1. Introduction

The mathematical analysis of a differential equation is usually based on the existence and uniqueness of a positive solution of the related Cauchy problem and on the dependence on the initial data (well-posed problem) [1,2]. The well-posed problem is analyzed under some (usually strongly) assumptions. However, if the differential equation is proposed as a general paradigm for the derivation of a mathematical model for a complex system [3,4], the definition of the assumptions is a delicate issue considering the restrictions that can be required on the system under consideration.

The present paper aims at generalizing the mathematical analysis of the discrete thermostatted kinetic theory framework recently proposed in [5,6] and employed in [7] for the modeling of the pedestrian dynamics into a metro station. The mathematical framework consists of a nonlinear differential equations system derived considering the balancing into the elementary volume of the microscopic states (space, velocity, strategy or activity) of the gain and loss particle-flows. The mathematical framework contains also a dissipative term, called thermostat, for balancing the action of an external force field which acts on the complex system, thus moving the system out-of-equilibrium [8,9]. The thermostat term allows for the existence and thus the modeling of the non-equilibrium stationary (possibly steady) states; see, among others, [10–15].

The mathematical analysis presented in [5] is based on the assumption that the discrete activity variable is greater than 1. In order to model complex systems, such as social and economical systems [16–20], the possibility for the activity variable to also attain negative values should be planned. Accordingly, this paper is devoted to a further generalization of the existence and uniqueness of the solution and of the non-equilibrium stationary solution for a real activity variable.

The present paper is organized as follows: after this introduction, Section 2 is devoted to the main definitions of the differential framework and the related non equilibrium stationary states; Section 3 deals with the new mathematical results.

2. The Mathematical Framework

Let $\mathbb{E}_p > 0$, $I_u = \{u_1, u_2, \ldots, u_n\}$, $u_i \in \mathbb{R}$, $\eta_{h,k} : I_u \times I_u \to \mathbb{R}^+$, $B^i_{hk} : I_u \times I_u \times I_u \to \mathbb{R}^+$, and $F_i : [0, +\infty[\to \mathbb{R}^+$, for $i, h, k \in \{1, 2, \ldots, n\}$.

This paper is devoted to the mathematical analysis of the solutions $f_i : [0, +\infty[\to \mathbb{R}^+$ of the following system of n nonlinear ordinary differential equations (called discrete thermostatted kinetic framework):

$$\frac{df_i}{dt}(t) = J_i[\mathbf{f}](t) + T_i[\mathbf{f}](t), \qquad i \in \{1, 2, \ldots, n\}, \tag{1}$$

where $\mathbf{f}(t) = (f_1(t), f_2(t), \ldots, f_n(t))$ is the vector solution, and $J_i[\mathbf{f}](t) := G_i[\mathbf{f}](t) - L_i[\mathbf{f}](t)$ and $T_i[\mathbf{f}](t)$ are the operators defined as follows:

$$G_i[\mathbf{f}] := \sum_{h=1}^{n} \sum_{k=1}^{n} B^i_{hk} \, \eta_{hk} \, f_h(t) f_k(t);$$

$$L_i[\mathbf{f}] := f_i(t) \sum_{k=1}^{n} \eta_{hk} \, f_k(t);$$

$$T_i[\mathbf{f}](t) := F_i - \left(\frac{\sum_{j=1}^{n} u_j^p \, (J_j[\mathbf{f}] + F_j)}{\mathbb{E}_p} \right) f_i(t).$$

Let $\mathbb{E}_p[\mathbf{f}](t)$ be the pth-order moment:

$$\mathbb{E}_p[\mathbf{f}](t) = \sum_{i=1}^{n} u_i^p \, f_i(t), \qquad p \in \mathbb{N},$$

and $\mathcal{R}^p$ the following function space:

$$\mathcal{R}^p := \left\{ \mathbf{f} \in C\left([0, +\infty[; (\mathbb{R}^+)^n\right) : \mathbb{E}_p[\mathbf{f}](t) = \mathbb{E}_p \right\}.$$

Let $\mathbf{f}^0 \in \mathcal{R}^p$, the Cauchy problem related to Equation (1) reads:

$$\begin{cases} \dfrac{d\mathbf{f}(t)}{dt} = \mathbf{J}[\mathbf{f}](t) + \mathbf{T}[\mathbf{f}](t) & t \in [0, +\infty[\\ \mathbf{f}(0) = \mathbf{f}^0, \end{cases} \tag{2}$$

where $\mathbf{J}[\mathbf{f}] = \mathbf{G}[\mathbf{f}] - \mathbf{L}[\mathbf{f}] = (J_1[\mathbf{f}], J_2[\mathbf{f}], \ldots, J_n[\mathbf{f}]) = (G_1[\mathbf{f}] - L_1[\mathbf{f}], G_2[\mathbf{f}] - L_2[\mathbf{f}], \ldots, G_n[\mathbf{f}] - L_n[\mathbf{f}])$ and $\mathbf{T}[\mathbf{f}] = (T_1[\mathbf{f}], T_2[\mathbf{f}], \ldots, T_n[\mathbf{f}])$.

The existence and uniqueness of the solution of the Cauchy problem (2) have been proved in [5], under the main assumption $u_i \geq 1$, for $i \in \{1, 2, \ldots, n\}$. This paper aims at generalizing the result of [5] when $u_i \in \mathbb{R}$.

A non-equilibrium stationary state of the framework (1) is a constant function f_i, for $i \in \{1, 2, \ldots, n\}$, solution of the following problem:

$$J_i[\mathbf{f}] - \left(\frac{\sum_{j=1}^{n} u_j^p \, (J_j[\mathbf{f}] + F_j)}{\mathbb{E}_p} \right) f_i = 0. \tag{3}$$

The existence and uniqueness of the non-equilibrium stationary state have been shown in [6] under the assumption $u_i \geq 1$, for all $i \in \{1, 2, \ldots, n\}$. This result can be relaxed as stated in Theorem 2.

Remark 1. *Nonlinear systems* (1) *are a mathematical framework proposed in* [5] *for the modeling of a complex system* $\mathcal{C}$ *homogeneous with respect to the* mechanical variables (space *and* velocity), *where* u, *called* activity, *models the states of the particles. The function* f_i, *for* $i \in \{1, 2, \ldots, n\}$, *denotes the* distribution function *of the ith functional subsystem.*

3. The Generalized Results

Let $\|x\|_p$ be the ℓ^p-norm on $\mathbb{R}^n$:

$$\|x\|_p := \left(\sum_{i=1}^n x_i^p \right)^{\frac{1}{p}},$$

and $\tilde{\mathbb{E}}_p$ the following number:

$$\tilde{\mathbb{E}}_p := \sup_{t>0} \left(\sum_{i=1}^n |u_i|^p f_i(t) \right). \tag{4}$$

It is worth stressing that, if p is even, then $\tilde{\mathbb{E}}_p = \mathbb{E}_p$; if p is odd, then $\mathbb{E}_p \leq \tilde{\mathbb{E}}_p$.

The main result of the paper follows.

Theorem 1. *Let* $p \in \mathbb{N}$, $\tilde{\mathbb{E}}_p < \infty$ *and* $\mathbf{f}^0 \in \mathcal{R}^p$. *Assume that*

- $u_i \in \mathbb{R} \setminus \{0\}$;
- $\sum_{i=1}^n B_{hk}^i = 1$, *for all* $h, k \in \{1, 2, \ldots, n\}$;
- *There exists a constant* $\eta > 0$ *such that* $\eta_{hk} \leq \eta$, *for all* $h, k \in \{1, 2, \ldots, n\}$;
- *There exists a constant* $F > 0$ *such that* $F_i(t) \leq F$, *for all* $i \in \{1, 2, \ldots, n\}$ *and* $t \geq 0$.

Then, there exists a unique positive function $\mathbf{f} \in \mathcal{R}^p$ *which is solution of the Cauchy problem* (2).

Proof. Let $\mathbf{f}, \mathbf{g} \in \mathcal{R}^p$. Since $\sum_{i=1}^n B_{hk}^i = 1$, for all $h, k \in \{1, 2, \ldots, n\}$, one has:

$$\begin{aligned}
\|\mathbf{G}[\mathbf{f}] - \mathbf{G}[\mathbf{g}]\|_1 &= \sum_{i=1}^n |G_i[\mathbf{f}] - G_i[\mathbf{g}]| \\
&= \sum_{i=1}^n \left| \sum_{h=1}^n \sum_{k=1}^n B_{hk}^i \eta_{hk} f_h(t) f_k(t) - \sum_{h=1}^n \sum_{k=1}^n B_{hk}^i \eta_{hk} g_h(t) g_k(t) \right| \\
&= \sum_{i=1}^n \left| \sum_{h=1}^n \sum_{k=1}^n B_{hk}^i \eta_{hk} (f_h(t) f_k(t) - g_h(t) g_k(t)) \right| \\
&\leq \eta \sum_{h=1}^n \sum_{k=1}^n |f_h(t) f_k(t) - g_h(t) g_k(t)| \tag{5} \\
&\leq \eta \sum_{h=1}^n \sum_{k=1}^n |f_h(t) f_k(t) - f_h(t) g_k(t) + f_h(t) g_k(t) - g_h(t) g_k(t)| \\
&\leq \eta \left| \sum_{h=1}^n f_h(t) + \sum_{h=1}^n g_h(t) \right| \sum_{k=1}^n |f_k(t) - g_k(t)| \\
&= \eta \left(\|\mathbf{f}\|_1 + \|\mathbf{g}\|_1 \right) \|\mathbf{f} - \mathbf{g}\|_1 \\
&= \eta \left| \mathbb{E}_0[\mathbf{f}](t) + \mathbb{E}_0[\mathbf{g}](t) \right| \|\mathbf{f} - \mathbf{g}\|_1,
\end{aligned}$$

and

$$\|\mathbf{L}[\mathbf{f}] - \mathbf{L}[\mathbf{g}]\|_1 = \sum_{i=1}^{n} \left| f_i(t) \sum_{k=1}^{n} \eta_{ik} f_k(t) - g_i(t) \sum_{k=1}^{n} \eta_{ik} g_i(t) \right| \qquad (6)$$
$$\leq \eta \left| \mathbb{E}_0[\mathbf{f}](t) + \mathbb{E}_0[\mathbf{g}](t) \right| \|\mathbf{f} - \mathbf{g}\|_1.$$

Since

$$|\mathbb{E}_0[\mathbf{f}]| = \left| \sum_{i=1}^{n} f_i(t) \right| = \left| \sum_{i=1}^{n} \frac{u_i^p}{u_i^p} f_i(t) \right|, \qquad (7)$$

if $L := \max_{0 \leq i \leq n} \left\{ \frac{1}{|u_i|^p} \right\}$, then, by Equation (7), one has:

$$|\mathbb{E}_0[\mathbf{f}]| \leq L \sum_{i=0}^{n} |u_i|^p f_i(t) \leq L \bar{\mathbb{E}}_p. \qquad (8)$$

By Equations (5), (6) and (8), one has:

$$\|\mathbf{J}[\mathbf{f}] - \mathbf{J}[\mathbf{g}]\|_1 \leq 2\eta \left| \mathbb{E}_0[\mathbf{f}] + \mathbb{E}_0[\mathbf{g}] \right| \|\mathbf{f} - \mathbf{g}\|_1 \qquad (9)$$
$$\leq 4 \eta L \bar{\mathbb{E}}_p \|\mathbf{f} - \mathbf{g}\|_1.$$

Moreover:

$$\|\mathbf{T}[\mathbf{f}] - \mathbf{T}[\mathbf{g}]\|_1 = \sum_{i=1}^{n} |T_i[\mathbf{f}] - T_i[\mathbf{g}]|$$
$$= \sum_{i=1}^{n} \left| \left(\frac{\sum_{j=1}^{n} u_j^p (J_j[\mathbf{f}] + F_j)}{\mathbb{E}_p} \right) f_i(t) - \left(\frac{\sum_{j=1}^{n} u_j^p (J_j[\mathbf{g}] + F_j)}{\mathbb{E}_p} \right) g_i(t) \right|$$
$$\leq \sum_{i=1}^{n} \left(\frac{\sum_{j=1}^{n} u_j^p F_j}{\mathbb{E}_p} \right) |f_i(t) - g_i(t)| + \sum_{i=1}^{n} \left| \left(\frac{\sum_{j=1}^{n} u_j^p J_j[\mathbf{f}]}{\mathbb{E}_p} \right) f_i(t) - \left(\frac{\sum_{j=1}^{n} u_j^p J_j[\mathbf{g}]}{\mathbb{E}_p} \right) g_i(t) \right| \qquad (10)$$
$$= \left(\frac{\sum_{j=1}^{n} u_j^p F_j}{\mathbb{E}_p} \right) \|\mathbf{f}(t) - \mathbf{g}(t)\|_1 + \sum_{i=1}^{n} \left| \left(\frac{\sum_{j=1}^{n} u_j^p J_j[\mathbf{f}]}{\mathbb{E}_p} \right) f_i(t) - \left(\frac{\sum_{j=1}^{n} u_j^p J_j[\mathbf{g}]}{\mathbb{E}_p} \right) g_i(t) \right|.$$

Bearing the expressions of the operator $\mathbf{J}[\mathbf{f}]$ in mind, one has:

$$\sum_{i=1}^{n} \left| \left(\frac{\sum_{j=1}^{n} u_j^p J_j[\mathbf{f}]}{\mathbb{E}_p} \right) f_i(t) - \left(\frac{\sum_{j=1}^{n} u_j^p J_j[\mathbf{g}]}{\mathbb{E}_p} \right) g_i(t) \right|$$

$$= \sum_{i=1}^{n} \left| \frac{1}{\mathbb{E}_p} \sum_{l=1}^{n} u_l^p J_l[\mathbf{f}] f_i(t) - \frac{1}{\mathbb{E}_p} \sum_{l=1}^{n} u_l^p J_l[\mathbf{g}] g_i(t) \right|$$

$$= \sum_{i=1}^{n} \left| \frac{1}{\mathbb{E}_p} \sum_{l=1}^{n} u_l^p \sum_{h=1}^{n} \sum_{k=1}^{n} B_{hk}^l \eta_{hk} f_h(t) f_k(t) f_i(t) \right.$$

$$- \frac{1}{\mathbb{E}_p} \sum_{l=1}^{n} u_l^p f_l(t) \sum_{k=1}^{n} \eta_{lk} f_k(t) f_i(t)$$

$$- \frac{1}{\mathbb{E}_p} \sum_{l=1}^{n} u_l^p \sum_{h=1}^{n} \sum_{k=1}^{n} B_{hk}^l \eta_{hk} g_h(t) g_k(t) g_i(t)$$

$$\left. + \frac{1}{\mathbb{E}_p} \sum_{l=1}^{n} u_l^p g_l(t) \sum_{k=1}^{n} \eta_{lk} g_k(t) g_i(t) \right| \quad (11)$$

$$\leq \sum_{i=1}^{n} \left| \frac{1}{\mathbb{E}_p} \sum_{l=1}^{n} u_l^p \sum_{h=1}^{n} \sum_{k=1}^{n} B_{hk}^l \eta_{hk} \left(f_h(t) f_k(t) f_i(t) - g_h(t) g_k(t) g_i(t) \right) \right|$$

$$+ \eta \sum_{i=1}^{n} \sum_{k=1}^{n} |f_k(t) f_i(t) - g_k(t) g_i(t)|$$

$$\leq \frac{\eta \sum_{j=1}^{n} u_j^p}{\mathbb{E}_p} \sum_{i=1}^{n} \sum_{h=1}^{n} \sum_{k=1}^{n} |f_h(t) f_k(t) f_i(t) - g_h(t) g_k(t) g_i(t)|$$

$$+ \eta \sum_{i=1}^{n} \sum_{k=1}^{n} |f_h(t) f_i(t) - g_k(t) g_i(t)|.$$

Since $\mathbf{f}$ and $\mathbf{g}$ belong to the space $\mathcal{R}^p$, by Equation (8), one has:

$$\sum_{i=1}^{n} \sum_{h=1}^{n} \sum_{k=1}^{n} |f_h(t) f_k(t) f_i(t) - g_h(t) g_k(t) g_i(t)|$$

$$= \sum_{i=1}^{n} \sum_{h=1}^{n} \sum_{k=1}^{n} \left| f_h(t) f_k(t) f_i(t) - f_h(t) f_k(t) g_i(t) + f_h(t) f_k(t) g_i(t) \right.$$

$$\left. - g_i(t) g_h(t) f_k(t) + g_i(t) g_h(t) f_k(t) - g_h(t) g_k(t) g_i(t) \right|$$

$$= \sum_{i=1}^{n} \sum_{h=1}^{n} \sum_{k=1}^{n} |f_h(t) f_k(t) (f_i(t) - g_i(t)) + g_i(t) f_k(t) (f_h(t) - g_h(t)) + g_i(t) g_h(t) (f_k(t) - g_k(t))| \quad (12)$$

$$\leq \sum_{h=1}^{n} f_h(t) \sum_{k=1}^{n} f_k(t) \sum_{i=1}^{n} |f_i(t) - g_i(t)| + \sum_{i=1}^{n} g_i(t) \sum_{k=1}^{n} f_k(t) \sum_{h=1}^{n} |f_h(t) - g_h(t)|$$

$$+ \sum_{i=1}^{n} g_i(t) \sum_{h=1}^{n} g_h(t) \sum_{k=1}^{n} |f_k(t) - g_k(t)|$$

$$\leq \|\mathbf{f} - \mathbf{g}\|_1 \left(\mathbb{E}_0^2[\mathbf{f}] + \mathbb{E}_0[\mathbf{f}] \mathbb{E}_0[\mathbf{g}] + \mathbb{E}_0^2[\mathbf{g}] \right)$$

$$\leq \|\mathbf{f} - \mathbf{g}\|_1 \left(\mathbb{E}_0[\mathbf{f}] + \mathbb{E}_0[\mathbf{g}] \right)^2$$

$$\leq \|\mathbf{f} - \mathbf{g}\|_1 4 L^2 \left(\bar{\mathbb{E}}_p \right)^2.$$

By Equation (8), one has:

$$\sum_{i=1}^{n}\sum_{k=1}^{n}|f_h(t)f_i(t) - g_h(t)g_i(t)| \leq (\mathbb{E}_0[\mathbf{f}] + \mathbb{E}_0[\mathbf{g}])\,\|\mathbf{f}-\mathbf{g}\|_1 \qquad (13)$$

$$\leq 2L\,\tilde{\mathbb{E}}_p\|\mathbf{f}-\mathbf{g}\|_1.$$

By Equations (10)–(13), one has:

$$\|\mathbf{T}[\mathbf{f}] - \mathbf{T}[\mathbf{g}]\|_1$$
$$\leq \left[\left(\frac{\sum_{j=1}^{n} u_j^p F_j}{\mathbb{E}_p}\right) + 4\eta \sum_{j=1}^{n} u_j^p L^2 \frac{(\tilde{\mathbb{E}}_p)^2}{\mathbb{E}_p} + 2\eta L\,\tilde{\mathbb{E}}_p\right]\|\mathbf{f}-\mathbf{g}\|_1. \qquad (14)$$

According to Equation (9) and Equation (14), the operators $\mathbf{J}[\mathbf{f}]$ and $\mathbf{T}[\mathbf{f}]$ are locally Lipschitz in $\mathbf{f}$, uniformly in t. Then, there exists a unique local solution of the Cauchy problem (2), and the solution $\mathbf{f}$ belongs to the space $\mathcal{R}^p$ (see Theorem 4.1 of [5]). The global existence of the solution is gained because $\mathbf{f}$ is globally bounded, for all $t > 0$, i.e., by Equation (7), one has:

$$\left|\sum_{i=1}^{n} f_i(t)\right| \leq L\,\tilde{\mathbb{E}}_p < +\infty, \quad \forall t > 0.$$

Then, the proof is gained. □

Remark 2. *If $u_i = 0$, Theorem 1 holds true if f_i is a bounded function, i.e., $\exists K > 0$ such that*

$$|f_i(t)| \leq K, \quad \forall t > 0.$$

Indeed, if l is such that $u_l = 0$, the estimates inequalities (8), (9), and (14) rewrite:

$$|\mathbb{E}_0[\mathbf{f}]| \leq L \sum_{i=1|u_i\neq 0}^{n} |u_i|^p f_i(t) + K \leq L\,\tilde{\mathbb{E}}_p + K,$$

$$|\mathbf{J}[\mathbf{f}] - \mathbf{J}[\mathbf{g}]\|_1 \leq 4M(L\tilde{\mathbb{E}}_p + K)^2\|\mathbf{f}-\mathbf{g}\|_1,$$

$$\|\mathbf{T}[\mathbf{f}] - \mathbf{T}[\mathbf{g}]\|_1 \leq$$
$$\leq \left[\left(\frac{\sum_{j=1}^{n} u_j^p F_j}{\mathbb{E}_p}\right) + \frac{4M\sum_{j=1}^{n} u_j^p}{\mathbb{E}_p}(L\tilde{\mathbb{E}}_p + K)^2 + 2M(L\tilde{\mathbb{E}}_p + K)\right]\|\mathbf{f}-\mathbf{g}\|_1.$$

Theorem 2. *Let $p \in \mathbb{N}$. Assume that*

- $u_i \in \mathbb{R} \setminus \{0\}$;
- $\sum_{i=1}^{n} B_{hk}^i = 1$, *for all $h,k \in \{1,2,\ldots,n\}$;*
- *There exists a constant $\eta > 0$ such that $\eta_{hk} \leq \eta$, for all $h,k \in \{1,2,\ldots,n\}$;*
- *There exists a constant $F > 0$ such that $F_i \leq F$, for all $i \in \{1,2,\ldots,n\}$;*
- *The following bound holds true:*

$$F > \eta\left[\frac{2\,\mathbb{E}_p^2}{\sum_{j=1}^{n} u_j^p} L + 4L^2\,\mathbb{E}_p^2\right], \qquad (15)$$

where $L := \max\limits_{u_i \neq 0} \left\{ \dfrac{1}{|u_i|^p} \right\}$.

Then, there exists a unique positive nonequilirbium stationary solution $\mathbf{f} = (f_1, f_2, \ldots, f_n) \in \mathcal{R}^p$ of the (3).

Proof. The non-equilibrium stationary problem (3) can be rewritten, for $i \in \{1, 2, \ldots, n\}$, as the following fixed point problem (see [6]):

$$f_i = S_i[\mathbf{f}] := \frac{\eta}{F}\left(\frac{\mathbb{E}_p}{\|\mathbf{U}^p\|_1} - f_i\right)\left(\sum_{h=1}^{n}\sum_{k=1}^{n} B_{hk}^i f_h f_k\right) + \frac{\mathbb{E}_p}{\sum_{j=1}^{n} u_j^p}. \tag{16}$$

By straightforward calculations, one has:

$$\|\mathbf{S}[\mathbf{f}] - \mathbf{S}[\mathbf{g}]\|_1 \leq \frac{\eta}{F}\left(\frac{\mathbb{E}_p}{\sum_{j=1}^{n} u_j^p}\sum_{h=1}^{n}\sum_{k=1}^{n}|f_h f_k - g_h g_k|\right) \\ + \frac{\eta}{F}\sum_{i=1}^{n}\left|(f_i - g_i)\sum_{h=1}^{n}\sum_{k=1}^{n} B_{hk}^i(f_h f_k - g_h g_k)\right|. \tag{17}$$

Furthermore, by the same arguments of the Theorem 1, one has:

$$\sum_{h=1}^{n}\sum_{k=1}^{n}|f_h f_k - g_h g_k| \leq \|\mathbf{f} - \mathbf{g}\|_1 \left(\mathbb{E}_0[\mathbf{f}] + \mathbb{E}_0[\mathbf{g}]\right) \tag{18}$$

$$\leq 2\|\mathbf{f} - \mathbf{g}\|_1 L \, \tilde{\mathbb{E}}_p.$$

Moreover,

$$\sum_{i=1}^{n}\left|(f_i - g_i)\sum_{h=1}^{n}\sum_{k=1}^{n} B_{hk}^i(f_h f_k - g_h g_k)\right| \\ \leq \|\mathbf{f} - \mathbf{g}\|_1 \left(\mathbb{E}_0^2[\mathbf{f}] + \mathbb{E}_0^2[\mathbf{g}]\right) \tag{19} \\ \leq 4\|\mathbf{f} - \mathbf{g}\|_1 L^2 \left(\tilde{\mathbb{E}}_p\right)^2.$$

Finally, by Equations (18) and (19), (17) rewrites:

$$\|\mathbf{S}[\mathbf{f}] - \mathbf{S}[\mathbf{g}]\|_1 \leq \frac{\eta}{F}\left[\frac{\mathbb{E}_p}{\sum_{j=1}^{n} u_j^p} 2L\tilde{\mathbb{E}}_p + 4L^2\left(\tilde{\mathbb{E}}_p\right)^2\right]\|\mathbf{f} - \mathbf{g}\|_1, \tag{20}$$

and, by assumption Equation (15), there exists a unique fixed point of the problem (16) (see [21]). Then, there exists a unique non-equilibrium stationary state for problem (3). □

Remark 3. *If $u_i = 0$, the Theorem 2 holds true if the ℓ_1-norm of $\mathbf{f}$ is bounded, i.e.,*

$$\|\mathbf{f}\|_1 \leq K.$$

Indeed, condition (15) rewrites:

$$F > \eta\left[\frac{\mathbb{E}_p}{\sum_{j=1}^{n} u_j^p} 2(L\tilde{\mathbb{E}}_p + K) + 4(L\tilde{\mathbb{E}}_p + K)^2\right].$$

Author Contributions: Conceptualization, C.B. and M.M.; methodology, C.B. and M.M.; formal analysis, C.B. and M.M.; investigation, C.B. and M.M.; resources, C.B. and M.M.; writing—original draft preparation, C.B. and M.M.; writing—review and editing, C.B. and M.M.; supervision, C.B. All authors have read and agreed to the published version of the manuscript.

Funding: This research received no external funding.

Conflicts of Interest: The authors declare no conflict of interest.

References

1. Coddington, E.A.; Levinson, N. *Theory of Ordinary Differential Equations*; Tata McGraw-Hill Education: New York, NY, USA, 1955.
2. Sattinger, D.H. *Topics in Stability and Bifurcation Theory*; Springer: Berlin, Germany, 2006; Volume 309.
3. Holland, J.H. Studying complex adaptive systems. *J. Syst. Sci. Complex.* **2006**, *19*, 1–8. [CrossRef]
4. Bar-Yam, Y. *Dynamics of Complex Systems*; CRC Press: Boca Raton, FL, USA, 2019.
5. Bianca, C.; Mogno, C. Qualitative analysis of a discrete thermostatted kinetic framework modeling complex adaptive systems. *Commun. Nonlinear Sci. Numer. Simul.* **2018**, *54*, 221–232. [CrossRef]
6. Bianca, C.; Menale, M. Existence and uniqueness of non-equilibrium stationary solutions in discrete thermostatted models. *Commun. Nonlinear Sci. Numer. Simul.* **2019**, *73*, 25–34. [CrossRef]
7. Bianca, C.; Mogno, C. Modelling pedestrian dynamics into a metro station by thermostatted kinetic theory methods. *Math. Comput. Model. Dyn. Syst.* **2018**, *24*, 207–235. [CrossRef]
8. Dickson, A.; Dinner, A.R. Enhanced sampling of non-equilibrium steady states. *Annu. Rev. Phys. Chem.* **2010**, *61*, 441–459. [CrossRef] [PubMed]
9. Wu, J. Non-equilibrium stationary states from the equation of motion of open systems. *New J. Phys.* **2010**, *12*, 083042. [CrossRef]
10. Eckmann, J.P.; Pillet, C.A.; Rey-Bellet, L. Non-equilibrium statistical mechanics of anharmonic chains coupled to two heat baths at different temperatures. *Commun. Math. Phys.* **1999**, *201*, 657–697. [CrossRef]
11. Wennberg, B.; Wondmagegne, Y. Stationary states for the Kac equation with a Gaussian thermostat. *Nonlinearity* **2004**, *17*, 633. [CrossRef]
12. Derrida, B. Non-equilibrium steady states: fluctuations and large deviations of the density and of the current. *J. Stat. Mech.* **2007**, *7*, P07023. [CrossRef]
13. Jepps, O.G.; Rondoni, L. Deterministic thermostats, theories of non-equilibrium systems and parallels with the ergodic condition. *J. Phys. Math. Theor.* **2010**, *43*, 133001. [CrossRef]
14. Tjhung, E.; Cates, M.E.D. Marenduzzo Non-equilibrium steady states in polar active fluids. *Soft Matter* **2011**, *7*, 7453–7464. [CrossRef]
15. Hurowitz, D.; Rahav, S.; Cohen, D. The non-equilibrium steady state of sparse systems with non-trivial topology. *EPL Europhys. Lett.* **2012**, *98*, 20002. [CrossRef]
16. Bronson, R.; Jacobson, C. Modeling the dynamics of social systems. *Comput. Math. Appl.* **1990**, *19*, 35–42. [CrossRef]
17. Carbonaro, B.; Serra, N. Towards mathematical models in psychology: A stochastic description of human feelings. *Math. Model. Methods Appl. Sci.* **2002**, *12*, 1453–1490. [CrossRef]
18. Karmeshu, V.P.J. Non-Linear Models of Social Systems. *Econ. Political Wkly.* **2003**, *38*, 3678–3685.
19. Giorno, V.; Spina, S. Rumor spreading models with random denials. *Phys. Stat. Mech. Its Appl.* **2016**, *461*, 569–576. [CrossRef]
20. Dobson, A.D.; de Lange, E.; Keane, A.; Ibbett, H.; Milner-Gulland, E.J. Integrating models of human behaviour between the individual and population levels to inform conservation interventions. *Philos. Trans. R. Soc.* **2019**, *374*, 20180053. [CrossRef] [PubMed]
21. Granas, A.; Dugundji, J. *Fixed Point Theory*; Springer Science & Business Media: Berlin, Germany, 2013.

© 2020 by the authors. Licensee MDPI, Basel, Switzerland. This article is an open access article distributed under the terms and conditions of the Creative Commons Attribution (CC BY) license (http://creativecommons.org/licenses/by/4.0/).

Article

A Convergence Theorem for the Nonequilibrium States in the Discrete Thermostatted Kinetic Theory

Carlo Bianca [1,2,*] and Marco Menale [1,3]

[1] Laboratoire Quartz EA 7393, École Supérieure d'Ingénieurs en Génie Électrique, Productique et Management Industriel, 95092 Cergy Pontoise CEDEX, France
[2] Laboratoire de Recherche en Eco-innovation Industrielle et Energétique, École Supérieure d'Ingénieurs en Génie Électrique, Productique et Management Industriel, 95092 Cergy Pontoise CEDEX, France
[3] Dipartimento di Matematica e Fisica, Università degli Studi della Campania "L. Vanvitelli", Viale Lincoln 5, I-81100 Caserta, Italy
* Correspondence: c.bianca@ecam-epmi.com

Received: 16 July 2019; Accepted: 25 July 2019; Published: 28 July 2019

Abstract: The existence and reaching of nonequilibrium stationary states are important issues that need to be taken into account in the development of mathematical modeling frameworks for far off equilibrium complex systems. The main result of this paper is the rigorous proof that the solution of the discrete thermostatted kinetic model catches the stationary solutions as time goes to infinity. The approach towards nonequilibrium stationary states is ensured by the presence of a dissipative term (thermostat) that counterbalances the action of an external force field. The main result is obtained by employing the Discrete Fourier Transform (DFT).

Keywords: thermostat; nonequilibrium stationary states; discrete Fourier transform; discrete kinetic theory; nonlinearity

1. Introduction

The modeling of a complex living system requires much attention considering the large number of components or active particles, the multiple nonlinear interactions, and the emerging collective behaviors [1–3]. The evolution of a complex system and the related global collective behaviors is usually driven by an external event (a predator for a swarm, an alert for a crowd of pedestrians, a vaccine for a tumor); see, among others, [4–6] and the references cited therein. Accordingly, a suitable modeling framework needs to take into account the nonequilibrium conditions under which a complex living system operates. Different modeling frameworks coming from the applied sciences have been developed [7–9], and in particular, the tools of nonequilibrium statistical mechanics have been proposed and employed in an attempt to follow the evolution of a complex system from the transient state to the stationary state; see [10–12].

Recently, the discrete thermostatted kinetic theory was proposed in [13,14] for the modeling and analysis of a far from equilibrium complex system; applications to biology [15,16] and crowd dynamics have been developed [17]. According to this theory, the complex system is divided into different functional subsystems composed by particles expressing the same task, which is usually a strategy. The strategy is modeled by introducing a scalar variable called activity; the interactions among the particles, called active particles, is modeled according to the stochastic game theory [18]. The nonequilibrium condition is modeled by introducing an external force field coupled to a dissipative term (called a thermostat, in analogy with the nonequilibrium statistical mechanics [19,20]), which allows the reaching of a nonequilibrium stationary state. Depending on the phenomenon under consideration, the activity variable can have a continuous or a discrete structure. Consequently, the mathematical structure reduces to a system of nonlinear partial integro-differential equations (continuous structure [21]) or nonlinear ordinary differential equations (discrete structure [13]).

The present paper is devoted to a further mathematical analysis of the discrete thermostatted kinetic theory framework proposed in [13]. The conditions of the existence and uniqueness of the nonequilibrium stationary solution were investigated in [14]. This paper provides the mathematical proof that the solutions of the discrete thermostatted kinetic theory framework approach the nonequilibrium stationary solutions when the time goes to infinity. The main result is obtained by employing the Discrete Fourier Transform (DFT).

It is worth noting that to the best of our knowledge, this is the first time that this proof has been presented for the discrete thermostatted kinetic theory proposed in [13]. However similar investigations have been addressed for the thermostatted Kac equation [22–25]. Further applications can be envisaged for biosystems [26,27], in medicine [28] and for complex systems [29].

The present paper is structured into three sections. In particular, Section 2 is devoted to the foundations and the main assumptions of the discrete thermostatted kinetic theory. The Discrete Fourier Transform (DFT) and the statement of the main result, concerning the convergence of the solutions of the discrete thermostatted kinetic theory to the related nonequilibrium stationary solutions, are presented in Section 3. The proof of the main result is detailed in Section 4.

2. The Discrete Thermostatted Framework

Let $I_u = \{u_1, u_2, \ldots, u_n\}$ be a discrete subset of $\mathbb{R}$, $F_i(t) \geq 0$ for $i \in \{1, 2, \ldots, n\}$ and $t > 0$, and $\mathbf{f}(t) = (f_1(t), f_2(t), \ldots, f_n(t))$, where, for $i \in \{1, 2, \ldots, n\}$,

$$f_i(t) := f(t, u_i) : [0, +\infty[\times I_u \to \mathbb{R}^+$$

is the solution of the following nonlinear ordinary differential equation:

$$\frac{df_i}{dt}(t) = J_i[\mathbf{f}](t) + F_i(t) - \sum_{i=1}^{n} \left(\frac{u_i^2 \left(J_i[\mathbf{f}] + F_i \right)}{\mathbb{E}_2[\mathbf{f}]} \right) f_i(t). \tag{1}$$

The operator $J_i[\mathbf{f}](t)$, for $i \in \{1, 2, \ldots, n\}$, is given by:

$$J_i[\mathbf{f}](t) = G_i[\mathbf{f}](t) - L_i[\mathbf{f}](t)$$
$$= \sum_{h=1}^{n} \sum_{k=1}^{n} \eta B_{hk}^i f_h(t) f_k(t) - f_i(t) \sum_{k=1}^{n} \eta f_k(t),$$

where $\eta > 0$ and, for $i, h, k \in \{1, 2, \ldots, n\}$, $B_{hk}^i : I_u \times I_u \times I_u \to \mathbb{R}^+$; the function $\mathbb{E}_2[\mathbf{f}](t)$ denotes the second order moment of $\mathbf{f}$:

$$\mathbb{E}_2[\mathbf{f}](t) = \sum_{i=1}^{n} u_i^2 f_i(t).$$

Let $\mathbb{E}_2 \in \mathbb{R}^+$ and $\mathcal{R}_\mathbf{f}^2$ denote the function space:

$$\mathcal{R}_\mathbf{f}^2 = \mathcal{R}_\mathbf{f}^2 \left(\mathbb{R}^+; \mathbb{E}_2 \right) = \left\{ \mathbf{f} \in C \left([0, +\infty]; \left(\mathbb{R}^+ \right)^n \right) : \mathbb{E}_2[\mathbf{f}] = \mathbb{E}_2 \right\}.$$

The existence and uniqueness of the solution of the related Cauchy problem has been proven in [13] under the following assumptions:

H1 The function B_{hk}^i is normalized with respect to i, namely for all $h, k \in \{1, 2, \ldots, n\}$, one has:

$$\sum_{i=1}^{n} B_{hk}^i = 1;$$

H2 $u_i \geq 1$, for all $i \in \{1, 2, \ldots, n\}$.

A nonequilibrium stationary solution of Equation (1) is a function f_i, for $i \in \{1, 2, \ldots, n\}$, which is the solution of the following equation:

$$J_i[\mathbf{f}] + F_i - \sum_{i=1}^{n} \left(\frac{u_i^2 (J_i[\mathbf{f}] + F_i)}{\mathbb{E}_2} \right) f_i = 0. \tag{2}$$

Let $\tilde{\mathcal{R}}_f^2$ be the following function space:

$$\tilde{\mathcal{R}}_f^2 \left(\mathbb{R}^+; \mathbb{E}_2 \right) = \left\{ \mathbf{f} \in \left(\mathbb{R}^+ \right)^n : \mathbb{E}_2[\mathbf{f}] = \mathbb{E}_2 \right\}.$$

The existence of nonequilibrium stationary solutions $\mathbf{g} \in \tilde{\mathcal{R}}_f^2$ was proven in [14], under Assumptions **H1–H2** and:

H3 $\sum_{i=1}^{n} u_i B_{hk}^i = 0$, for all $h, k \in \{1, 2, \ldots, n\}$;

H4 $\sum_{i=1}^{n} u_i^2 B_{hk}^i = u_h^2$, for all $h, k \in \{1, 2, \ldots, n\}$;

H5 $\mathbb{E}_0[\mathbf{f}] := \sum_{i=1}^{n} f_i = \mathbb{E}_2[\mathbf{f}] = 1$.

In particular, in [14], it was proven that the nonequilibrium stationary solution is unique if the following constraint on the force field $\mathbf{F}(t) = (F, F, \ldots, F)$ holds true:

$$F \geq 2\eta \mathbb{E}_2^2 \left(1 + \frac{1}{\|u\|_2^2} \right).$$

Proposition 1 ([14]). *Assume that Assumptions **H1–H5** hold true.*
Then:

1. *The evolution equation of $\mathbb{E}_1[\mathbf{f}](t) = \sum_{i=1}^{n} u_i f_i(t)$ reads:*

$$\mathbb{E}_1'[\mathbf{f}](t) + \left(\eta + \sum_{i=1}^{n} u_i^2 f_i \right) \mathbb{E}_1[\mathbf{f}](t) - \sum_{i=1}^{n} u_i F_i = 0; \tag{3}$$

2. *The first-order moment converges to a constant, which depends on the parameters of the system, as t goes to infinity, i.e.,*

$$\mathbb{E}_1[\mathbf{f}](t) \to K := \frac{\sum_{i=1}^{n} u_i F_i}{\eta + \sum_{i=1}^{n} u_i^2 F_i}; \tag{4}$$

3. *Let $\mathbf{f}_0$ be the initial data of the Cauchy problem related to (1), then:*

$$|\mathbb{E}_1[\mathbf{f}](t) - K| \leq c\, e^{-\left(\eta + \sum_{i=1}^{n} u_i^2 F_i \right) t}, \tag{5}$$

where c is a constant that depends on the system.

Remark 1. *Equation (1) was proposed in [13] for the modeling of a complex system, which is assumed to be composed of n subsystems (called **functional subsystems**). In particular:*

- *The function $f_i(t)$, for $i \in \{1, 2, \ldots, n\}$, denotes the distribution function of the ith functional subsystem;*

- The function $\mathbf{F}(t) = (F_1(t), F_2(t), \ldots, F_n(t))$ is the external force field acting on the whole system;
- The term:

$$\alpha := \sum_{i=1}^{n} \left(\frac{u_i^2 \left(J_i[\mathbf{f}] + F_i \right)}{\mathbb{E}_2} \right)$$

represents the thermostat term, which allows keeping the quantity $\mathbb{E}_2[\mathbf{f}](t)$ constant;
- The term η_{hk} is the interaction rate related to the encounters between the functional subsystem h and the functional subsystem k, for $h, k \in \{1, 2, \ldots, n\}$;
- The function B_{hk}^i denotes the transition probability density that the functional subsystem h falls into i after interacting with the functional subsystem k, for $i, h, k \in \{1, 2, \ldots, n\}$;
- The operator $J_i[\mathbf{f}](t)$, for $i \in \{1, 2, \ldots, n\}$, models the net flux related to the ith functional subsystem; $G_i[\mathbf{f}](t)$ denotes the gain term operator and $L_i[\mathbf{f}](t)$ the loss term operator.

Remark 2. Let $p \in \mathbb{N}$. Equation (1) can be further generalized as follows:

$$\frac{df_i}{dt}(t) = J_i[\mathbf{f}](t) + F_i(t) - \sum_{j=1}^{n} \left(\frac{u_j^p \left(J_j[\mathbf{f}] + F_j \right)}{\mathbb{E}_p[\mathbf{f}]} \right) f_i(t).$$

The above framework allows keeping the following pth order moment constant:

$$\mathbb{E}_p[\mathbf{f}](t) = \sum_{i=1}^{n} u_i^p f_i(t).$$

3. Convergence to the Stationary State

The main result of this paper is the proof that the function $\mathbf{f}(t)$, the solution of Equation (1), converges to the nonequilibrium stationary solution $\mathbf{g}$ of (2). The proof is based on the Discrete Fourier Transform (DFT).

Let $\mathbf{x} = (x_1, x_2, \ldots, x_n) \in \mathbb{R}^n$; the DFT is defined as follows:

$$\hat{x}_m = \sum_{l=1}^{n} x_l e^{-\frac{2\pi \iota}{n} m(l-1)}, \quad m \in \{1, 2, \ldots, n\},$$

where ι denotes the imaginary unit.

Let $\hat{f}_m(t)$ and $\hat{g}_m$, the DFT of the solution $\mathbf{f}(t)$ of (1) and of the solution $\mathbf{g}$ of (2), respectively, be defined as follows:

$$\hat{f}_m(t) = \sum_{l=1}^{n} f_l(t) e^{-\frac{2\pi \iota}{n} m(l-1)}, \quad \hat{g}_m = \sum_{l=1}^{n} g_l e^{-\frac{2\pi \iota}{n} m(l-1)}, \tag{6}$$

for $m \in \{1, 2, \ldots, n\}$.

Theorem 1. Let $\mathbf{f}(t)$ be the solution of Equation (1) and $\mathbf{g}$ the solution of Equation (2). If Assumptions **H1–H4** hold true, then $\mathbf{f}(t)$ converges to $\mathbf{g}$, as $t \to +\infty$.

4. Proof of the Main Result

Proof of Theorem 1. The first step is the derivation of the DFT of the discrete thermostatted Equation (1).

Multiplying both sides of (1) by $e^{-\frac{2\pi \iota}{n} m(l-1)}$ and summing for l from 1–n, one has:

$$\sum_{l=1}^{n} \frac{df_l}{dt}(t) e^{-\frac{2\pi \iota}{n} m(l-1)} = \sum_{l=1}^{n} \left(J_l[\mathbf{f}](t) + F_l(t) - \alpha f_l(t) \right) e^{-\frac{2\pi \iota}{n} m(l-1)}, \tag{7}$$

for $m \in \{1, 2, \ldots, n\}$.

The left-hand side of (7) is written as:

$$\sum_{l=1}^{n} \frac{df_l}{dt}(t) e^{-\frac{2\pi i}{n} m(l-1)} = \frac{d}{dt}\left(\sum_{l=1}^{n} f_l(t) e^{-\frac{2\pi i}{n} m(l-1)}\right) \tag{8}$$

$$= \frac{d\hat{f}_m}{dt}(t).$$

The first term in the right-hand side of the (7), using the property of operators $G_i[\mathbf{f}](t)$ and $L_i[\mathbf{f}](t)$, is written as:

$$\sum_{l=1}^{n} J_l[\mathbf{f}](t) e^{-\frac{2\pi i}{n} m(l-1)} = \sum_{l=1}^{n} (G_l[\mathbf{f}](t) - L_l[\mathbf{f}](t)) e^{-\frac{2\pi i}{n} m(l-1)}$$

$$= \sum_{l=1}^{n} \left(\sum_{h=1}^{n} \sum_{k=1}^{n} \eta B_{hk}^{l} f_h(t) f_k(t)\right) e^{-\frac{2\pi i}{n} m(l-1)} \tag{9}$$

$$- \sum_{l=1}^{n} \left(f_l(t) \sum_{k=1}^{n} \eta f_k(t)\right) e^{-\frac{2\pi i}{n} m(l-1)},$$

where:

$$\sum_{l=1}^{n} \left(\sum_{h=1}^{n} \sum_{k=1}^{n} \eta B_{hk}^{l} f_h(t) f_k(t)\right) e^{-\frac{2\pi i}{n} m(l-1)}$$

$$= \eta \sum_{h=1}^{n} \sum_{k=1}^{n} f_h(t) f_k(t) \left(\sum_{l=1}^{n} B_{hk}^{l} e^{-\frac{2\pi i}{n} m(l-1)}\right). \tag{10}$$

For Assumption **H5**, one has:

$$\sum_{l=1}^{n} \left(f_l(t) \sum_{k=1}^{n} \eta f_k(t)\right) e^{-\frac{2\pi i}{n} m(l-1)} = \eta \sum_{l=1}^{n} f_l(t) e^{-\frac{2\pi i}{n} m(l-1)} \tag{11}$$

$$= \eta \hat{f}_m(t).$$

The second term of (7) reads:

$$\sum_{l=1}^{n} F_l e^{-\frac{2\pi i}{n} m(l-1)} = \hat{F}_m. \tag{12}$$

For Assumptions **H1**, **H3**, **H4**, and **H5**, the third term of the right-hand side of the (7) is written as:

$$\sum_{l=1}^{n} \left(\sum_{l=1}^{n} u_l^2 \left(J_l[\mathbf{f}](t) + F_l\right)\right) f_l(t) e^{-\frac{2\pi i}{n} m(l-1)}$$

$$= \sum_{l=1}^{n} \left(\sum_{l=1}^{n} u_l^2 F_l\right) f_l(t) e^{-\frac{2\pi i}{n} m(l-1)} \tag{13}$$

$$= \left(\sum_{l=1}^{n} u_l^2 F_l\right) \hat{f}_m(t).$$

By (8)–(13), (7) is rewritten, for $m \in \{1, 2, \ldots, n\}$, as follows:

$$\frac{d\hat{f}_m}{dt}(t) + \hat{f}_m(t)\left(\sum_{l=1}^{n} u_l^2 F_l\right) + \eta \hat{f}_m(t) - \hat{F}_m$$
$$= \eta \sum_{h=1}^{n} \sum_{k=1}^{n} f_h(t) f_k(t) \left(\sum_{l=1}^{n} B_{hk}^l e^{-\frac{2\pi i}{n} m(l-1)}\right). \tag{14}$$

Let $\mathbf{f}_1(t) = (f_{1l}(t))_l$ and $\mathbf{f}_2(t) = (f_{2l}(t))_l$, for $l \in \{1, 2, \ldots, n\}$, two different solutions of the framework (1), and:

$$v_m(t) := \left(\hat{f}_{1m}(t) - \hat{f}_{2m}(t)\right) + m\left(\mathbb{E}_1[\mathbf{f}_1](t) - \mathbb{E}_1[\mathbf{f}_2](t)\right), \quad m \in \{1, 2, \ldots, n\}. \tag{15}$$

By (15) and (14) and straightforward calculations, one has:

$$\frac{dv_m}{dt}(t) + \eta v_m(t) = \frac{d\hat{f}_{1m}}{dt}(t) - \frac{d\hat{f}_{2m}}{dt}(t) + \eta \hat{f}_{1m} - \eta \hat{f}_{2m}$$
$$+ \eta m \left(\mathbb{E}_1[\mathbf{f}_1](t) - \mathbb{E}_1[\mathbf{f}_2](t)\right) + m \left(\mathbb{E}_1'[\mathbf{f}_1](t) - \mathbb{E}_1'[\mathbf{f}_2](t)\right)$$
$$= -\hat{f}_{1m}\left(\sum_{l=1}^{n} u_l^2 F_l\right) + \hat{f}_{2m}\left(\sum_{l=1}^{n} u_l^2 F_l\right)$$
$$+ \eta \left(\sum_{h=1}^{n} \sum_{k=1}^{n} f_{1h}(t) f_{1k}(t) \left(\sum_{l=1}^{n} B_{hk}^l e^{-\frac{2\pi i}{n} m(l-1)}\right)\right) \tag{16}$$
$$- \eta \left(\sum_{h=1}^{n} \sum_{k=1}^{n} f_{2h}(t) f_{2k}(t) \left(\sum_{l=1}^{n} B_{hk}^l e^{-\frac{2\pi i}{n} m(l-1)}\right)\right)$$
$$+ \eta m \left(\mathbb{E}_1[\mathbf{f}_1](t) - \mathbb{E}_1[\mathbf{f}_2](t)\right) + m \left(\mathbb{E}_1'[\mathbf{f}_1](t) - \mathbb{E}_1'[\mathbf{f}_2](t)\right).$$

By using the (3), one has:

$$\mathbb{E}_1'[\mathbf{f}_1](t) - \mathbb{E}_1'[\mathbf{f}_2](t) = -\mathbb{E}_1[\mathbf{f}_1](t)\left(\eta + \sum_{l=1}^{n} u_l^2 f_l\right) + \mathbb{E}_1[\mathbf{f}_2](t)\left(\eta + \sum_{l=1}^{n} u_l^2 f_l\right)$$
$$= \left(\eta + \sum_{l=1}^{n} u_l^2 F_l\right)\left(\mathbb{E}_1[\mathbf{f}_2](t) - \mathbb{E}_1[\mathbf{f}_1](t)\right). \tag{17}$$

Bearing (17) in mind, Equation (16) reads:

$$\frac{dv_m}{dt}(t) + \eta v_m(t) = \left(\hat{f}_{2m}(t) - \hat{f}_{1m}(t)\right)\left(\sum_{l=1}^{n} u_l^2 F_l\right)$$
$$+ \eta m \left(\mathbb{E}_1[\mathbf{f}_1](t) - \mathbb{E}_1[\mathbf{f}_2](t)\right)$$
$$+ \left(\eta + \sum_{l=1}^{n} u_l^2 F_l\right) m \left(\mathbb{E}_1[\mathbf{f}_2](t) - \mathbb{E}_1[\mathbf{f}_1](t)\right) \tag{18}$$
$$+ \eta \left(\sum_{h=1}^{n} \sum_{k=1}^{n} f_{1h}(t) f_{1k}(t) \left(\sum_{l=1}^{n} B_{hk}^l e^{-\frac{2\pi i}{n} m(l-1)}\right)\right)$$
$$- \eta \left(\sum_{h=1}^{n} \sum_{k=1}^{n} f_{2h}(t) f_{2k}(t) \left(\sum_{l=1}^{n} B_{hk}^l e^{-\frac{2\pi i}{n} m(l-1)}\right)\right).$$

By straightforward calculations, (18) is rewritten as:

$$\begin{aligned}
\frac{dv_m}{dt}(t) + \eta v_m(t) = & -\left(\hat{f}_{1m}(t) - \hat{f}_{2m}(t)\right)\left(\sum_{l=1}^{n} u_l^2 F_l\right) \\
& + \left(\sum_{l=1}^{n} u_l^2 F_l\right) m \left(\mathbb{E}_1[\mathbf{f}_1](t) - \mathbb{E}_1[\mathbf{f}_2](t)\right) \\
& - \left(\sum_{l=1}^{n} u_l^2 F_l\right) m \left(\mathbb{E}_1[\mathbf{f}_1](t) - \mathbb{E}_1[\mathbf{f}_2](t)\right) \\
& + \eta \, m \left(\mathbb{E}_1[\mathbf{f}_1](t) - \mathbb{E}_2[\mathbf{f}_2](t)\right) \\
& + \left(\eta + \sum_{l=1}^{n} u_l^2 F_l\right) m \left(\mathbb{E}_1[\mathbf{f}_2](t) - \mathbb{E}_1[\mathbf{f}_1](t)\right) \\
& + \eta \left(\sum_{h=1}^{n}\sum_{k=1}^{n} f_{1h}(t) f_{1k}(t) \left(\sum_{l=1}^{n} B_{hk}^l e^{-\frac{2\pi i}{n} m(l-1)}\right)\right) \\
& - \eta \left(\sum_{h=1}^{n}\sum_{k=1}^{n} f_{2h}(t) f_{2k}(t) \left(\sum_{l=1}^{n} B_{hk}^l e^{-\frac{2\pi i}{n} m(l-1)}\right)\right) \\
= & -\left(\sum_{l=1}^{n} u_l^2 F_l\right) v_m(t) \\
& + \eta \left(\sum_{h=1}^{n}\sum_{k=1}^{n} f_{1h}(t) f_{1k}(t) \left(\sum_{l=1}^{n} B_{hk}^l e^{-\frac{2\pi i}{n} m(l-1)}\right)\right) \\
& - \eta \left(\sum_{h=1}^{n}\sum_{k=1}^{n} f_{2h}(t) f_{2k}(t) \left(\sum_{l=1}^{n} B_{hk}^l e^{-\frac{2\pi i}{n} m(l-1)}\right)\right).
\end{aligned} \quad (19)$$

By straightforward calculations, the second and the third terms of the right-hand side of (19) are written as:

$$\begin{aligned}
& \eta \left(\sum_{h=1}^{n}\sum_{k=1}^{n} f_{1h}(t) f_{1k}(t) \left(\sum_{l=1}^{n} B_{hk}^l e^{-\frac{2\pi i}{n} m(l-1)}\right)\right) \\
& - \eta \left(\sum_{h=1}^{n}\sum_{k=1}^{n} f_{2h}(t) f_{2k}(t) \left(\sum_{l=1}^{n} B_{hk}^l e^{-\frac{2\pi i}{n} m(l-1)}\right)\right) \\
= & \, \eta \sum_{h=1}^{n}\sum_{k=1}^{n} (f_{1h}(t) f_{1k}(t) - f_{2h}(t) f_{2k}(t)) \left(\sum_{l=1}^{n} B_{hk}^l e^{-\frac{2\pi i}{n} m(l-1)}\right) \\
= & \, \eta \sum_{h=1}^{n}\sum_{k=1}^{n} (f_{1h}(t) f_{1k}(t) - f_{2h}(t) f_{1k}(t) + f_{2h}(t) f_{1k}(t) - f_{2h}(t) f_{2k}(t)) \\
& \cdot \left(\sum_{l=1}^{n} B_{hk}^l e^{-\frac{2\pi i}{n} m(l-1)}\right) \\
= & \, \eta \sum_{h=1}^{n}\sum_{k=1}^{n} (f_{1k}(t)(f_{1h}(t) - f_{2h}(t)) + f_{2h}(t)(f_{1k}(t) - f_{2k}(t))) \\
& \cdot \left(\sum_{l=1}^{n} B_{hk}^l e^{-\frac{2\pi i}{n} m(l-1)}\right).
\end{aligned} \quad (20)$$

By summing and subtracting $m f_{1k}(t) \left(\mathbb{E}_1[\mathbf{f}_1](t) - \mathbb{E}_1[\mathbf{f}_2](t)\right)$ and $m f_{2h}(t) \left(\mathbb{E}_1[\mathbf{f}_1](t) - \mathbb{E}_1[\mathbf{f}_2](t)\right)$, one has:

$$
\begin{aligned}
&\eta \sum_{h=1}^{n}\sum_{k=1}^{n}(f_{1h}(t)-f_{2h}(t))\,f_{1k}(t)\sum_{l=1}^{n}B_{hk}^{l}e^{-\frac{2\pi i}{n}m(l-1)}\\
&-\eta \sum_{h=1}^{n}\sum_{k=1}^{n}(f_{1k}(t)-f_{2k}(t))\,f_{2h}(t)\sum_{l=1}^{n}B_{hk}^{l}e^{-\frac{2\pi i}{n}m(l-1)}\\
&=\eta \sum_{h=1}^{n}\sum_{k=1}^{n}\Big(f_{1k}(t)\left(f_{1h}(t)-f_{2h}(t)\right)+m f_{1k}(t)\left(\mathbb{E}_{1}[\mathbf{f}_{1}](t)-\mathbb{E}_{1}[\mathbf{f}_{2}](t)\right)\\
&\quad -m f_{1k}(t)\left(\mathbb{E}_{1}[\mathbf{f}_{1}](t)-\mathbb{E}_{1}[\mathbf{f}_{2}](t)\right)\Big)\sum_{l=1}^{n}B_{hk}^{l}e^{-\frac{2\pi i}{n}m(l-1)}\\
&\quad +\eta \sum_{h=1}^{n}\sum_{k=1}^{n}\Big(f_{2h}(t)\left(f_{1k}(t)-f_{2k}(t)\right)+m f_{2h}(t)\left(\mathbb{E}_{1}[\mathbf{f}_{1}](t)-\mathbb{E}_{1}[\mathbf{f}_{2}](t)\right)\\
&\quad -m f_{2h}(t)\left(\mathbb{E}_{1}[\mathbf{f}_{1}](t)-\mathbb{E}_{1}[\mathbf{f}_{2}](t)\right)\Big)\sum_{l=1}^{n}B_{hk}^{l}e^{-\frac{2\pi i}{n}m(l-1)}.
\end{aligned}
\tag{21}
$$

Rearranging into (21), one has:

$$
\begin{aligned}
&\eta\left(\sum_{h=1}^{n}\sum_{k=1}^{n}f_{1h}(t)f_{1k}(t)\left(\sum_{l=1}^{n}B_{hk}^{l}e^{-\frac{2\pi i}{n}m(l-1)}\right)\right)\\
&-\eta\left(\sum_{h=1}^{n}\sum_{k=1}^{n}f_{2h}(t)f_{2k}(t)\left(\sum_{l=1}^{n}B_{hk}^{l}e^{-\frac{2\pi i}{n}m(l-1)}\right)\right)\\
&=\eta\sum_{h=1}^{n}\sum_{k=1}^{n}\left((f_{1h}(t)-f_{2h}(t))+m\left(\mathbb{E}_{1}[\mathbf{f}_{1}](t)-\mathbb{E}_{1}[\mathbf{f}_{2}](t)\right)\right)\\
&\quad \cdot f_{1k}(t)\sum_{l=1}^{n}B_{hk}^{l}e^{-\frac{2\pi i}{n}m(l-1)}\\
&\quad -\eta m\left(\mathbb{E}_{1}[\mathbf{f}_{1}](t)-\mathbb{E}_{1}[\mathbf{f}_{2}](t)\right)\sum_{h=1}^{n}\sum_{k=1}^{n}f_{1k}(t)\sum_{l=1}^{n}B_{hk}^{l}e^{-\frac{2\pi i}{n}m(l-1)}\\
&\quad -\eta\sum_{h=1}^{n}\sum_{k=1}^{n}\left((f_{1k}(t)-f_{2k}(t))+m\left(\mathbb{E}_{1}[\mathbf{f}_{1}](t)-\mathbb{E}_{1}[\mathbf{f}_{2}](t)\right)\right)\\
&\quad \cdot f_{2h}(t)\sum_{l=1}^{n}B_{hk}^{l}e^{-\frac{2\pi i}{n}m(l-1)}\\
&\quad +\eta m\left(\mathbb{E}_{1}[\mathbf{f}_{1}](t)-\mathbb{E}_{1}[\mathbf{f}_{2}](t)\right)\sum_{h=1}^{n}\sum_{k=1}^{n}f_{2h}(t)\sum_{l=1}^{n}B_{hk}^{l}e^{-\frac{2\pi i}{n}m(l-1)}.
\end{aligned}
\tag{22}
$$

By (22), the (19) is rewritten, for $m \in \{1, 2, \ldots, n\}$, as follows:

$$\frac{dv_m}{dt}(t) + \eta v_m(t) = -v_m(t)\left(\sum_{l=1}^{n} u_l^2 F_l\right)$$
$$-\eta m \left(\mathbb{E}_1[\mathbf{f}_1](t) - \mathbb{E}_1[\mathbf{f}_2](t)\right) \sum_{h=1}^{n}\sum_{k=1}^{n} f_{1k}(t) \sum_{l=1}^{n} B_{hk}^l e^{-\frac{2\pi i}{n}m(l-1)}$$
$$+\eta m \left(\mathbb{E}_1[\mathbf{f}_1](t) - \mathbb{E}_1[\mathbf{f}_2](t)\right) \sum_{h=1}^{n}\sum_{k=1}^{n} f_{2h}(t) \sum_{l=1}^{n} B_{hk}^l e^{-\frac{2\pi i}{n}m(l-1)}$$
$$+\eta \sum_{h=1}^{n}\sum_{k=1}^{n} \left((f_{1h}(t) - f_{2h}(t)) + m\left(\mathbb{E}_1[\mathbf{f}_1](t) - \mathbb{E}_1[\mathbf{f}_2](t)\right)\right) \quad (23)$$
$$\cdot f_{1k}(t) \sum_{l=1}^{n} B_{hk}^l e^{-\frac{2\pi i}{n}m(l-1)}$$
$$-\eta \sum_{h=1}^{n}\sum_{k=1}^{n} \left((f_{1k}(t) - f_{2k}(t)) + m\left(\mathbb{E}_1[\mathbf{f}_1](t) - \mathbb{E}_1[\mathbf{f}_2](t)\right)\right)$$
$$\cdot f_{2h}(t) \sum_{l=1}^{n} B_{hk}^l e^{-\frac{2\pi i}{n}m(l-1)}.$$

Finally, (23) can be written as follows:

$$\frac{dv_m}{dt}(t) + \eta v_m(t) + \left(\sum_{i=1}^{n} u_i^2 F_i\right) v_m(t) =$$
$$\left(\mathbb{E}_1[\mathbf{f}_1](t) - \mathbb{E}_1[\mathbf{f}_2](t)\right)\left(-\eta m \sum_{h=1}^{n}\sum_{k=1}^{n} f_{1k}(t) \sum_{l=1}^{n} B_{hk}^l e^{-\frac{2\pi i}{n}m(l-1)}\right.$$
$$\left.+\eta m \sum_{h=1}^{n}\sum_{k=1}^{n} f_{2h}(t) \sum_{l=1}^{n} B_{hk}^l e^{-\frac{2\pi i}{n}m(l-1)}\right)$$
$$+\eta \sum_{h=1}^{n}\sum_{k=1}^{n} \left((f_{1h}(t) - f_{2h}(t)) + m\left(\mathbb{E}_1[\mathbf{f}_1](t) - \mathbb{E}_1[\mathbf{f}_2](t)\right)\right) \quad (24)$$
$$\cdot f_{1k}(t) \sum_{l=1}^{n} B_{hk}^l e^{-\frac{2\pi i}{n}m(l-1)}$$
$$-\eta \sum_{h=1}^{n}\sum_{k=1}^{n} \left((f_{1k}(t) - f_{2k}(t)) + m\left(\mathbb{E}_1[\mathbf{f}_1](t) - \mathbb{E}_1[\mathbf{f}_2](t)\right)\right)$$
$$\cdot f_{2h}(t) \sum_{l=1}^{n} B_{hk}^l e^{-\frac{2\pi i}{n}m(l-1)}.$$

Let $\mathbf{f}_1(t) = \mathbf{f}(t)$ and $\mathbf{f}_2(t) = \mathbf{g}$. By (4), (24), for $m \in \{1, 2, \ldots, n\}$, reads:

$$\frac{dv_m}{dt}(t) + \eta v_m(t) + \left(\sum_{i=1}^{n} u_i^2 F_i\right) v_m(t) =$$

$$(\mathbb{E}_1[\mathbf{f}](t) - K)\left(-\eta m \sum_{h=1}^{n}\sum_{k=1}^{n} f_k(t) \sum_{l=1}^{n} B_{hk}^l e^{-\frac{2\pi i}{n} m(l-1)}\right.$$

$$\left. + \eta m \sum_{h=1}^{n}\sum_{k=1}^{n} g_h \sum_{l=1}^{n} B_{hk}^l e^{-\frac{2\pi i}{n} m(l-1)}\right)$$

$$+ \eta \sum_{h=1}^{n}\sum_{k=1}^{n} \left((f_h(t) - g_h(t)) + m\left(\mathbb{E}_1[\mathbf{f}](t) - K\right)\right) \quad (25)$$

$$\cdot f_k(t) \sum_{l=1}^{n} B_{hk}^l e^{-\frac{2\pi i}{n} m(l-1)}$$

$$- \eta \sum_{h=1}^{n}\sum_{k=1}^{n} \left((f_k(t) - g_k) + m\left(\mathbb{E}_1[\mathbf{f}_1](t) - K\right)\right)$$

$$\cdot g_h(t) \sum_{l=1}^{n} B_{hk}^l e^{-\frac{2\pi i}{n} m(l-1)}.$$

Let $\underline{v}_m(t) := e^{\eta t} v_m(t)$, then:

$$\frac{d\underline{v}_m}{dt}(t) = \eta e^{\eta t} v_m(t) + e^{\eta t}\frac{dv_m}{dt}(t). \quad (26)$$

Bearing (26) in mind and multiplying by $e^{\eta t}$ both sides of (25), one has:

$$\frac{d\underline{v}_m}{dt}(t) + \left(\sum_{i=1}^{n} u_i^2 F_i\right) \underline{v}_m(t) = e^{\eta t} \left(\mathbb{E}_1[\mathbf{f}] - K\right)$$

$$\cdot \left(-\eta m \sum_{h=1}^{n}\sum_{k=1}^{n} f_k(t) \sum_{l=1}^{n} B_{hk}^l e^{-\frac{2\pi i}{n} m(l-1)}\right.$$

$$\left. + \eta m \sum_{h=1}^{n}\sum_{k=1}^{n} g_h \sum_{l=1}^{n} B_{hk}^l e^{-\frac{2\pi i}{n} m(l-1)}\right)$$

$$+ \eta \sum_{h=1}^{n} e^{\eta t} \left((f_h(t) - g_h) + m\left(\mathbb{E}_1[\mathbf{f}] - K\right)\right) \sum_{k=1}^{n} f_k(t) \quad (27)$$

$$\cdot \sum_{l=1}^{n} B_{hk}^l e^{-\frac{2\pi i}{n} m(l-1)}$$

$$+ \eta \sum_{k=1}^{n} e^{\eta t} \left((f_k(t) - g_k) + m\left(\mathbb{E}_1[\mathbf{f}] - K\right)\right) \sum_{h=1}^{n} g_h$$

$$\cdot \sum_{l=1}^{n} B_{hk}^l e^{-\frac{2\pi i}{n} m(l-1)}.$$

By (5) and straightforward calculations, one has:

$$\left| e^{\eta t} \left(\mathbb{E}_1[\mathbf{f}] - K \right) \cdot \left(-\eta m \sum_{h=1}^{n} \sum_{k=1}^{n} f_k(t) \sum_{l=1}^{n} B_{hk}^l e^{-\frac{2\pi i}{n} m(l-1)} \right. \right.$$
$$\left. \left. + \eta m \sum_{h=1}^{n} \sum_{k=1}^{n} \hat{g}_h \sum_{l=1}^{n} B_{hk}^l e^{-\frac{2\pi i}{n} m(l-1)} \right) \right| \leq c e^{\eta t} e^{-\left(\eta + \sum_{i=1}^{n} u_i^2 F_i \right) t}, \quad (28)$$

where c is a constant that depends on the parameters of the system. By using (27) and (28), one has:

$$\frac{d|\underline{v}|_m(t)}{dt}(t) \leq c e^{\eta t} e^{-\left(\eta + \sum_{i=1}^{n} u_i^2 F_i \right) t} + \lambda \underline{v}_m(t), \quad m \in \{1, 2, \ldots, n\}. \quad (29)$$

By integrating (29) between zero and t and using (29), one has:

$$|\underline{v}_m(t)| \leq |\underline{v}_m(0)| + c \left| \int_0^t e^{\eta \tau} e^{-\left(\eta + \sum_{i=1}^{n} u_i^2 F_i \right) \tau} d\tau \right| + \lambda \left| \int_0^t \underline{v}_m(\tau) \, d\tau \right|$$
$$\leq |\underline{v}_m(0)| + c \int_0^t e^{-\sum_{i=1}^{n} u_i^2 F_i \tau} d\tau + \lambda \int_0^t |\underline{v}_m(0)| \, d\tau. \quad (30)$$

By the integral Gronwall inequality [30] and (30), one has:

$$|\underline{v}_m(t)| \leq |\underline{v}_m(0)| e^{\lambda t} + c e^{-\sum_{i=1}^{n} u_i^2 F_i t}. \quad (31)$$

By dividing (31) by $e^{\eta t}$ and bearing (26) in mind, one has:

$$|\underline{v}_m(t)| \leq |\underline{v}_m(0)| e^{(\lambda - \eta) t} + c e^{-\left(\eta + \sum_{i=1}^{n} u_i^2 F_i t \right)}. \quad (32)$$

Finally, by (32) and (15), for $m \in \{1, 2, \ldots, n\}$, one has:

$$\left| \hat{f}_m(t) - \hat{g}_m \right| \leq (c + c_0) e^{-\left(\eta + \sum_{i=1}^{n} u_i^2 F_i t \right)}, \quad (33)$$

where c is a constant that depends on the parameters of the system and c_0 on the initial data of the related problem.

It is possible to conclude that:

$$\hat{f}_m(t) \xrightarrow{t \to +\infty} \hat{g}_m,$$

for $m \in \{1, 2, \ldots, n\}$. The proof is thus gained. □

Author Contributions: Conceptualization, C.B. and M.M.; methodology, C.B. and M.M.; formal analysis, C.B. and M.M.; investigation, C.B. and M.M.; resources, C.B. and M.M.; writing-original draft preparation, C.B. and M.M.; writing-review and editing, C.B. and M.M.; supervision, C.B.

Funding: This research received no external funding.

Conflicts of Interest: The authors declare no conflict of interest.

References

1. Holland, J.H. Studying complex adaptive systems. *J. Syst. Sci. Complex.* **2006**, *19*, 1–8. [CrossRef]
2. Bar-Yam, Y. *Dynamics of Complex Systems, Studies in Nonlinearity*; Westview Press: Boulder, CO, USA, 2003.
3. Zöttl, A.; Stark, H. Emergent behavior in active colloids. *J. Phys. Condens. Matter* **2016**, *28*, 253001. [CrossRef]
4. Schaerf, T.M.; Dillingham, P.W.; Ward, A.J.W. The effects of external cues on individual and collective behavior of shoaling fish. *Sci. Adv.* **2017**, *23*, e1603201. [CrossRef]
5. Nagy, M.; Ákos, Z.; Biro, D.; Vicsek, T. Hierarchical group dynamics in pigeon flocks. *Nature* **2010**, *464*, 890–899. [CrossRef] [PubMed]
6. Fetecau, C.; Guo, A. A mathematical model for flight guidance in honeybee swarms. *Bull. Math. Biol.* **2012**, *74*, 2600–2621. [CrossRef] [PubMed]
7. Bao, L.; Fritchman, J.C. Information of complex systems and applications in agent based modeling. *Sci. Rep.* **2018**, *8*, 6177. [CrossRef] [PubMed]
8. Bisi, M.; Spiga, G.; Toscani, G. Kinetic models of conservative economies with wealth redistribution. *Comm. Math. Sci.* **2009**, *7*, 901–916. [CrossRef]
9. Aletti, G.; Naldi, G.; Toscani, G. First-order continuous models of opinion formation. *SIAM J. Appl. Math.* **2007**, *67*, 837–853. [CrossRef]
10. Eckmann, J.P.; Pillet, C.A.; Rey-Bellet, L. Non-equilibrium statistical mechanics of an harmonic chains coupled to two heat baths at different temperatures. *Commun. Math. Phys.* **1999**, *201*, 657–697. [CrossRef]
11. Kwon, C.; Ao, P. Nonequilibrium steady state of a stochastic system driven by a nonlinear drift force. *Phys. Rev. E Stat. Nonlinear Soft Matter Phys.* **2001**, *84*, 061106. [CrossRef]
12. Tjhung, E.; Cates, M.E.; Marenduzzo, D. Nonequilibrium steady states in polar active uids. *Soft Matter* **2011**, *7*, 7453–7464. [CrossRef]
13. Bianca, C.; Mogno, C. Qualitative analysis of a discrete thermostatted kinetic framework modeling complex adaptive systems. *Commun. Nonlinear Sci. Numer. Simul.* **2018**, *54*, 221–232. [CrossRef]
14. Bianca, C.; Menale, M. Existence and Uniqueness of Nonequilibrium Stationary Solutions in Discrete Thermostatted Models. *Commun. Nonlinear Sci. Numer. Simul.* **2019**, *73*, 25–34. [CrossRef]
15. Bianca, C.; Riposo, J. Mimic therapeutic actions against keloid by thermostatted kinetic theory methods. *Eur. Phys. J. Plus* **2015**, *130*, 159. [CrossRef]
16. Bianca, C.; Brézin, L. Modeling the antigen recognition by B-cell and T-cell receptors through thermostatted kinetic theory methods. *Int. J. Biomath.* **2017**, *10*, 1750072. [CrossRef]
17. Bianca, C.; Mogno, C. A thermostatted kinetic theory model for event-driven pedestrian dynamics. *Eur. Phys. J. Plus* **2018**, *133*, 213. [CrossRef]
18. Myerson, R.B. *Game Theory: Analysis of Conflict*; Harvard University Press: Cambridge, MA, USA, 1997.
19. Morriss, G.P.; Dettmann, C.P. Thermostats: Analysis and application. *Chaos* **1998**, *8*, 321–336. [CrossRef]
20. Jepps, O.G.; Rondoni, L. Deterministic thermostats, theories of nonequilibrium systems and parallels with the ergodic condition. *J. Phys. A* **2010**, *43*, 133001. [CrossRef]
21. Bianca, C. Kinetic theory for active particles modelling coupled to Gaussian thermostats. *Appl. Math. Sci.* **2012**, *6*, 651–660.
22. Wennberg, B.; Wondmagegne, Y. Stationary states for the Kac equation with a Gaussian thermostat. *Nonlinearity* **2004**, *17*, 633. [CrossRef]
23. Wennberg, B.; Wondmagegne, Y. The Kac equation with a thermostatted force field. *J. Stat. Phys.* **2006**, *124*, 859–880. [CrossRef]
24. Bagland, V.; Wennberg, B.; Wondmagegne, Y. Stationary states for the noncutoff Kac equation with a Gaussian thermostat. *Nonlinearity* **2007**, *20*, 583–604. [CrossRef]
25. Bagland, V. Well-posedness and large time behaviour for the non-cutoff Kac equation with a Gaussian thermostat. *J. Stat. Phys.* **2010**, *138*, 838–875. [CrossRef]

26. Deisboeck, T.S.; Wang, Z.; Macklin, P.; Cristini, V. Multiscale cancer modeling. *Annu. Rev. Biomed. Eng.* **2011**, *13*, 127–155. [CrossRef]
27. Lucia, U.; Grisolia, G.; Ponzetto, A.; Deisboeck, T.S. Thermodynamic considerations on the role of heat and mass transfer in biochemical causes of carcinogenesis. *Phys. A Stat. Mech. Appl.* **2018**, *490*, 1164–1170. [CrossRef]
28. Silvagni, A.J.; Levy, L.A.; Mcfee, R.B. Educating health professionals, first responders, and the community about bioterrorism and weapons of mass destruction. *J. Am. Osteopath. Assoc.* **2002**, *102*, 491–498.
29. Bejan, A.; Errera, M.R. Complexity, organization, evolution, and constructal law. *J. Appl. Phys.* **2016**, *119*, 074901. [CrossRef]
30. Walter, W. *Differential and Integral Inequalities*; Springer Science & Business Media: Berlin, Germany, 2012.

© 2019 by the authors. Licensee MDPI, Basel, Switzerland. This article is an open access article distributed under the terms and conditions of the Creative Commons Attribution (CC BY) license (http://creativecommons.org/licenses/by/4.0/).

Article
Dependence on the Initial Data for the Continuous Thermostatted Framework

Bruno Carbonaro [1] and Marco Menale [1,2,*]

[1] Dipartimento di Matematica e Fisica, Università degli Studi della Campania "L. Vanvitelli", Viale Lincoln 5, I-81100 Caserta, Italy

[2] Laboratoire Quartz EA 7393, École Supérieure d'Ingénieurs en Génie Électrique, Productique et Management Industriel, 95092 Cergy Pontoise Cedex, France

* Correspondence: marco.menale@unicampania.it

Received: 6 June 2019; Accepted: 4 July 2019; Published: 6 July 2019

Abstract: The paper deals with the problem of continuous dependence on initial data of solutions to the equation describing the evolution of a complex system in the presence of an external force acting on the system and of a thermostat, simply identified with the condition that the second order moment of the activity variable (see Section 1) is a constant. We are able to prove that these solutions are stable with respect to the initial conditions in the Hadamard's sense. In this connection, two remarks spontaneously arise and must be carefully considered: first, one could complain the lack of information about the "distance" between solutions at any time $t \in [0, +\infty)$; next, one cannot expect any more complete information without taking into account the possible distribution of the transition probabiliy densities and the interaction rates (see Section 1 again). This work must be viewed as a first step of a research which will require many more steps to give a sufficiently complete picture of the relations between solutions (see Section 5).

Keywords: kinetic theory; integro-differential equations; complex systems; stability; evolution equations

1. Introduction

The aim of the present paper is to contribute a first result about the stability (and, consequently, the uniqueness) of the solutions to the equation describing the evolution of a thermostatted complex system. As is well known, a complex system is a set of a very large number of objects that, in connection with the physical origin of the notion, can be called "particles", but—in view of its present applications—should perhaps be identified by the more general name of "individuals". These objects, of course, enjoy a number of empirical properties, that define their "state": in a purely mechanical framework, these properties are position and velocity. In principle, the state of the whole system should be considered as completely known when the states of all of its particles are known. But, since these states are modified by the mutual interactions of the particles, and these interactions in turn depend on the states of the particles, also in view of the extremely large number of interactions to be taken into account, it is readily seen that a complete knowledge of the state of the system is impossible [1] at any time. This gave rise to statistical mechanics and to Boltzmann's Kinetic Theory of Gases, that aimed to describe the states of any such system and their evolution in terms of average states of the particles and of probability of their interactions.

But the "complexity" of a system is more than the large number of its particles. In the mechanical framework, each interaction between two particles p and q is assumed to be independent of the presence of the remaining particles of the system, so that the result of simultaneous interactions of p with $q_1, q_2, \ldots, q_n$ is simply the sum of the results of single interaction. When this assumption is given up, and each particle is allowed to interact—under suitable conditions—not only with other particles but also with their interactions, then the system is "complex" in the full meaning of the term. So, not only the states of a complex systems and their evolution can be described only by using a statistical (or probabilistic) language and statistical (or probabilistic) methods, but also the interactions between its particles. Accordingly, all what we may assume to be able to know about complex systems at each instant are (a) a probability (or relative frequency) distribution f on the set D_u of all possible states of all of its particles; (b) for any pair (u_*, u^*) of possible states, an average frequency $\eta(u_*, u^*)$ of interactions between particles that are in the state u_* and particles that are in the state u^*; (c) for any pair (u_*, u^*) of possible states, a probability (or relative frequency) distribution on the possible effects of interactions (see Section 2 below).

This is a generalized form of Boltzmann's framework [1,2]. As a matter of fact, the introduction of the notion of "complex systems", in addition to the methodological choice to work on probability distributions in the same way as in the Kinetic Theory of Gases, corresponds to the need, hence to the attempt, to go beyond the limits of purely mechanical applications of the farmework, in order to give some contributions to a formal statement, and if possible to the solution, of many problems arising in the life of large sets of individuals of many different kinds (cell populations [3], with particular respect to the interaction between cancer cells and cells of immune systems [4], animal populations, social and economic communities like human nations [5,6], problems about vehicular traffic [7] and about the behaviour of swarms [8], crowds [9,10] and pedestrians [11], opinion formation [12], etc.). Of course, this attempt started about thirty years ago, was quite succesful and gave origin to a wide literature, investigating both the mathematical problems posed by the framework and its possible new applications [5,6,13]. The mathematical scheme provided by the equations proposed in this framework (see Section 2 below) could perhaps be considered as the most versatile among the ones from time to time constructed to formulate and solve problems arising in medicine, natural sciences, economics and even in behavioural sciences.

In this perspective, in order to widen the "world" of applications of the model, the notion of thermostat was recently introduced [14–17] to describe e.g. a crowd in a place hall compelled to go out by an emergency (the external force) but calmed down and organized to move in a suitable order by a security service (the thermostat) [18,19]. Of course, this extended model presents the same mathematical problems of all the models of the same kind (existence and uniqueness of solutions and their continuous dependence on initial and boundary conditions). Existence and uniqueness results were given in [18,20] (for the case in which the variable u representing the state of any particle is continuous) and [21,22] (for the discrete case). This paper offers a result about contiunous dependence for the continuous model.

The matter of the paper is distributed as follows: in Section 2 a short description of the problem is given, with the explicit statement of the equation governing the evolution of a general thermostatted complex system; the statement of our result about the continuous dependence of solutions to the evolution equation on the initial data is given in Section 3, and its proof in Section 4; finally, Section 5 is devoted to point out the reasons why the result stated and proved in the previous sections must be just considered as the first step of a further research, as well as to draw the perspectives of such reasearch.

2. The Framework

Let $\mathcal{C}$ be a complex system [13,23–25] composed by a large number of objects, called active particles [13], which can interact with each other. The system is assumed to be homogeneous with respect to mechanical variables (position and velocity). The state of each of its particles will be then described, at any time $t \geq 0$, by only one scalar variable $u \in D_u \subseteq \mathbb{R}$, called activity variable, which will be denoted by u. According to the statistical (or stochastic) scheme adopted to describe complex systems (see Introduction), the states of single particles cannot be known at any time, and only a probability (or relative frequency) distribution on the set D_u of all possible states (values of u) can be considered. This distribution function is

$$f(t, u) \in [0, +\infty[\times D_u \to \mathbb{R}^+.$$

Interactions between particles are described by the following two functions:

- a function $\eta(u_*, u^*) : D_u \times D_u \to \mathbb{R}^+$, the interaction rate between particles in the state u_* and particles in the state u^*;
- a function $\mathcal{A}(u_*, u^*, u) : D_u \times D_u \times D_u \to \mathbb{R}^+$, the transition probability density that a particle in the state u_* falls into the state u after interacting with a particle in the state u^*, such that for all $u_*, u^* \in D_u$:

$$\int_{D_u} \mathcal{A}(u_*, u^*, u) \, du = 1.$$

Let now $p \in \mathbb{N}$. The p-th order moment of the system, a time $t > 0$, is:

$$\mathbb{E}_p[f](t) := \int_{D_u} u^p f(t, u) \, du.$$

We assume that the system $\mathcal{C}$ is subject to an external force field acting on it, described by a function $F(u) : D_u \to \mathbb{R}^+$. Moreover, a thermostat is applied on $\mathcal{C}$, in order to keep the second order moment

$$\mathbb{E}_2[f](t) = \int_{D_u} u^2 f(t, u) \, du$$

constant [18].

Accordingly, the evolution equation for the thermostatted framework of the system takes the form

$$\partial_t f(t, u) + \partial_u \left(F(u) \left(1 - u \mathbb{E}_1[f](t) \right) f(t, u) \right) = J[f, f](t, u), \tag{1}$$

where the operator $J[f, f](t, u)$ can be split as follows:

$$J[f, f](t, u) = G[f, f](t, u) - L[f, f](t, u)$$
$$= \int_{D_u \times D_u} \eta(u_*, u^*) \, \mathcal{A}(u_*, u^*, u) \, f(t, u_*) \, f(t, u^*) \, du_* du^*$$
$$- f(t, u) \int_{D_u} \eta(u, u^*) \, f(t, u^*) \, du^*.$$

The term $G[f, f,](t, u)$ is called the gain operator, and the term $L[f, f](t, u)$ is the loss operator.

Given an initial datum $f^0(u)$, from Equation (1) we obtain the Cauchy problem related to the thermostatted:

$$\begin{cases} \partial_t f(t, u) + \partial_u \left(F(u) \left(1 - u \mathbb{E}_1[f](t) \right) f(t, u) \right) = J[f, f](t, u) & [0, +\infty[\times D_u \\ f(0, u) = f^0(u) & D_u \end{cases}. \tag{2}$$

Let now

$$\mathcal{K}(D_u) := \{ f(t, u) \in [0, +\infty[\times D_u \to \mathbb{R}^+ : \mathbb{E}_0[f](t) = \mathbb{E}_2[f](t) = 1 \}$$

and assume that

Assumption 1 (H1): for all $u_*, u^* \in D_u$:

$$\int_{D_u} \mathcal{A}(u_*, u^*, u) \, u^2 \, du = u_*^2;$$

Assumption 2 (H2): the interaction function is constant, i.e., there exists an $\eta > 0$ such that $\eta(u_*, u^*) = \eta$, for all $u_*, u^* \in D_u$;

Assumption 3 (H3): the distribution function vanishes on the boundary of D_u, i.e. $f(t, u) = 0$ for $u \in \partial D_u$;

Assumption 4 (H4): there exists $F > 0$ such that $F(t) = F$, for all $t > 0$;

Assumption 5 (H5): the initial data verifies the conditions $\mathbb{E}_0[f^0] = \mathbb{E}_2[f^0] = 1$.

If assumptions H1–H5 are satisfied, then there exists a unique solution $f(t,u) \in C\left((0,+\infty); L^1(D_u)\right) \cap \mathcal{K}(D_u)$ to the Cauchy problem (2) [18].

Lemma 1. *[18] Under assumptions H1–H5, the relations*

1.
$$G[f,f] - G[g,g] = G[f-g,f] + G[g,f-g],$$

2.
$$\int_{D_u} G[f,f](t,u)\, du = \eta,$$

3.
$$\int_{D_u} u\, G[f,f](t,u)\, du = 0,$$

4.
$$\int_{D_u} u^2\, G[f,f](t,u)\, du = \eta$$

are true.

Now, the evolution equation of $\mathbb{E}_1[f](t)$ reads, [18]:

$$\mathbb{E}_1'[f](t) = F(1 - (\mathbb{E}_1[f](t))^2) - \eta \mathbb{E}_1[f](t).$$

Lemma 2. *[18] Under the assumptions H1–H5:*

$$\mathbb{E}_1[f](t) = \frac{\mathbb{E}_1^+(\mathbb{E}_1^- - \mathbb{E}_1^0) - \mathbb{E}_1^-(\mathbb{E}_1^+ - \mathbb{E}_1^0) e^{-\frac{\sqrt{\eta^2+4F^2}}{F}t}}{(\mathbb{E}_1^- - \mathbb{E}_1^0) - (\mathbb{E}_1^+ - \mathbb{E}_1^0) e^{-\frac{\sqrt{\eta^2+4F^2}}{F}t}},$$

where $\mathbb{E}_1^0 := \mathbb{E}_1[f^0]$ and $\mathbb{E}_1^\pm := \frac{-\eta \pm \sqrt{\eta^2+4F^2}}{2F}$.

Furthermore, the following result holds.

Lemma 3. *If assumptions H1–H5 are satisfied, then*

$$\mathbb{E}_1[f](t) \to \mathbb{E}_1^+, \text{ as } t \to +\infty,$$

and

$$|\mathbb{E}_1[f](t) - \mathbb{E}_1^+| \leq C e^{-\frac{\sqrt{\eta^2+4F^2}}{F}t},$$

where $C = C(\eta, F)$ is a constant depending on the system.

Remark 1. *Assumptions H2 and H4 can be relaxed as follows: there exist $\eta > 0$ and $F > 0$ such that*

$$\eta(u_*, u^*) \leq \eta,$$

for all $u_*, u^* \in D_u$, and

$$F(t) \leq F,$$

for all $t > 0$.

3. Dependence on the Initial Data

The main aim of this paper is to prove a result about the dependence on the initial data of the continuous thermostatted framework (1) that defines the related Cauchy problem (2).

Let $f_1(t,u)$ and $f_2(t,u)$ be two solutions to the Cauchy problem (2) corresponding to the initial data $f_1^0(u)$ and $f_2^0(u)$ respectively. If assumptions H1–H5 hold true, then $f_1(t,u)$, $f_2(t,u) \in C\left((0,+\infty); L^1(D_u)\right) \cap \mathcal{K}(D_u)$. Suppose that $f_1^0(u)$ and $f_2^0(u)$ belong to $L^1(D_u)$.

Theorem 1 obtains an estimate of the norm

$$\|f_1(t,u) - f_2(t,u)\|_{C\left((0,+\infty); L^1(D_u)\right) \cap \mathcal{K}(D_u)}$$
$$= \max_{t \in [0,T]} \left(\int_{D_u} |f_1(t,u) - f_2(t,u)| \, du \right),$$

when

$$\left\| f_1^0(u) - f_2^0(u) \right\|_{L^1(D_u)} = \int_{D_u} \left| f_1^0(u) - f_2^0(u) \right| du \leq \delta,$$

where $\delta > 0$. Precisely,

Theorem 1. *Assume that conditions **H1–H5** are verified. If*

$$\left\| f_1^0(u) - f_2^0(u) \right\|_{L^1(D_u)} \leq \delta,$$

for $\delta > 0$, then, for all $T > 0$

$$\|f_1(t,u) - f_2(t,u)\|_{C\left([0,T]; L^1(D_u)\right) \cap \mathcal{K}(D_u)}$$
$$\leq \delta e^{\tilde{C} T},$$

where $\tilde{C} := \left(3\eta + 2F\left(C(\eta, F) + \mathbb{E}_1^+\right)\right)$, and $C(\eta, F)$ is the constant of Lemma 1.

This result ensures that the dependence of solution to the continuous thermostatted framework (2) on the initial data is continuous, so that any solution to the Cauchy problem (2) is stable in the Hadamard sense.

4. Proof of the Result

The aim of this section is to give the proof of Theorem 1.

Proof of Theorem 1. Let $f_1(t,u)$ and $f_2(t,u)$ be two solutions to the Cauchy problem (2) belonging to the space $C\left([0,T]; L^1(D_u)\right) \cap \mathcal{K}(D_u)$, corresponding to the initial data $f_1^0(u)$ and $f_2^0(u)$ in the space $L^1(D_u)$, respectively.

Then, Equation (1) can be written:

$$\partial_t f(t,u) = J[f,f](t,u) - F \partial_u \left((1 - u \mathbb{E}_1[f](t)) f(t,u) \right). \tag{3}$$

Integrating Equation (3) between 0 and t, for $t > 0$, one has:

$$\int_0^t \partial_t f(\tau, u) \, d\tau = \int_0^t J[f, f](\tau, u) \, d\tau - \int_0^t F \partial_u \left((1 - u\mathbb{E}_1[f](\tau))f(\tau, u)\right) d\tau,$$

and

$$f(t, u) = f^0(u) + \int_0^t J[f, f](\tau, u) \, d\tau - F \int_0^t \partial_u \left((1 - u\mathbb{E}_1[f](\tau))f(\tau, u)\right) d\tau. \tag{4}$$

Bearing in mind the initial data $f_1^0(u)$ and $f_2^0(u)$ and using relation (4), we obtain

$$f_1(t, u) = f_1^0(u) + \int_0^t J[f_1, f_1](\tau, u) \, d\tau - F \int_0^t \partial_u \left((1 - u\mathbb{E}_1[f_1](\tau))f_1(\tau, u)\right) d\tau, \tag{5}$$

and

$$f_2(t, u) = f_2^0(u) + \int_0^t J[f_2, f_2](\tau, u) \, d\tau - F \int_0^t \partial_u \left((1 - u\mathbb{E}_1[f_2](\tau))f_2(\tau, u)\right) d\tau, \tag{6}$$

so that, subtracting the (5) and (6), we find

$$\begin{aligned} f_1(t, u) - f_2(t, u) &= \left(f_1^0(u) - f_2^0(u)\right) \\ &+ \int_0^t \left(J[f_1, f_1](\tau, u) - J[f_2, f_2](\tau, u)\right) d\tau \\ &+ F \int_0^t \partial_u \left((1 - u\mathbb{E}_1[f_2](\tau))f_2(\tau, u)\right) d\tau \\ &- F \int_0^t \partial_u \left((1 - u\mathbb{E}_1[f_1](\tau))f_1(\tau, u)\right), d\tau, \end{aligned} \tag{7}$$

In virtue of Lemma 1:

$$\begin{aligned} J[f_1, f_1](\tau, u) - J[f_2, f_2](\tau, u) &= G[f_1, f_1](\tau, u) - \eta f_1(\tau, u) \\ &\quad - G[f_2, f_2](\tau, u) + \eta f_2(\tau, u) \\ &= G[f_1 - f_2, f_1](\tau, u) \\ &\quad + G[f_2, f_1 - f_2](\tau, u) \\ &\quad + \eta \left(f_2(\tau, u) - f_1(\tau, u)\right). \end{aligned} \tag{8}$$

Furthermore:

$$\begin{aligned} &\partial_u \left((1 - u\mathbb{E}_1[f_2](\tau))f_2(\tau) - (1 - u\mathbb{E}_1[f_1](\tau))f_1(\tau)\right) \\ &= \partial_u \left((f_2(\tau, u) - f_1(\tau, u)) - u\mathbb{E}_1[f_2](\tau)f_2(\tau, u) + u\mathbb{E}_1[f_1](\tau)f_1(\tau, u)\right) \\ &= \partial_u (f_2(\tau, u) - f_1(\tau, u)) - \partial_u \left(u(\mathbb{E}_1[f_2](\tau)f_2(\tau, u) - \mathbb{E}_1[f_1](\tau)f_1(\tau, u))\right). \end{aligned} \tag{9}$$

In virtue of relations (8) and (9), Equation (7) may be rewritten in the form

$$f_1(t,u) - f_2(t,u) = \left(f_1^0(u) - f_2^0(u)\right)$$
$$+ \int_0^t \left(G[f_1 - f_2, f_1](\tau, u) + G[f_2, f_1 - f_2](\tau, u)\right) d\tau \quad (10)$$
$$+ \eta \int_0^t (f_2(\tau, u) - f_1(\tau, u)) \, d\tau + F \int_0^t \partial_u (f_2(\tau, u) - f_1(\tau, u)) \, d\tau$$
$$- F \int_0^t \partial_u \left(u(\mathbb{E}_1[f_2](\tau) f_2(\tau, u) - \mathbb{E}_1[f_1](\tau) f_1(\tau, u))\right) d\tau,$$

and the application of the triangle inequality to Equation (10) and an integration on D_u lead to

$$\int_{D_u} |f_1(t,u) - f_2(t,u)| \, du \leq \int_{D_u} \left|f_1^0(u) - f_2^0(u)\right| du$$
$$+ \int_{D_u} \left|\int_0^t G[f_1 - f_2, f_1](\tau, u) + G[f_2, f_1 - f_2](\tau, u) \, d\tau\right| du$$
$$+ \eta \int_0^t \left(\int_{D_u} |f_2(\tau, u) - f_1(\tau, u)| \, du\right) d\tau \quad (11)$$
$$+ F \int_{D_u} \left|\int_0^t \partial_u (f_2(\tau, u) - f_1(\tau, u)) \, d\tau\right| du$$
$$+ F \int_{D_u} \left|\int_0^t \partial_u (u(\mathbb{E}_1[f_2](\tau) f_2(\tau, u) - \mathbb{E}_1[f_1](\tau) f_1(\tau, u))) \, d\tau\right| du.$$

To conclude the proof, we only need now to estimate the five terms at the right hand side of Equation (11).

First of all, by the assumptions of the theorem,

$$\int_{D_u} \left|f_1^0(u) - f_2^0(u)\right| du \leq \delta, \quad (12)$$

so that, by straightforward calculations, we obtain for the second term at the right hand side of inequality (11) the following estimate:

$$\int_{D_u} \left|\int_0^t G[f_1 - f_2, f_1](\tau, u) + G[f_2, f_1 - f_2](\tau, u) \, d\tau\right| du$$
$$\leq \int_{D_u} \int_0^t \int_{D_u \times D_u} |\eta \mathcal{A}(u_*, u^*, u) (f_1(\tau, u_*) - f_2(\tau, u_*)) f_1(\tau, u^*)| \, du_* \, du^* \, du \, d\tau$$
$$+ \int_{D_u} \int_0^t \int_{D_u \times D_u} |\eta \mathcal{A}(u_*, u^*, u) (f_1(\tau, u^*) - f_2(\tau, u^*)) f_2(\tau, u_*)| \, du_* \, du^* \, du \, d\tau \quad (13)$$
$$\leq 2\eta \int_{D_u} \left(\int_0^t |f_1(\tau, u) - f_2(\tau, u)| \, d\tau\right) du$$
$$\leq 2\eta \int_0^t \left(\int_{D_u} |f_2(\tau, u) - f_1(\tau, u)| \, du\right) d\tau.$$

Next, since

$$F \int_{D_u} \left|\int_0^t \partial_u (f_2(\tau, u) - f_1(\tau, u)) \, d\tau\right| du$$
$$\leq F \int_0^t \left(\int_{D_u} |\partial_u (f_2(\tau, u) - f_1(\tau, u))| \, du\right) d\tau,$$

and $f(t, u) = 0$ for $u \in \partial D_u$, the fourth term at the right hand side of inequality (11) vanishes.

Furthermore, the fifth term at the right hand side of relation (11) is easily estimated as follows:

$$F \int_{D_u} \left| \int_0^t \partial_u \left(u(\mathbb{E}_1[f_2](\tau) f_2(\tau, u) - \mathbb{E}_1[f_1](\tau) f_1(\tau, u)) \right) d\tau \right| du$$
$$\leq F \int_{D_u} \int_0^t |\mathbb{E}_1[f_2](\tau) f_2(\tau, u) - \mathbb{E}_1[f_1](\tau) f_1(\tau, u)| \, d\tau \, du \quad (14)$$
$$+ F \int_{D_u} \int_0^t |u \, \partial_u \left(\mathbb{E}_1[f_2](\tau) f_2(\tau, u) - \mathbb{E}_1[f_1](\tau) f_1(\tau, u) \right)| \, d\tau \, du.$$

Now, integrating by parts the second term of the right hand side of relation (14) and bearing in mind that $f(t, u) = 0$ for $u \in \partial D_u$, one has

$$F \int_{D_u} \left| \int_0^t \partial_u \left(u(\mathbb{E}_1[f_2](\tau) f_2(\tau, u) - \mathbb{E}_1[f_1](\tau) f_1(\tau, u)) \right) d\tau \right| du$$
$$\leq 2F \int_{D_u} \int_0^t |\mathbb{E}_1[f_2](\tau) f_2(\tau, u) - \mathbb{E}_1[f_1](\tau) f_1(\tau, u)| \, d\tau \, du. \quad (15)$$

By adding and subtracting $\mathbb{E}_1^+ f_2(\tau, u)$ and $\mathbb{E}_1^+ f_1(\tau, u)$ at the right hand side of relation (15) and using Lemma (2), it follows that

$$F \int_{D_u} \left| \int_0^t \partial_u \left(u(\mathbb{E}_1[f_2](\tau) f_2(\tau, u) - \mathbb{E}_1[f_1](\tau) f_1(\tau, u)) \right) d\tau \right| du$$
$$\leq 2F \left[\int_{D_u} \int_0^t \left| f_2(\tau, u) \left(\mathbb{E}_1[f_2](\tau) - \mathbb{E}_1^+ \right) \right. \right.$$
$$\left. - f_1(\tau, u) \left(\mathbb{E}_1[f_1](\tau) - \mathbb{E}_1^+ \right) \right| d\tau \, du$$
$$+ \mathbb{E}_1^+ \int_0^t \left(\int_{D_u} |f_2(\tau, u) - f_1(\tau, u)| \, du \right) d\tau \right] \quad (16)$$
$$\leq 2F \left[C(\eta, F) \int_{D_u} \int_0^t e^{-\frac{\sqrt{\eta^2 + 4F^2}}{F} t} |f_2(\tau, u) - f_1(\tau, u)| \, d\tau \, du \right.$$
$$\left. + \mathbb{E}_1^+ \int_0^t \left(\int_{D_u} |f_2(\tau, u) - f_1(\tau, u)| \, du \right) d\tau \right]$$
$$\leq 2F \left(C(\eta, F) + \mathbb{E}_1^+ \right) \int_0^t \left(\int_{D_u} |f_2(\tau, u) - f_1(\tau, u)| \, du \right) d\tau.$$

By relations (12), (13), (16) and the condition $f(t, u) = 0$ for $u \in \partial D_u$, inequality (11) becomes

$$\int_{D_u} |f_1(t, u) - f_2(t, u)| \, du \leq \delta$$
$$+ 2\eta \int_0^t \left(\int_{D_u} |f_2(\tau, u) - f_1(\tau, u)| \, du \right) d\tau$$
$$+ \eta \int_0^t \left(\int_{D_u} |f_2(\tau, u) - f_1(\tau, u)| \, du \right) d\tau \quad (17)$$
$$+ 2F \left(C(\eta, F) + \mathbb{E}_1^+ \right) \int_0^t \left(\int_{D_u} |f_2(\tau, u) - f_1(\tau, u)| \, du \right) d\tau,$$

and, by reordering the terms of this last relation and using the constant

$$\bar{C} := \left(3\eta + 2F \left(C(\eta, F) + \mathbb{E}_1^+ \right) \right),$$

we get

$$\int_{D_u} |f_1(t,u) - f_2(t,u)|\, du \tag{18}$$
$$\leq \delta + \bar{C} \int_0^t \left(\int_{D_u} |f_2(\tau,u) - f_1(\tau,u)|\, du \right) d\tau.$$

Finally, the Gronwall inequality [26] and inequality (15) yield

$$\int_{D_u} |f_1(t,u) - f_2(t,u)|\, du \leq \delta\, e^{\bar{C} t}, \tag{19}$$

for all $t > 0$, so that, for $T > 0$,

$$\max_{t \in [0,T]} \left(\int_{D_u} |f_1(t,u) - f_2(t,u)|\, du \right) \leq \delta\, e^{\bar{C} t},$$

and our claim is proved. □

5. Concluding Remarks and Research Perspectives

As already pointed out at the end of Section 3, the result proved in the foregoing section can be expressed by the statement that the solutions to the Cauchy problem (2) are stable in the Hadamard sense. As it stands, this result could seem to be not quite satisfactory, and the question spontaneously arises on whether these solutions can be proved to be stable or unstable in the whole interval $(0, +\infty)$, and whether the answer to such a question could be find by some suitable technical refinements of the method of proof used in the previous section. Several observations witness for a negative answer to this last question, but a simple heuristic reasoning—based on the acknowledged versatility of the scheme expressed by problem (2)—will be sufficient. As said in the Introduction, and shown in the references quoted there, the scheme applies to many different kinds of systems, in particular to social and economic systems. Our experience shows that slightly different distributions of wealth can evolve in time in very similar or very different ways depending on the behaviour of the particles of each system, i.e. on the terms in $J[f,f](t,u)$ describing their interaction rates and the probable effects of such interactions. The relations between these terms should play an important role in producing stability or instability: for instance, frequent interactions could produce instability when joined with strongly asymmetric transition probability distributions, but even uniform stability in the whole interval $(0, +\infty)$ when the transition probabilities are symmetric for any pair of initial states of interacting individuals. Thus, one of the first goals of subsequent researches about the problem considered in the present paper should be to check whether different assumptions on the forms of $\eta(u_*, u^*)$ and $\mathcal{A}(u_*, u^*, u)$ give stronger but different results about stability or instability.

In this connection, a very important step of subsequent research will be the numerical formulation and some numerical simulations of its solutions. In particular, problem (2) should be formulated in particular contexts, like the one of socio–economic problems, and the interaction rate $\eta(u_*, u^*)$ and the probability distribution $\mathcal{A}(u_*, u^*, u)$ on the effects of interactions should be assigned according to the results of previous statistical researches and the laws of micro-economy.

Furthermore, the choice of the distance

$$d(f_1, f_2) = \int_{D_u} |f_1(t,u) - f_2(t,u)|\, du$$

in the space of solutions of the Cauchy problem (2) can be improved, as it is not so significant from a statistical viewpoint. For instance, if f_1 is a uniform probability density, it tells us that the deviations of f_2 from uniformity are, in some sense, "small", but says nothing about the symmetry (or asymmetry) of these deviations. From a statistical viewpoint, we are interested to know whether such deviations are "large" but concentrated in a small subset of D_u, or "small" and spread over D_u. So, the analysis must be concentrated on a comparison of the derivatives $\partial_u f_1$ and $\partial_u f_2$.

Finally, it is of the greatest relevance to study continuous dependence on initial data in the discrete case, mainly in order to check it also by means of numerical simulations. This must be a preliminary step to the statistical studies proposed above.

Author Contributions: Conceptualization, B.C. and M.M.; methodology, B.C. and M.M.; formal analysis, B.C. and M.M.; investigation, B.C. and M.M.; resources, B.C. and M.M.; writing—original draft preparation, B.C. and M.M.; writing—review and editing, B.C. and M.M.; supervision, M.M.

Funding: This research received no external funding

Conflicts of Interest: The authors declare no conflict of interest.

References

1. Cercignani, C. *The Boltzmann Equation and Its Applications*; Springer: Berlin/Heidelberg, Germany, 1988.
2. Arkeryd, L. On the Boltzmann equation. *Arch. Ration. Mech. Anal.* **1972**, *45*, 1–16. [CrossRef]
3. Bianca, C.; Brézin, L. Modeling the antigen recognition by B-cell and T-cell receptors through thermostatted kinetic theory methods. *Int. J. Biomath.* **2017**, *10*, 1750072. [CrossRef]
4. Masurel, L.; Bianca, C.; Lemarchand, A. On the learning control effects in the cancer-immune system competition. *Phys. A Stat. Mech. Appl.* **2018**, *506*, 462–475. [CrossRef]
5. Bertotti, M.L. Modelling taxation and redistribution: A discrete active particle kinetic approach. *Appl. Math. Comput.* **2010**, *217*, 752–762. [CrossRef]
6. Bertotti, M.L.; Modanese, G. Microscopic Models for Welfare Measures addressing a Reduction of Economic Inequality. *Complexity* **2016**, *21*, 89–98. [CrossRef]
7. Iannini, M.L.L.; Dickman, R. Kinetic theory of vehicular traffic. *Am. J. Phys.* **2016**, *84*, 135–145. [CrossRef]
8. Carrillo, J.A.; D'Orsogna, M.R.; Panferov, V. Double milling in self-propelled swarms from kinetic theory. *Kinet. Relat. Mod.* **2009**, *2* 363–378. [CrossRef]
9. Bellomo, N.; Dogbe, C. On the modelling crowd dynamics from scaling to hyperbolic macroscopic models. *Math. Models Methods Appl. Sci.* **2008**, *18* (Suppl. 1), 1317–1345. [CrossRef]
10. Cristiani, E.; Piccoli, B.; Tosin, A. Multiscale modeling of granular flows with application to crowd dynamics. *Multiscale Model. Simul.* **2011**, *9* 155–182. [CrossRef]
11. Bianca, C.; Mogno, C. Modelling pedestrian dynamics into a metro station by thermostatted kinetic theory methods. *Math. Comput. Model. Dyn. Syst.* **2018**, *24*, 207–235. [CrossRef]
12. Kacperski, K. Opinion formation model with strong leader and external impact: A mean field approach. *Phys. A Stat. Mech. Appl.* **1999**, *269*, 511–526. [CrossRef]
13. Bellomo, N.; Carbonaro, B. Toward a mathematical theory of living systems focusing on developmental biology and evolution: A review and perspectives. *Phys. Life Rev.* **2011**, *8*, 1–18. [CrossRef] [PubMed]
14. Bianca, C. Thermostatted kinetic equations as models for complex systems in physics and life sciences. *Phys. Life Rev.* **2012**, *9*, 359–399. [CrossRef]
15. Evans, D.J.; Morriss, G.P. *Statistical Mechanics of Nonequilibrium Liquids*; Cambridge University Press: Cambridge, UK, 2008.
16. Morriss, G.P.; Dettmann, C.P. Thermostats: Analysis and application. *Chaos* **1998**, *8*, 321–336. [CrossRef]
17. Wennberg, B.; Wondmagegne, Y. The Kac equation with a thermostatted force field. *J. Stat. Phys.* **2006**, *124*, 859–880. [CrossRef]
18. Bianca, C. An existence and uniqueness theorem to the Cauchy problem for thermostatted-KTAP models. *Int. J. Math. Anal.* **2012**, *6*, 813–824.
19. Bianca, C. Kinetic theory for active particles modelling coupled to Gaussian thermostats. *Appl. Math. Sci.* **2012**, *6*, 651–660.
20. Bianca, C. Existence of stationary solutions in kinetic models with Gaussian thermostats. *Math. Methods Appl. Sci.* **2013**, *36*, 1768–1775. [CrossRef]
21. Bianca, C.; Mogno, C. Qualitative analysis of a discrete thermostatted kinetic framework modeling complex adaptive systems. *Commun. Nonlinear Sci. Numer. Simul.* **2018**, *54*, 221–232. [CrossRef]

22. Bianca, C.; Menale, M. Existence and uniqueness of nonequilibrium stationary solutions in discrete thermostatted models. *Commun. Nonlinear Sci. Numer. Simul.* **2019**, *73*, 25–34. [CrossRef]
23. Bar-Yam, Y. *Dynamics of Complex Systems*; Addison-Wesley: Reading, MA, USA, 1997; Volume 213.
24. Holland, J.H. Studying complex adaptive systems. *J. Syst. Sci. Complex.* **2006**, *19*, 1–8. [CrossRef]
25. Ladyman, J.; Lambert, J.; Wiesner, K. What is a complex system? *Eur. J. Philos. Sci.* **2013**, *3*, 33–67. [CrossRef]
26. Walter, W. *Differential and Integral Inequalities*; Springer Science & Business Media: Berlin/Heidelberg, Germany, 2012; Volume 55.

© 2019 by the authors. Licensee MDPI, Basel, Switzerland. This article is an open access article distributed under the terms and conditions of the Creative Commons Attribution (CC BY) license (http://creativecommons.org/licenses/by/4.0/).

Article

Fixed-Point Results for a Generalized Almost (s,q)−Jaggi F-Contraction-Type on b−Metric-Like Spaces

Hasanen A. Hammad [1] and Manuel De la Sen [2],*

1. Department of Mathematics, Faculty of Science, Sohag University, 82524 Sohag, Egypt; h_elmagd89@yahoo.com
2. Institute of Research and Development of Processes, University of the Basque Country, 48940 Leioa (Bizkaia), Spain
* Correspondence: manuel.delasen@ehu.eus; Tel.: +34-94-601-2548

Received: 15 October 2019; Accepted: 17 December 2019; Published: 2 January 2020

Abstract: The purpose of this article is to present a new generalized almost (s,q)−Jaggi F−contraction-type and a generalized almost (s,q)−Jaggi F−Suzuki contraction-type and some results in related fixed point on it in the context of b−metric-like spaces are discussed. Also, we support our theoretical results with non-trivial examples. Finally, applications to find a solution for the electric circuit equation and second-order differential equations are presented and an strong example is given here to support the first application.

Keywords: electric circuit equations; wardoski contraction; almost (s,q)−Jaggi-type; b−metric-like spaces; second-order differential equations

MSC: 47H10; 54H25

1. Introduction

Mathematical models can take many forms, including dynamical systems, statistical models, differential equations and game theoretic models and real world problems. In various branches of mathematics, The existence of solution for these matters has been checked, for example, differential equations, integral equations, functional analysis, etc. Fixed point technique is one of these methods to find the solution of these problems. So this technique has many applications not only is limited to mathematics but also occurs in various sciences, such that, economics, biology, chemistry, computer science, physics, etc. More clearly, for example, In economics, this technique is applied to find the solution of the equilibrium problem in game theory.

Problems in the nonlinear analysis are solved by a popular tool called Banach contraction principle. This principle appeared in Banach's thesis [1], where it was used in proving the existence and uniqueness of solution of integral equations, it stated as: A nonlinear self mapping Γ on a metric space (Ω, d) is called a Banach contraction if there exists $\delta \in [0, 1)$ such that

$$d(\Gamma \kappa, \Gamma \mu) \leq \delta d(\kappa, \mu), \forall \kappa, \mu \in \Omega. \tag{1}$$

Notice that the contractive condition (1) is satisfied for all $\kappa, \mu \in \Omega$ which forces the mapping Γ to be continuous, while it is not applicable in case of discontinuity. In view of the applicability of contraction principle this is the major draw-back of this principle. Many authors attempted to overcome this drawback (see, for example [2–4]).

In 1989, one of the interesting generalizations of this basic principle was given by Bakhtin [5] (and also Czerwik [6], 1993) by introducing the concept of $b-$metric spaces. For fixed point results in $b-$metric spaces. See [7–14].

In 2010, the concept of a $b-$metric-like initiated by Alghamdi et al. [9] as an extension of a $b-$metric. They studied some related fixed point consequences concerning with this space. Recently, many contributions on fixed points results via certain contractive conditions in mentioned spaces are made (for example, see [15–20]).

In 2012, a new contraction called $F-$ contraction-type is presented by Wardowski [21], where $F : R^+ \to R$. By this style, recent fixed point results and strong examples to obtain a different type of contractions are discussed.

Definition 1 ([21]). *A mapping $\Gamma : \Omega \longrightarrow \Omega$ defined on a metric space (Ω, d), is called an $F-$contraction if there is $F \in \Sigma$ and $\tau > 0$ such that*

$$d(\Gamma\kappa, \Gamma\mu) > 0 \text{ implies } \tau + F(d(\Gamma\kappa, \Gamma\mu)) \leq F(d(\kappa, \mu)), \forall \kappa, \mu \in \Omega, \quad (2)$$

where Σ is the set of functions $F : (0, +\infty) \to \mathbb{R}$ satisfying the following assumptions:
($\Im_1$) *F is strictly increasing, i.e., for all $a, b \in \mathbb{R}^+$ such that $a < b$, $F(a) < F(b)$;*
($\Im_2$) *For every sequence $\{a_n\}_{n \in \mathbb{N}}$ of positive numbers, $\lim_{n \to \infty} a_n = 0$ iff $\lim_{n \to \infty} F(a_n) = -\infty$;*
($\Im_3$) *There exists $\mu \in (0, 1)$ such that $\lim_{a \to 0^+} a^\mu F(a) = 0$.*

The following functions $F_a : (0, +\infty) \longrightarrow \mathbb{R}$ for $a \in \{1, 2, 3, 4\}$, are the elements of Σ. Furthermore, substituting these functions in (2), Wardowski obtained the following contractions:

(1) $F_1(\theta) = \ln(\theta)$, $\frac{d(\Gamma\kappa, \Gamma\mu)}{d(\kappa, \mu)} \leq e^{-\tau}$,

(2) $F_2(\theta) = \ln(\theta) + \theta$, $\frac{d(\Gamma\kappa, \Gamma\mu)}{d(\kappa, \mu)} \leq e^{-\tau + d(\kappa, \mu) - d(\Gamma\kappa, \Gamma\mu)}$,

(3) $F_3(\theta) = \frac{-1}{\sqrt{\theta}}$, $\frac{d(\Gamma\kappa, \Gamma\mu)}{d(\kappa, \mu)} \leq \frac{1}{\left(1 + \tau\sqrt{d(\kappa,\mu)}\right)^2}$,

(4) $F_4(\theta) = \ln(\theta^2 + \theta)$, $\frac{d(\Gamma\kappa, \Gamma\mu)(1 + d(\Gamma\kappa, \Gamma\mu))}{d(\kappa, \mu)(1 + d(\kappa, \mu))} \leq e^{-\tau}$.

for all $\kappa, \mu \in \Omega$ with $\theta > 0$ and $\Gamma\kappa \neq \Gamma\mu$.

Remark 1. *It follows from (2) that*

$$d(\Gamma\kappa, \Gamma\mu) < d(\kappa, \mu), \text{ for all } \kappa, \mu \in \Omega,$$

this means that Γ is contractive with $\Gamma\kappa \neq \Gamma\mu$. Hence, if the mapping is $F-$contraction, then it continuous.

Remark 2 ([22]). *For $p > 1$ and $\theta > 0$, the function $F(\theta) = \frac{-1}{\sqrt[p]{\theta}}$ belong to Σ.*

In a different way to generalize the Banach contraction principle, Wardowski [21] established the following theorem:

Theorem 1 ([21]). *Suppose that (Ω, d) is a complete metric space and Γ is a self-mapping on it satisfying the condition (2). Then there exists a unique fixed point κ^* of Γ. As well as, the sequence $\{\Gamma^n \kappa_\circ\}_{n \in \mathbb{N}}$ is convergent to κ^*, for any $\kappa_\circ \in \Omega$.*

The Wardowski-contraction is extended by many authors such as Abbas et al. [23] to give certain fixed point results, Batra et al. [24,25], to generalize it on graphs and alter distances, and Cosentino and Vetro [26] to introduce some fixed point consequences for Hardy-Rogers-type self-mappings in ordered and complete metric spaces.

In 2014, some fixed point consequences proved via the notion of an $F-$Suzuki contraction by Piri and Kumam [27]. This concept is stated as follows:

Definition 2 ([27]). *Let (Ω, d) be a complete metric space and a mapping $\Gamma: \Omega \to \Omega$ is called $F-$Suzuki contraction if there exists $F \in \Sigma$, and $\tau > 0$ such that*

$$\frac{1}{2}d(\kappa, \Gamma\kappa) < d(\kappa, \mu) \Rightarrow \tau + F(d(\Gamma\kappa, \Gamma\mu)) \leq F(d(\kappa, \mu)), \forall \kappa, \mu \in \Omega,$$

with $\Gamma\kappa \neq \Gamma\mu$.

In 1975, Jaggi [28] defined the concept of a generalized Banach contraction principle as follows:

Definition 3. *Let (Ω, d) be a complete metric space. A continuous self-mapping Γ on a set Ω is called Jaggi contraction-type if*

$$d(\Gamma\kappa, \Gamma\mu) \leq \alpha \frac{d(\kappa, \Gamma\kappa).d(\mu, \Gamma\mu)}{d(\kappa, \mu)} + \beta d(\kappa, \mu),$$

for all $\kappa, \mu \in \Omega$, $\kappa \neq \mu$ and for some $\alpha, \beta \in [0, 1)$ with $\alpha + \beta < 1$.

Recently, the same author [29], extended his above result on $b-$metric-like spaces as follows:

Definition 4. *Let (Ω, ϖ) be a $b-$metric-like space with parameter $s \geq 1$. A nonlinear self-mapping Γ on a set Ω is called $(s, q)-$Jaggi contraction type if it satisfies the following condition*

$$s^q \varpi(\Gamma\kappa, \Gamma\mu) \leq \alpha \frac{\varpi(\kappa, \Gamma\kappa).\varpi(\mu, \Gamma\mu)}{\varpi(\kappa, \mu)} + \beta \varpi(\kappa, \mu),$$

for all $\kappa, \mu \in \Omega$, whenever $\varpi(\kappa, \mu) \neq 0$, where $\alpha, \beta \in [0, 1)$ with $\alpha + \beta < 1$ and for some $q \geq 1$.

In addition, Berinde [30] introduced the notion of almost contraction by generalized the Zamfirescu fixed point theorem, his result incorporated as follows:

Definition 5. *Let Γ be a nonlinear self-mapping on a complete metric space (Ω, d). Then it called Ciric almost contraction, if there exists $\delta \in [0, 1)$ and $\exists L \geq 0$ such that*

$$d(\Gamma\kappa, \Gamma\mu) \leq \delta d(\kappa, \mu) + Ld(\mu, \Gamma\kappa), \tag{3}$$

for all $\kappa, \mu \in \Omega$, $\kappa \neq \mu$.

After that, the same author [31] extended the contraction (3) and obtained some related fixed point results on complete metric spaces as follows:

Theorem 2. *Let (Ω, d) be a complete metric space and a self-mapping Γ on the set Ω be a Ciric almost contraction, if there exist $\alpha \in [0, 1)$ and $\exists L \geq 0$ such that*

$$d(\Gamma\kappa, \Gamma\mu) \leq \alpha M(\kappa, \mu) + Ld(\mu, \Gamma\kappa) \text{ for all } \kappa, \mu \in \Omega,$$

where $M(\kappa, \mu) = \max\{d(\kappa, \mu), d(\kappa, \Gamma\kappa), d(\mu, \Gamma\mu), d(\kappa, \Gamma\mu), d(\mu, \Gamma\kappa)\}$. Then,

i. there is a non-empty fixed point of the mapping Γ, i.e., $Fix(\Gamma) \neq \emptyset$;

ii. for any $\kappa_\circ = \kappa \in \Omega$, $n \geq 0$ the Picard iteration $\kappa_{n+1} = \Gamma \kappa_n$ converges to $\kappa^* \in Fix(\Gamma)$;

iii. The following estimate holds

$$d(\kappa_n, \kappa^*) \leq \frac{\alpha^n}{(1-\alpha)^2} d(\kappa, \Gamma\kappa),$$

for all $n \geq 1$.

Inspired by Definitions 1, 4 and 5, we introduce a new generalized (s,q)–Jaggi F–contraction-type on the context of b–metric-like spaces as the following:

Definition 6. *Let Γ be a self-mapping on a b–metric-like space (Ω, ϖ) with parameter $s \geq 1$. Then the mapping Γ is said to be generalized (s,q)–Jaggi F contraction-type if there is $F \in \Sigma$ and $\tau > 0$ such that*

$$\varpi(\Gamma\kappa, \Gamma\mu) > 0 \Rightarrow \tau + F(s^q \varpi(\Gamma\kappa, \Gamma\mu)) \leq F\left(\alpha \frac{\varpi(\kappa, \Gamma\kappa) \cdot \varpi(\mu, \Gamma\mu)}{\varpi(\kappa, \mu)} + \beta \varpi(\kappa, \mu) + \gamma \varpi(\mu, \Gamma\kappa)\right), \quad (4)$$

for all $\kappa, \mu \in \Omega$ and $\alpha, \beta, \gamma \geq 0$ with $\alpha + \beta + 2\gamma s < 1$, and for some $q > 1$.

To support our definition, we state the following example:

Example 1. *Let $\Omega = [0, +\infty)$ and $\varpi(\kappa, \mu) = \kappa^2 + \mu^2 + |\kappa - \mu|^2 \, \forall \kappa, \mu \in \Omega$. It's obvious that ϖ is a b–metric like on Ω, with coefficient $s = 2$. Define a nonlinear self-mapping $\Gamma : \Omega \to \Omega$ by $\Gamma\kappa = \frac{1}{16}\ln(1 + \frac{\kappa}{2})$, for all $\kappa \in \Omega$, and the function $F(\theta) = \ln(\theta)$. Consider the constants $q = 2$, $\tau = \ln(2)$, $\alpha = \gamma = 0$ and $\beta = \frac{1}{4}$. So $\alpha + \beta + 2s\gamma = \frac{1}{4} < 1$. Since $t \geq \ln(1+t)$ for each $t \in [0, \infty)$, for all $\kappa, \mu \in \Omega$, we have*

$$\begin{aligned}
\tau + F(s^q \varpi(\Gamma\kappa, \Gamma\mu)) &= \tau + F(s^2 \varpi(\Gamma\kappa, \Gamma\mu)) = \tau + F\left(s^2 \left(T^2\kappa + T^2\mu - |T\kappa - T\mu|^2\right)\right) \\
&= \ln(2) + F\left(4\left(\left(\frac{\ln(1+\frac{\kappa}{2})}{16}\right)^2 + \left(\frac{\ln(1+\frac{\mu}{2})}{16}\right)^2 + \left|\frac{\ln(1+\frac{\kappa}{2})}{16} - \frac{\ln(1+\frac{\mu}{2})}{16}\right|^2\right)\right) \\
&\leq \ln(2) + F\left(\frac{1}{16}\left(\frac{\kappa^2}{4} + \frac{\mu^2}{4} + \left|\frac{\kappa}{4} - \frac{\mu}{4}\right|^2\right)\right) \\
&= \ln(2) + \ln\left(\frac{1}{16}\left(\frac{\kappa^2}{4} + \frac{\mu^2}{4} + \left|\frac{\kappa}{4} - \frac{\mu}{4}\right|^2\right)\right) \\
&\leq \ln\left(\frac{1}{4}\left(\kappa^2 + \mu^2 + |\kappa - \mu|^2\right)\right) \\
&= F\left(\alpha \frac{\varpi(\kappa, \Gamma\kappa) \cdot \varpi(\mu, \Gamma\mu)}{\varpi(\kappa, \mu)} + \beta \varpi(\kappa, \mu) + \gamma \varpi(\mu, \Gamma\kappa)\right).
\end{aligned}$$

Therefore the mapping Γ is a generalized almost (s,q)–Jaggi F–contraction-type.

In this article, we present some related fixed point results for a generalized almost (s,q)–Jaggi F-contraction-type and generalized almost (s,q)–Jaggi F–Suzuki contraction-type on b–metric-like spaces. Also, we give some examples to illustrate these main results. Moreover, applications to find solutions of electric circuit equations and second-order differential equations are discussed and we justify the first application with an example.

2. Preliminaries and Known Results

In the context of this paper, we will use the following notations: $\mathbb{N}, \mathbb{R}, \mathbb{R}^+$ and $\mathbb{Q}$ denotes the set of positive integers, real numbers, nonnegative real numbers and rational numbers, respectively. We begin this part with backgrounds about metric-like and b–metric-like spaces.

Definition 7 ([9]). *Let Ω be a nonempty set. A mapping $\omega : \Omega \times \Omega \to \mathbb{R}^+$ is said to be dislocated (metric-like) if the following three conditions hold for all $\kappa, \mu, \tau \in \Omega$:*
(ω_1) $\omega(\kappa, \mu) = 0 \Rightarrow \kappa = \mu$;
(ω_2) $\omega(\kappa, \mu) = \omega(\mu, \kappa)$;
(ω_3) $\omega(\kappa, \tau) \leq \omega(\kappa, \mu) + \omega(\mu, \tau)$.
In this case, the pair (Ω, ω) is called a dislocated (metric-like) space.

Definition 8 ([32]). *A $b-$dislocated on a nonempty set Ω is a function $\varpi : \Omega \times \Omega \to \mathbb{R}^+$ such that for all $\kappa, \mu, \tau \in \Omega$ and a constant $s \geq 1$, the following three conditions are satisfied:*
(ϖ_1) $\varpi(\kappa, \mu) = 0 \Rightarrow \kappa = \mu$;
(ϖ_2) $\varpi(\kappa, \mu) = \varpi(\mu, \kappa)$;
(ϖ_3) $\varpi(\kappa, \tau) \leq s[\varpi(\kappa, \mu) + \varpi(\mu, \tau)]$.

In this case, the pair (Ω, ϖ) is called a $b-$dislocated (metric-like) space (with constant s).

It should be noted that the class of $b-$metric-like spaces is larger than the class of metric-like spaces, since a $b-$metric-like is a metric-like with $s = 1$.

For new examples in metric-like and $b-$metric-like spaces (see [33,34]).

A $b-$metric-like on Ω satisfies all of the conditions of a metric except that $\varpi(\kappa, \mu)$ may be positive for $\kappa \in \Omega$, so each $b-$metric-like ϖ on Ω generates a topology $\Re_\varpi$ on Ω whose base is the family of open $\varpi-$balls

$$\Re_\varpi(\kappa, \epsilon) = \{\mu \in \Omega : |\varpi(\kappa, \mu) - \varpi(\kappa, \kappa)| < \frac{\epsilon}{s}\},$$

for all $\kappa \in \Omega, s \geq 1$ and $\epsilon > 0$.
According to a topology $\Re_\varpi$, we can present the following results:

Definition 9. *Let (Ω, ϖ) be a $b-$metric-like space and χ be a subset of Ω. We say that χ is a $\varpi-$open subset of Ω, if for all $\kappa \in \Omega$ there exists $\epsilon > 0$ such that $\Re_\varpi(\kappa, \epsilon) \subseteq \chi$. Also $\sigma \subseteq \Omega$ is a $\varpi-$closed subset of Ω if $\Omega \setminus \chi$ is a $\varpi-$open subset in Ω.*

Lemma 1. *Let (Ω, ϖ) be a $b-$metric-like space and σ be a $\varpi-$closed subset of Ω. Let $\{\kappa_n\}$ be a sequence in σ such that $\lim_{n \to \infty} \kappa_n = \kappa$. Then $\kappa \in \sigma$.*

Proof. Let $\kappa \notin \sigma$ by Definition 9, $(\Omega \setminus \sigma)$ is a $\varpi-$open set. Then there exists $\epsilon > 0$ such that $\Re_\varpi(\kappa, \epsilon) \subseteq \Omega \setminus \sigma$. On the other hand, we have $\lim_{n \to \infty} |\varpi(\kappa_n, \kappa) - \varpi(\kappa, \kappa)| = 0$ since $\lim_{n \to \infty} \kappa_n = \kappa$. Hence, there exists $\iota_\circ \in N$ such that

$$|\varpi(\kappa_n, \kappa) - \varpi(\kappa, \kappa)| < \frac{\epsilon}{s},$$

for all $\iota \geq \iota_\circ$. So, we conclude that $\{\kappa_n\} \subseteq \Re_\varpi(\kappa, \epsilon) \subseteq \Omega \setminus \sigma$ for all $\iota \geq \iota_\circ$. This is a contradiction since $\{\kappa_n\} \subseteq \sigma$ for all $\iota_\circ \in N$. □

In a $b-$metric-like space (Ω, ϖ), if $\kappa, \mu \in \Omega$ and $\varpi(\kappa, \mu) = 0$, then $\kappa = \mu$, but the converse is not true in general.

Example 2. *Let $\Omega = \{0, 1, 2, 3, 4\}$ and let*

$$\varpi(\kappa, \mu) = \begin{cases} 5, & \kappa = \mu = 0, \\ \frac{1}{5}, & otherwise. \end{cases}$$

Then (Ω, ϖ) is a $b-$metric-like space with the constant $s = 5$.

Definition 10. Let $\{\kappa_n\}$ be a sequence on a $b-$metric-like space (Ω,ϖ) with a coefficient s. Then
1. If $\lim_{m,n\to\infty}\varpi(\kappa_n,\kappa) = \varpi(\kappa,\kappa)$, then the sequence $\{\kappa_n\}$ is said to be convergent to κ. $\{\kappa_n\}$ is said to be a Cauchy sequence if $\lim_{m,n\to\infty}\varpi(\kappa_m,\kappa_n)$ exists and is finite. The pair (Ω,ϖ) is said to be a complete $b-$metric-like space if for every Cauchy sequence $\{\kappa_n\}$ in Ω, there exists a $\kappa \in \Omega$, such that

$$\lim_{m,n\to\infty}\varpi(\kappa_m,\kappa_n) = \varpi(\kappa,\kappa) = \lim_{n\to\infty}\varpi(\kappa_n,\kappa).$$

2. $\{\kappa_n\}$ in Ω is called a $0-\varpi-$Cauchy sequence if $\lim_{m,n\to\infty}\varpi(\kappa_m,\kappa_n) = 0$. The space (Ω,ϖ) is said to be $0-\varpi-$complete if every $0-\varpi-$Cauchy sequence in Ω converges with respect to $\Re_\varpi$ to a point $\kappa \in \Omega$ such that $\varpi(\kappa,\kappa) = 0$.
3. A nonlinear mapping Γ is continuous on the set Ω, if the following limits

$$\lim_{n\to\infty}\varpi(\kappa_n,\kappa), \quad \lim_{n\to\infty}\varpi(\Gamma\kappa_n,\Gamma\kappa).$$

Existing and equal.

The following example elucidates every $\varpi-$complete $b-$metric-like space is $0-\varpi-$complete but the converse is not true.

Example 3. Let $\Omega = [0,1) \cap \mathbb{Q}$ and $\varpi : \Omega \times \Omega \to [0,+\infty)$ be a function defined by

$$\varpi(\kappa,\mu) = \begin{cases} 2\kappa & \text{if } \kappa = \mu, \\ (\max\{\kappa,\mu\})^2 & \text{otherwise.} \end{cases}$$

$\forall \kappa,\mu \in \Omega$. Then (Ω,ϖ) is a $b-$metric like spaces with a coefficient $s = 2$. Also, if we take a Cauchy sequence $\{\kappa_n\} = \left(\frac{1}{n}\right)_{n\in\mathbb{N}}$, then $\lim_{m,n\to\infty}\varpi(\kappa_m,\kappa_n) = 0$. So $\{\kappa_n\}$ is a $0-\varpi-$Cauchy sequence converges to a point $0 \in \Omega$. Therefore the pair (Ω,ϖ) is a $0-\varpi-$complete $b-$metric-like space, while, if we consider $\{\kappa_n\} = \left(\frac{n}{n+1}\right)_{n\in\mathbb{N}}$, then $\lim_{m,n\to\infty}\varpi(\kappa_m,\kappa_n)$ exists and is finite but converges to a point $2 \notin \Omega$, so, the pair (Ω,ϖ) is not a $\varpi-$complete $b-$metric-like space.

Remark 3. In a $b-$metric-like space the limit of a sequence need not be unique and a convergent sequence need not be a Cauchy sequence.

To show this remark, we gave the following example:

Example 4. Let $\Omega = [0,+\infty)$. Define a function $\varpi : [0,+\infty) \times [0,+\infty) \to [0,+\infty)$ by $\varpi(\kappa,\mu) = (\max\{\kappa,\mu\})^2$. Then (Ω,ϖ) is a $b-$metric-like space with a coefficient $s = 2$. Suppose that

$$\{\kappa_n\} = \begin{cases} 0 & \text{when } n \text{ is odd} \\ 1 & \text{when } n \text{ is even} \end{cases}.$$

For $\kappa \in \Omega$, $\lim_{n\to\infty}\varpi(\kappa_n,\kappa) = \lim_{n\to\infty}(\max\{\kappa_n,\kappa\})^2 = \kappa^2 = \varpi(\kappa,\kappa)$. Therefore, it is a convergent sequence and $\kappa_n \to \kappa$. Now if we take $\kappa = 0$, therefore,
if n is an odd number, we have $\lim_{n\to\infty}\varpi(\kappa_n,\kappa) = \lim_{n\to\infty}(\max\{0,0\})^2 = 0$,
if n is an even number, we get $\lim_{n\to\infty}\varpi(\kappa_n,\kappa) = \lim_{n\to\infty}(\max\{1,0\})^2 = 1$.
That is, the sequence has not limit although it has two subsequences (for odd n and for even n) both having a limit with both limits being distinct.

3. New Fixed Point Results

This section is devoted to present some new fixed point results for a generalized almost (s,q)–Jaggi F–contraction-type and almost (s,q)–Jaggi F–Suzuki-type contraction on the context of $b-$ metric-like spaces.

We begin with the first main result.

Theorem 3. *Let (Ω, ϖ) be a $0-\varpi-$complete $b-$metric-like space with a coefficient $s \geq 1$ and Γ be a self mapping satisfying a generalized almost (s,q)–Jaggi F–contraction-type (4). Then, Γ has a unique fixed point whenever F or Γ is continuous.*

Proof. Let $\kappa_\circ$ be an arbitrary point of Ω. Define a sequence $\{\kappa_n\}_{n \in \mathbb{N}}$ by $\kappa_{n+1} = \Gamma \kappa_n$. If $\Gamma \kappa_\circ = \kappa_\circ$, then the proof is finished. Again, if there exists $i_\circ \in \{1, 2, ..\}$ the right hand side of (4) is 0 for $\kappa = \kappa_{i_\circ - 1}$ and $\mu = \kappa_{i_\circ}$ so the proof is stopped. So, without loss of generality we may assume that $\kappa_{n+1} \neq \kappa_n$ for all $n \geq 1$ and $\varpi_n = \varpi(\kappa_{n+1}, \kappa_n)$. Then $\varpi_n > 0$. On the other hand, Γ is a generalized almost (s,q)–Jaggi F–contraction-type, hence we get

$$\begin{aligned}
\tau + F(\varpi_n) &\leq \tau + F(s^q \varpi_n) = \tau + F(s^q \varpi(\kappa_{n+1}, \kappa_n)) \\
&= \tau + F(s^q \varpi(\Gamma \kappa_n, \Gamma \kappa_{n-1})) \\
&\leq F\left(\alpha \frac{\varpi(\kappa_n, \Gamma \kappa_n).\varpi(\kappa_{n-1}, \Gamma \kappa_{n-1})}{\varpi(\kappa_n, \kappa_{n-1})} + \beta \varpi(\kappa_n, \kappa_{n-1}) + \gamma \varpi(\kappa_{n-1}, \Gamma \kappa_n) \right) \\
&= F\left(\alpha \varpi(\kappa_n, \kappa_{n+1}) + \beta \varpi(\kappa_n, \kappa_{n-1}) + \gamma \varpi(\kappa_{n-1}, \kappa_{n+1}) \right).
\end{aligned} \quad (5)$$

By condition (ϖ_3), we have

$$\varpi(\kappa_{n-1}, \kappa_{n+1}) \leq s[\varpi(\kappa_{n-1}, \kappa_n) + \varpi(\kappa_n, \kappa_{n+1})]. \quad (6)$$

Applying (6) in (5), one can write

$$\tau + F(\varpi_n) \leq F\left(\alpha \varpi(\kappa_n, \kappa_{n+1}) + \beta \varpi(\kappa_n, \kappa_{n-1}) + \gamma s \varpi(\kappa_{n-1}, \kappa_n) + \gamma s \varpi(\kappa_n, \kappa_{n+1}) \right).$$

Since F is strictly increasing, then

$$\varpi_n < \alpha \varpi_n + \beta \varpi_{n-1} + \gamma s \varpi_{n-1} + \gamma s \varpi_n,$$

this leads to

$$(1 - \alpha - \gamma s) \varpi_n < (\beta + \gamma s) \varpi_{n-1} \text{ for all } n \geq 1.$$

Since $\alpha + \beta + 2\gamma s < 1$, we deduce that $1 - \alpha - \gamma s > 0$, and thus

$$\varpi_n < \frac{\beta + \gamma s}{1 - \alpha - \gamma s} \varpi_{n-1} < \varpi_{n-1}.$$

Consequently,

$$\tau + F(\varpi_n) \leq F(\varpi_{n-1}).$$

By the same method, we can prove that

$$\begin{aligned}
F(\varpi_n) &\leq F(\varpi_{n-1}) - \tau \\
&\leq F(\varpi_{n-2}) - 2\tau \\
&\quad \vdots \\
&\leq F(\varpi_\circ) - n\tau \text{ for all } n \geq 1.
\end{aligned} \quad (7)$$

Passing the limit as $n \to \infty$ in (7), we can get

$$\lim_{n \to \infty} F(\varpi_n) = -\infty.$$

So, by $(\Im_2)$, we obtain

$$\lim_{n \to \infty} \varpi_n = 0. \tag{8}$$

Apply $(\Im_3)$, there exists $\lambda \in (0,1)$ such that

$$\lim_{n \to \infty} \varpi_n^\lambda F(\varpi_n) = 0. \tag{9}$$

By (7), for all $n \geq 1$, yields

$$\varpi_n^\lambda \left(F(\varpi_n) - F(\varpi_\circ) \right) \leq -n\tau \varpi_n^\lambda \leq 0. \tag{10}$$

Considering (8), (9) and passing $n \to \infty$ in (10), we get

$$\lim_{n \to \infty} n \varpi_n^\lambda = 0. \tag{11}$$

By (11), there exists $n_1 \in \mathbb{N}$ such that $n \varpi_n^\lambda \leq 1$ for all $n \geq n_1$, or

$$\varpi_n \leq \frac{1}{n^{\frac{1}{\lambda}}} \ \forall n \geq n_1. \tag{12}$$

Now, we shall prove that $\{\kappa_n\}$ is $0 - \varpi-$Cauchy sequence, let $m, n \in \mathbb{N}$ such that $m > n \geq n_1$. By (12) and the assumption (ϖ_3), we have

$$\begin{aligned}
\varpi(\kappa_n, \kappa_m) &\leq s\varpi(\kappa_n, \kappa_{n+1}) + s\varpi(\kappa_{n+1}, \kappa_m) \\
&\leq s\varpi(\kappa_n, \kappa_{n+1}) + s^2 \varpi(\kappa_{n+1}, \kappa_{n+2}) + s^2 \varpi(\kappa_{n+2}, \kappa_m) \\
&\vdots \\
&\leq s\varpi(\kappa_n, \kappa_{n+1}) + s^2 \varpi(\kappa_{n+1}, \kappa_{n+2}) + s^3 \varpi(\kappa_{n+2}, \kappa_{n+3}) + \ldots \\
&\quad + s^{m-n-1} \varpi(\kappa_{m-2}, \kappa_{m-1}) + s^{m-n} \varpi(\kappa_{m-1}, \kappa_m) \\
&= \sum_{i=n}^{m-1} s^{i-n+1} \varpi_i \\
&\leq \sum_{i=n}^{\infty} s^{i-n+1} \left(\frac{1}{i^{\frac{1}{\lambda}}} \right).
\end{aligned}$$

Since the series $\sum_{i=n}^{\infty} \frac{1}{i^{\frac{1}{\lambda}}}$ is converges, as $i \to \infty$ and since multiply a scalar number in a convergent series gives a convergent series, so, $\varpi(\kappa_n, \kappa_m) \to 0$. Therefore $\{\kappa_n\}$ is $0 - \varpi-$ Cauchy sequence in (Ω, ϖ). Since Ω is $0- \varpi-$complete $b-$metric-like space, there exists $\kappa^* \in \Omega$ such that $\kappa_n \to \kappa^*$ or equivalently,

$$\lim_{n,m \to \infty} \varpi(\kappa_n, \kappa_m) = \lim_{n \to \infty} \varpi(\kappa_n, \kappa^*) = \varpi(\kappa^*, \kappa^*) = 0. \tag{13}$$

If Γ is ϖ continuous, it follows from (12) that

$$\lim_{n \to \infty} \varpi(\Gamma \kappa_n, \Gamma \kappa^*) = \lim_{n \to \infty} \varpi(\kappa_{n+1}, \Gamma \kappa^*) = \varpi(\kappa^*, \Gamma \kappa^*) = 0.$$

this implies that

$$\kappa^* = \Gamma \kappa^*.$$

Furthermore, suppose that F is continuous, we prove that κ^* is a fixed point of Γ by contrary, suppose $\kappa^* \neq \Gamma\kappa^*$, so there exist an $n_\circ \in \mathbb{N}$ and a subsequence $\{\kappa_{n_i}\}$ of $\{\kappa_n\}$ such that $\omega(\kappa_{n_i+1}, \Gamma\kappa^*) > 0$ for all $n_i \geq n_\circ$ (otherwise, there exists $n_1 \in N$ such that $\kappa_n = \Gamma\kappa^* \ \forall n \geq n_1$, which implies that $\kappa_n \to \kappa^*$. That is a contradiction, with $\kappa^* \neq \Gamma\kappa^*$). Since $\omega(\kappa_{n_i+1}, \Gamma\kappa^*) > 0 \ \forall n_i \geq n_\circ$, then by (4), we have

$$\begin{aligned}
\tau + F(s^q\omega(\kappa_{n_i+1}, \Gamma\kappa^*)) & \\
= \tau + F(s^q\omega(\Gamma\kappa_{n_i}, \Gamma\kappa^*)) & \\
\leq F\left(\alpha\frac{\omega(\kappa_{n_i}, \Gamma\kappa_{n_i}).\omega(\kappa^*, \Gamma\kappa^*)}{\omega(\kappa_{n_i}, \kappa^*)} + \beta\omega(\kappa_{n_i}, \kappa^*) + \gamma\omega(\kappa^*, \Gamma\kappa_{n_i})\right) & \\
= F\left(\alpha\frac{\omega(\kappa_{n_i}, \kappa_{n_i+1}).\omega(\kappa^*, \Gamma\kappa^*)}{\omega(\kappa_{n_i}, \kappa^*)} + \beta\omega(\kappa_{n_i}, \kappa^*) + \gamma\omega(\kappa^*, \kappa_{n_i+1})\right). & \quad (14)
\end{aligned}$$

Letting $n \to \infty$ in (14) and since F is continuous, we can get

$$\begin{aligned}
\tau + F(s^q\omega(\kappa^*, \Gamma\kappa^*)) &\leq F(\alpha\omega(\kappa^*, \Gamma\kappa^*)) \\
&< F(\omega(\kappa^*, \Gamma\kappa^*)),
\end{aligned}$$

the above inequality say that $s^q < 1$ for some $q > 1$. This a contradiction. Hence $\kappa^* = \Gamma\kappa^*$.

For uniqueness. Suppose that κ_1^* and κ_2^* are two distinct fixed points of a mapping Γ, hence $F(\omega(\Gamma\kappa_1^*, \Gamma\kappa_2^*)) = \omega(\kappa_1^*, \kappa_2^*) > 0$, which implies by (4) that

$$\begin{aligned}
F(s^q\omega(\kappa_1^*, \kappa_2^*)) &= F(s^q\omega(\Gamma\kappa_1^*, \Gamma\kappa_2^*)) \\
&< \tau + F(s^q\omega(\Gamma\kappa_1^*, \Gamma\kappa_2^*)) \\
&\leq F\left(\alpha\frac{\omega(\kappa_1^*, \Gamma\kappa_1^*).\omega(\kappa_2^*, \Gamma\kappa_2^*)}{\omega(\kappa_1^*, \kappa_2^*)} + \beta\omega(\kappa_1^*, \kappa_2^*) + \gamma\omega(\kappa_2^*, \Gamma\kappa_1^*)\right) \\
&= F((\beta+\gamma)\omega(\kappa_1^*, \kappa_2^*)) \\
&< F(\omega(\kappa_1^*, \kappa_2^*)),
\end{aligned}$$

a contradiction again. Hence, the fixed point is unique. The proof is finished. □

Remark 4. *In the real, we can obtain some classical results of our new contraction (4) if we take the following considerations on a complete metric space (Ω, d).*
- *Put $\alpha = \gamma = 0$ and $\beta = 1$, we have Wardowski contraction [21].*
- *Take $\alpha = \gamma = 0$, $\beta \in [0,1)$, $F(\theta) = \ln(\theta)$ with $\theta > 0$, we get Banach contraction [1].*
- *Consider $\gamma = 0$ and $F(\theta) = \ln(\theta)$ with $\theta > 0$, we have Jaggi-contraction [29].*
- *Let $\gamma = 0$, $F(\theta) = \ln(\theta)$ with $\theta > 0$, we have Jaggi-contraction [28].*
- *Set $\alpha = 0$, $\beta \in [0,1)$, $\gamma \geq 0$, $F(\theta) = \ln(\theta)$ with $\theta > 0$, we get Ciric almost contraction [30].*

Now, we present the following example to discuss the validity results of Theorem 3.

Example 5. *Let $\Omega = [0,1) \cap \mathbb{Q}$ and $\omega : \Omega \times \Omega \to \mathbb{R}^+$ be a function defined by*

$$\omega(\kappa, \mu) = \begin{cases} 2\sqrt[3]{\kappa}, & \text{if } \kappa = \mu \\ \left(\max\{2\sqrt[3]{\kappa}, 2\sqrt[3]{\mu}\}\right)^2 & \text{otherwise} \end{cases}$$

for all $\kappa, \mu \in \Omega$. Suppose that $\{\kappa_n\} \in \Omega$. If $\lim_{n\to\infty} \varpi(\kappa_n, \kappa) = (\kappa, \kappa) = 0$, then $\lim_{n\to\infty} \varpi(\kappa_n, \mu) = (\kappa, \mu)$ for all $\mu \in \Omega$. By the condition (ϖ_3), we can write

$$\varpi(\kappa_n, \kappa_m) \leq 2(\varpi(\kappa_n, \kappa) + \varpi(\kappa, \kappa_m)). \tag{15}$$

Passing limit as $n, m \to \infty$ in (15), we obtain $\lim_{n,m\to\infty} \varpi(\kappa_n, \kappa_m) = 0$. Thus (Ω, ϖ) is $0 - \omega$-complete b-metric-like space with a coefficient $s = 2$. Note, here (Ω, ϖ) is not a complete b- metric like space. Indeed, consider the sequence $\{\kappa_n\} = \frac{1}{n}$ for $n \in \mathbb{N}$ in Ω. then $\lim_{n\to\infty} \kappa_n = 0$. Let $\lim_{n\to\infty} \varpi(\kappa_n, \kappa) = \varpi(0, \kappa)$ for all $\kappa \in \Omega$.

If $\kappa = 0$, then $\varpi(0, \kappa) = 0$. So $\lim_{n,m\to\infty} \varpi(\kappa_n, \kappa_m) = \varpi(0, \kappa) = \varpi(\kappa_n, \kappa)$.

If $\kappa \neq 0$, then $\varpi(0, \kappa) = 2\sqrt[3]{\kappa}$. So $\lim_{n,m\to\infty} \varpi(\kappa_n, \kappa_m) \neq \varpi(0, \kappa) = \varpi(\kappa_n, \kappa)$.

Therefore, (Ω, ϖ) is a $0 - \omega$-complete b-metric-like space, which is not a ϖ-complete b-metric-like space. Define a nonlinear mapping Γ by $\Gamma\kappa = \frac{\kappa}{512}$. Take $F(\theta) = \ln(\theta) + \theta$, $q = 3$ and $\tau = \ln(2)$. We shall prove that Γ satisfy the condition (4) with $\alpha = \gamma = 0$ and $\beta = \frac{1}{4}$. So $\alpha + \beta + 2\gamma s = \frac{1}{4} < 1$. Then for $\kappa < \mu$,

$$\varpi(\Gamma\kappa, \Gamma\mu) = \varpi\left(\frac{\kappa}{512}, \frac{\mu}{512}\right) = \left(\frac{2\sqrt[3]{\mu}}{8}\right)^2 = \frac{(\sqrt[3]{\mu})^2}{16} > 0,$$

by simple calculations, we can get

$$\varpi(\kappa, \Gamma\kappa) = \varpi\left(\kappa, \frac{\kappa}{512}\right) = 4\left(\sqrt[3]{\kappa}\right)^2, \quad \varpi(\mu, \Gamma\mu) = \varpi\left(\mu, \frac{\mu}{512}\right) = 4\left(\sqrt[3]{\mu}\right)^2,$$
$$\varpi(\kappa, \mu) = 4\left(\sqrt[3]{\mu}\right)^2, \qquad \varpi(\mu, \Gamma\kappa) = \varpi\left(\mu, \frac{\kappa}{512}\right) = 4\left(\sqrt[3]{\mu}\right)^2,$$

and

$$\alpha\frac{\varpi(\kappa, \Gamma\kappa).\varpi(\mu, \Gamma\mu)}{\varpi(\kappa, \mu)} + \beta\varpi(\kappa, \mu) + \gamma\varpi(\mu, \Gamma\kappa)$$
$$= \left(\alpha \times 4\left(\sqrt[3]{\kappa}\right)^2\right) + \left(\beta \times 4\left(\sqrt[3]{\mu}\right)^2\right) + \left(\gamma \times 4\left(\sqrt[3]{\mu}\right)^2\right) = \left(\sqrt[3]{\mu}\right)^2.$$

Hence

$$\begin{aligned}
\tau + F(s^q(\varpi(\Gamma\kappa, \Gamma\mu))) &= \ln(2) + F\left(8 \times \frac{(\sqrt[3]{\mu})^2}{16}\right) \\
&= \ln(2) + F\left(\frac{(\sqrt[3]{\mu})^2}{2}\right) \\
&\leq \ln(2) + \ln\left(\frac{(\sqrt[3]{\mu})^2}{2}\right) + \frac{(\sqrt[3]{\mu})^2}{2} \\
&\leq \ln\left((\sqrt[3]{\mu})^2\right) + (\sqrt[3]{\mu})^2 \\
&= F\left((\sqrt[3]{\mu})^2\right) \\
&= F(\alpha\frac{\varpi(\kappa, \Gamma\kappa) \times \varpi(\mu, \Gamma\mu)}{\varpi(\kappa, \mu)} + \beta\varpi(\kappa, \mu) + \gamma\varpi(\mu, \Gamma\kappa)).
\end{aligned}$$

By the same manner, for $\kappa = \mu \neq 0$, one gets that

$$\begin{aligned}
\tau + F(s^q(\varpi(\Gamma\kappa,\Gamma\mu))) &= \ln(2) + F(8 \times \frac{(\sqrt[3]{\kappa})^2}{16}) \\
&= \ln(2) + F\left(\frac{(\sqrt[3]{\kappa})^2}{2}\right) \\
&\leq \ln(2) + \ln\left(\frac{(\sqrt[3]{\kappa})^2}{2}\right) + \frac{(\sqrt[3]{\kappa})^2}{2} \\
&\leq \ln\left((\sqrt[3]{\kappa})^2\right) + (\sqrt[3]{\kappa})^2 \\
&= F\left((\sqrt[3]{\kappa})^2\right) \\
&= F(\alpha\frac{\varpi(\kappa,\Gamma\kappa).\varpi(\mu,\Gamma\mu)}{\varpi(\kappa,\mu)} + \beta\varpi(\kappa,\mu) + \gamma\varpi(\mu,\Gamma\kappa)).
\end{aligned}$$

So, all required hypotheses of Theorem 3 are verified and the point $0 \in \Omega$ is a unique fixed point of Γ.

The second result of this section is to introduce the notion of a generalized almost (s,q)–Jaggi F–Suzuki contraction-type in the context of b–metric-like spaces and study some related fixed point results in this direction.

Definition 11. *Let Γ be a self-mapping on a b–metric-like space (Ω, ϖ) with parameter $s \geq 1$. Then the mapping Γ is said to be a generalized almost (s,q)–Jaggi F–Suzuki contraction-type if there exists $F \in \Sigma$ and $\tau \in (0, +\infty)$ such that*

$$\frac{1}{2s}\varpi(\kappa,\Gamma\kappa) < \varpi(\kappa,\mu) \Rightarrow \tau + F\left(s^q \varpi(\Gamma\kappa,\Gamma\mu)\right) \leq F\left(\alpha \frac{\varpi(\kappa,\Gamma\kappa).\varpi(\mu,\Gamma\mu)}{\varpi(\kappa,\mu)} + \beta\varpi(\kappa,\mu) + \gamma\varpi(\mu,\Gamma\kappa)\right), \quad (16)$$

for all $\kappa, \mu \in \Omega$ and $\alpha, \beta, \gamma \geq 0$ with $\alpha + \beta + 2\gamma s < 1$, for some $q > 1$ and satisfying $\varpi(\Gamma\kappa, \Gamma\mu) > 0$.

Theorem 4. *Let (Ω, ϖ) be a $0 - \varpi$–complete b–metric-like space with a coefficient $s \geq 1$ and Γ be a self mapping satisfying a generalized almost (s,q)–Jaggi-type F–Suzuki contraction (16). Then, Γ has a unique fixed point whenever F or Γ is continuous.*

Proof. Let $\kappa_\circ \in \Omega$ and $\{\kappa_n\}_{n=1}^\infty$ defined by $\kappa_{n+1} = \Gamma\kappa_n = \Gamma^{n+1}\kappa_\circ$. If there exists $n \in \mathbb{N}$ such that $\varpi(\kappa_n, \Gamma\kappa_n) = 0$, thus the proof is completed. So, suppose that $0 < \varpi(\kappa_n, \Gamma\kappa_n) = \varpi(\kappa_n, \kappa_{n+1}) = \varpi_n$, therefore for all $n \in \mathbb{N}$

$$\frac{1}{2s}\varpi(\kappa_n, \Gamma\kappa_n) < \varpi(\kappa_n, \Gamma\kappa_n),$$

yields

$$\begin{aligned}
\tau + F\left(\varpi(\Gamma\kappa_n, \Gamma^2\kappa_n)\right) &\leq \tau + F\left(s^q \varpi(\Gamma\kappa_n, \Gamma^2\kappa_n)\right) = \tau + F\left(s^q \varpi(\Gamma\kappa_n, \Gamma(\Gamma\kappa_n))\right) \\
&\leq F\left(\alpha \frac{\varpi(\kappa_n, \Gamma\kappa_n).\varpi(\Gamma\kappa_n, \Gamma^2\kappa_n)}{\varpi(\kappa_n, \Gamma\kappa_n)} + \beta\varpi(\kappa_n, \Gamma\kappa_n) + \gamma\varpi(\Gamma\kappa_n, \Gamma\kappa_n)\right) \\
&= F\left(\alpha\varpi(\Gamma\kappa_n, \Gamma^2\kappa_n) + \beta\varpi(\kappa_n, \Gamma\kappa_n) + \gamma\varpi(\Gamma\kappa_n, \Gamma\kappa_n)\right) \\
&\leq F\left(\alpha\varpi(\Gamma\kappa_n, \Gamma^2\kappa_n) + \beta\varpi(\kappa_n, \Gamma\kappa_n) + 2s\gamma\varpi(\kappa_n, \Gamma\kappa_n)\right).
\end{aligned}$$

Since F is strictly increasing, then

$$\varpi(\Gamma\kappa_n, \Gamma^2\kappa_n) < \alpha\varpi(\Gamma\kappa_n, \Gamma^2\kappa_n) + \beta\varpi(\kappa_n, \Gamma\kappa_n) + 2s\gamma\varpi(\kappa_n, \Gamma\kappa_n). \quad (17)$$

Since $1 - \alpha > 0$, and $\alpha + \beta + 2\gamma s < 1$ then we can write

$$\varpi(\Gamma\kappa_n, \Gamma^2\kappa_n) < \left(\frac{\beta + 2s\gamma}{1 - \alpha}\right)\varpi(\kappa_n, \Gamma\kappa_n) < \varpi(\kappa_n, \Gamma\kappa_n). \tag{18}$$

Consequently,

$$\tau + F(\varpi(\Gamma\kappa_n, \Gamma^2\kappa_n)) \leq F(\varpi(\kappa_n, \Gamma\kappa_n)),$$

or,

$$F(\varpi(\Gamma\kappa_n, \Gamma^2\kappa_n)) \leq F(\varpi(\kappa_n, \Gamma\kappa_n)) - \tau,$$

By the same method, we can deduce that

$$\begin{aligned} F(\varpi_n) &= F(\varpi(\kappa_n, \Gamma\kappa_n)) = F(\varpi(\Gamma\kappa_{n-1}, \Gamma^2\kappa_{n-1})) \\ &\leq F(\varpi(\kappa_{n-1}, \Gamma\kappa_{n-1})) - \tau \\ &\leq F(\varpi(\kappa_{n-2}, \Gamma\kappa_{n-2})) - 2\tau \\ &\quad \vdots \\ &\leq F(\varpi(\kappa_\circ, \Gamma\kappa_\circ)) - n\tau \text{ for all } n \geq 1. \end{aligned}$$

By the same manner of Theorem 3, we deduce that $\{\kappa_n\}$ is $0 - \varpi$-Cauchy sequence in (Ω, ϖ). Since Ω is $0 - \varpi$-complete b-metric-like space, there exists $\kappa^* \in \Omega$ such that $\kappa_n \to \kappa^*$ or equivalently,

$$\lim_{n,m \to \infty} \varpi(\kappa_n, \kappa_m) = \lim_{n \to \infty} \varpi(\kappa_n, \kappa^*) = \varpi(\kappa^*, \kappa^*) = 0. \tag{19}$$

Now, we shall prove that

$$\frac{1}{2s}\varpi(\kappa_n, \Gamma\kappa_n) < \varpi(\kappa_n, \kappa^*) \text{ or } \frac{1}{2s}\varpi(\Gamma\kappa_n, \Gamma^2\kappa_n) < \varpi(\Gamma\kappa_n, \kappa^*). \tag{20}$$

Assuming the opposite, that there is $m \in \mathbb{N}$ such that

$$\frac{1}{2s}\varpi(\kappa_m, \Gamma\kappa_m) \geq \varpi(\kappa_m, \kappa^*) \text{ or } \frac{1}{2s}\varpi(\Gamma\kappa_m, \Gamma^2\kappa_m) \geq \varpi(\Gamma\kappa_m, \kappa^*). \tag{21}$$

Hence

$$2s\varpi(\kappa_m, \kappa^*) \leq \varpi(\kappa_m, \Gamma\kappa_m) \leq s[\varpi(\kappa_m, \kappa^*) + \varpi(\kappa^*, \Gamma\kappa_m)],$$

which leads to

$$s\varpi(\kappa_m, \kappa^*) \leq s\varpi(\kappa^*, \Gamma\kappa_m),$$

or

$$\varpi(\kappa_m, \kappa^*) \leq \varpi(\kappa^*, \Gamma\kappa_m), \tag{22}$$

From (21) and (22), we have

$$\varpi(\kappa_m, \kappa^*) \leq \varpi(\kappa^*, \Gamma\kappa_m) \leq \frac{1}{2s}\varpi(\Gamma\kappa_m, \Gamma^2\kappa_m).$$

Since $\frac{1}{2s}\varpi(\kappa_m, \Gamma\kappa_m) < \varpi(\kappa_m, \Gamma\kappa_m)$ and using (16), we can get

$$\begin{aligned}
\tau + F\left(\varpi(\Gamma\kappa_m, \Gamma^2\kappa_m)\right) &\leq \tau + F\left(s^q\varpi(\Gamma\kappa_m, \Gamma^2\kappa_m)\right) \\
&\leq F\left(\alpha\frac{\varpi(\kappa_m, \Gamma\kappa_m)}{\varpi(\kappa_m, \Gamma\kappa_m)} + \beta\varpi(\kappa_m, \Gamma\kappa_m) + \gamma\varpi(\Gamma\kappa_m, \Gamma\kappa_m)\right) \\
&= F\left(\alpha\varpi(\Gamma\kappa_m, \Gamma^2\kappa_m) + \beta\varpi(\kappa_m, \Gamma\kappa_m) + \gamma\varpi(\Gamma\kappa_m, \Gamma\kappa_m)\right) \\
&\leq F\left(\alpha\varpi(\Gamma\kappa_m, \Gamma^2\kappa_m) + \beta\varpi(\kappa_m, \Gamma\kappa_m) + 2s\gamma\varpi(\kappa_m, \Gamma\kappa_m)\right).
\end{aligned}$$

Replace n with m in the inequalities (17) and (18) and apply the same above steps, we can write

$$\varpi(\Gamma\kappa_m, \Gamma^2\kappa_m) < \varpi(\kappa_m, \Gamma\kappa_m).$$

It follows from (21) and (22), that

$$\begin{aligned}
\varpi(\Gamma\kappa_m, \Gamma^2\kappa_m) &< \varpi(\kappa_m, \Gamma\kappa_m) \\
&\leq s[\varpi(\kappa_m, \kappa^*) + \varpi(\kappa^*, \Gamma\kappa_m)] \\
&\leq s[\varpi(\kappa^*, \Gamma\kappa_m) + \frac{1}{2s}\varpi(\Gamma\kappa_m, \Gamma^2\kappa_m)] \\
&\leq s[\frac{1}{2s}\varpi(\Gamma\kappa_m, \Gamma^2\kappa_m) + \frac{1}{2s}\varpi(\Gamma\kappa_m, \Gamma^2\kappa_m)] \\
&= \varpi(\Gamma\kappa_m, \Gamma^2\kappa_m).
\end{aligned}$$

A contradiction, so (20) holds for all $n \in \mathbb{N}$, this leads to

$$\begin{aligned}
\tau + F\left(2s^2\varpi(\Gamma\kappa_n, \Gamma\kappa^*)\right) &\leq \tau + F(2s^q\varpi(\Gamma\kappa_n, \Gamma\kappa^*)) \\
&\leq F\left(\alpha\frac{\varpi(\kappa_n, \Gamma\kappa_n).\varpi(\kappa^*, \Gamma\kappa^*)}{\varpi(\kappa_n, \kappa^*)} + \beta\varpi(\kappa_n, \kappa^*) + \gamma\varpi(\kappa^*, \Gamma\kappa_n)\right) \\
&< F\left(\alpha\frac{2s\varpi(\kappa_n, \kappa^*).\varpi(\kappa^*, \Gamma\kappa^*)}{\varpi(\kappa_n, \kappa^*)} + \beta\varpi(\kappa_n, \kappa^*) + \gamma\varpi(\kappa^*, \Gamma\kappa_n)\right) \\
&= F(2s\alpha\varpi(\kappa^*, \Gamma\kappa^*) + \beta\varpi(\kappa_n, \kappa^*) + \gamma\varpi(\kappa^*, \Gamma\kappa_n)) \\
&\leq F\left(\begin{array}{c} 2s^2\alpha\varpi(\kappa^*, \kappa_n) + 2s^3\alpha\varpi(\kappa_n, \Gamma\kappa_n) + 2s^3\alpha\varpi(\Gamma\kappa_n, \Gamma\kappa^*) \\ +\beta\varpi(\kappa_n, \kappa^*) + \gamma s\varpi(\kappa^*, \kappa_n) + \gamma s\varpi(\kappa_n, \Gamma\kappa_n) \end{array}\right),
\end{aligned}$$

a gain, since $\varpi(\kappa_n, \Gamma\kappa_n) < 2s\varpi(\kappa_n, \kappa^*)$, $\tau > 0$ and F is strictly increasing, this yields,

$$2s^2\varpi(\Gamma\kappa_n, \Gamma\kappa^*) < (2s^2\alpha + 4s^4\alpha + \beta + \gamma s + 2s^2\gamma)\varpi(\kappa_n, \kappa^*) + 2s^3\alpha\varpi(\Gamma\kappa_n, \Gamma\kappa^*),$$

or,

$$(1 - 2\alpha s)\varpi(\Gamma\kappa_n, \Gamma\kappa^*) < (\alpha + 2s^2\alpha + \beta + 2\gamma)\varpi(\kappa_n, \kappa^*),$$

Thus,

$$\begin{aligned}
\tau + F((1 - 2\alpha s)\varpi(\Gamma\kappa_n, \Gamma\kappa^*)) &\leq F((\alpha + 2s^2\alpha + \beta + 2\gamma)\varpi(\kappa_n, \kappa^*)) \\
&= F(\alpha + 2s^2\alpha + \beta + 2\gamma)\varpi(\Gamma\kappa_{n+1}, \kappa^*)). \quad (23)
\end{aligned}$$

Passing the limit as $n \to \infty$ in (23), using $(\Im_2)$ and (19) we can get

$$\lim_{n\to\infty} \varpi(\Gamma\kappa_n, \Gamma\kappa^*) = 0.$$

Therefore, $\varpi(\kappa^*, \Gamma\kappa^*) = \lim_{n\to\infty} \varpi(\kappa_n, \Gamma\kappa^*) = \lim_{n\to\infty} \varpi(\Gamma\kappa_{n+1}, \Gamma\kappa^*) = 0.$

Similarly, by (20) for all $n \in \mathbb{N}$, one can write

$$\begin{aligned}
\tau + F\left(2s^2\varpi(\Gamma^2\kappa_n, \Gamma\kappa^*)\right) &\leq \tau + F\left(2s^q\varpi(\Gamma^2\kappa_n, \Gamma\kappa^*)\right) \\
&\leq F\left(\alpha\frac{\varpi(\Gamma\kappa_n, \Gamma^2\kappa_n).\varpi(\kappa^*, \Gamma\kappa^*)}{\varpi(\Gamma\kappa_n, \kappa^*)} + \beta\varpi(\Gamma\kappa_n, \kappa^*) + \gamma\varpi(\kappa^*, \Gamma^2\kappa_n)\right) \\
&< F\left(\alpha\frac{2s\varpi(\Gamma\kappa_n, \kappa^*).\varpi(\kappa^*, \Gamma\kappa^*)}{\varpi(\Gamma\kappa_n, \kappa^*)} + \beta\varpi(\Gamma\kappa_n, \kappa^*) + \gamma\varpi(\kappa^*, \Gamma^2\kappa_n)\right) \\
&= F\left(2s\alpha\varpi(\kappa^*, \Gamma\kappa^*) + \beta\varpi(\Gamma\kappa_n, \kappa^*) + \gamma\varpi(\kappa^*, \Gamma^2\kappa_n)\right) \\
&\leq F\left(\begin{array}{c} 2s^2\alpha\varpi(\kappa^*, \Gamma\kappa_n) + 2s^3\alpha\varpi(\Gamma\kappa_n, \Gamma^2\kappa_n) + 2s^3\alpha\varpi(\Gamma^2\kappa_n, \Gamma\kappa^*) \\ +\beta\varpi(\Gamma\kappa_n, \kappa^*) + \gamma s\varpi(\kappa^*, \Gamma\kappa_n) + \gamma s\varpi(\Gamma\kappa_n, \Gamma^2\kappa_n) \end{array}\right),
\end{aligned}$$

since $\varpi(\Gamma\kappa_n, \Gamma^2\kappa_n) < 2s\varpi(\Gamma\kappa_n, \kappa^*)$, $\tau > 0$ and F is strictly increasing, this leads to

$$2s^2\varpi(\Gamma^2\kappa_n, \Gamma\kappa^*) < (2s^2\alpha + 4s^4\alpha + \beta + \gamma s + 2\gamma s^2)\varpi(\kappa^*, \Gamma\kappa_n) + 2s^3\alpha\varpi(\Gamma^2\kappa_n, \Gamma\kappa^*),$$

or,

$$(1 - 2\alpha s)\varpi(\Gamma^2\kappa_n, \Gamma\kappa^*) < (\alpha + 2s^2\alpha + \beta + 2\gamma)\varpi(\kappa^*, \Gamma\kappa_n).$$

Thus,

$$\begin{aligned}
\tau + F((1-2\alpha s)\varpi(\Gamma^2\kappa_n, \Gamma\kappa^*)) &\leq F((\alpha + 2s^2\alpha + \beta + 2\gamma)\varpi(\kappa^*, \Gamma\kappa_n)) \\
&= F(\alpha + 2s^2\alpha + \beta + 2\gamma)\varpi(\kappa_{n+1}, \kappa^*)).
\end{aligned} \qquad (24)$$

Taking the limit as $n \to \infty$ in (24), using $(\Im_2)$ and (19) we can get

$$\lim_{n\to\infty} \varpi(\Gamma^2\kappa_n, \Gamma\kappa^*) = 0.$$

Therefore, $\varpi(\kappa^*, \Gamma\kappa^*) = \lim_{n\to\infty} \varpi(\kappa_{n+2}, \Gamma\kappa^*) = \lim_{n\to\infty} \varpi(\Gamma^2\kappa_n, \Gamma\kappa^*) = 0$. Hence, $\kappa^* = \Gamma\kappa^*$.

The uniqueness follows immediately from the proof of Theorem 3, and this completes the proof. □

Example 6. *By taking all assumptions of Example 5, if $\kappa < \mu$, we have*

$$\frac{1}{2s}\varpi(\kappa, \Gamma\kappa) = \frac{1}{4} \times 4\left(\sqrt[3]{\kappa}\right)^2 = \left(\sqrt[3]{\kappa}\right)^2 < 4\left(\sqrt[3]{\mu}\right)^2 = \varpi(\kappa, \mu),$$

also, if $\kappa = \mu \neq 0$, we deduce that

$$\frac{1}{2s}\varpi(\kappa, \Gamma\kappa) = \frac{1}{4} \times 4\left(\sqrt[3]{\kappa}\right)^2 = \left(\sqrt[3]{\kappa}\right)^2 < 4\left(\sqrt[3]{\kappa}\right)^2 = \varpi(\kappa, \mu).$$

Therefore all required hypotheses of Theorem 4 are satisfied and a mapping Γ has a unique fixed point $0 \in \Omega$.

4. Solution of Electric Circuit Equation

Fixed point theory is involved in physical applications especially the solution of the the electric circuit equation, which was presented in [35,36]. The authors applied their theorems obtained to solve this equation under $F-$contraction mapping. In this part, we present the solution of electric circuit equation, which is in the form of second-order differential equation. It contains a resistor R, an electromotive force E, a capacitor C, an inductor L and a voltage V in series as Figure 1.

If the rate of change of charge q with respect to time t denoted by the current I, i.e., $I = \frac{dq}{dt}$. We get the following relations:

- $V = IR$;
- $V = \frac{q}{C}$;
- $V = L\frac{dI}{dt}$.

The sum of these voltage drops is equal to the supplied voltage (law of Kirchhoff voltage), i.e.,

$$IR + \frac{q}{C} + L\frac{dI}{dt} = V(t),$$

or

$$L\frac{d^2q}{dt^2} + R\frac{dq}{dt} + \frac{q}{C} = V(t), \ q(0) = 0, \ q'(0) = 0, \quad (25)$$

where $C = \frac{4L}{R^2}$ and $\tau = \frac{R}{2L}$, this case is said to be the resonance solution in a Physics context. Then, the Green function associated with (25) is given by

$$\Lambda(t, \varrho) = \begin{cases} -\varrho e^{-\tau(\varrho-t)} & \text{if } 0 \leq \varrho \leq t \leq 1 \\ -t e^{-\tau(\varrho-t)} & \text{if } 0 \leq t \leq \varrho \leq 1 \end{cases}.$$

Using Green function, problem (25) is equivalent to the following nonlinear integral equation

$$\kappa(t) = \int_0^t \Lambda(t, \varrho) \chi(\varrho, \kappa(\varrho)) d\varrho, \quad (26)$$

where $t \in [0, 1]$.

Let $\Omega = C([0, 1])$ be the set of all continuous functions defined on $[0, 1]$, endowed with

$$\omega(\kappa, \mu) = (\|\kappa\|_\infty + \|\mu\|_\infty)^m \text{ for all } \kappa, \mu \in \Omega,$$

where $\|\kappa\|_\infty = \sup_{t \in [0,1]} \{|\kappa(t)| e^{-2t\tau m}\}$ and $m > 1$. It is clear that (Ω, ω) is a complete b-metric-like space with parameter $s = 2^{m-1}$.

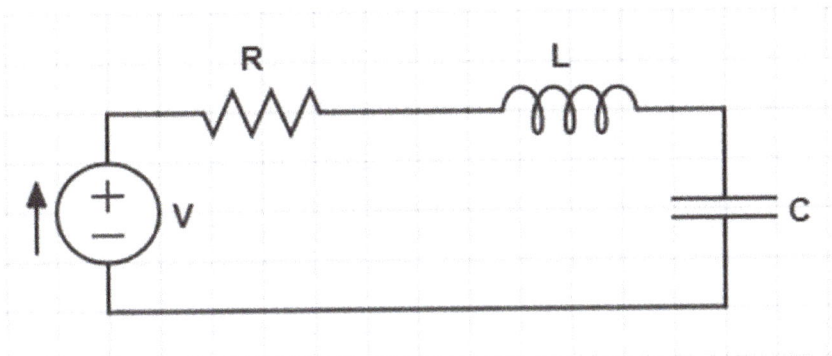

Figure 1. Electric circuit.

Now, we state and prove the main theorem of this section.

Theorem 5. *Let Γ be a nonlinear self mapping on Ω of a b-metric-like space (Ω, ω), such that the following conditions hold*

(i) $\Lambda : [0, 1] \times [0, 1] \to [0, \infty)$ is a continuous function;

(ii) $\chi : [0,1] \times \mathbb{R} \to \mathbb{R}$, where $\chi(\varrho, .)$ is monotone nondecreasing mapping for all $\varrho \in [0,1]$;
(iii) there exists a constant $\tau \in \mathbb{R}^+$ such that for all $(t, \varrho) \in [0,1]^2$ and $\kappa, \mu \in \mathbb{R}^+$,

$$|\chi(t,\kappa) + \chi(t,\mu)| \leq \tau^2 \left(\frac{e^{-\tau}}{s^2}\right)^{\frac{1}{m}} B^{\frac{1}{m}}(\kappa, \mu),$$

where

$$B(\kappa, \mu) = \alpha \frac{\varpi(\kappa, \Gamma\kappa).\varpi(\mu, \Gamma\mu)}{\varpi(\kappa, \mu)} + \beta\varpi(\kappa, \mu) + \gamma\varpi(\mu, \Gamma\kappa),$$

for all, $t \in [0,1]$ and $\alpha, \beta, \gamma \geq 0$ such that $\alpha + \beta + 2\gamma s < 1$. Then the Equation (25) has a unique solution.

Proof. Define a nonlinear self-mapping $\Gamma : \Omega \to \Omega$ by

$$\Gamma\kappa(t) = \int_0^t \Lambda(t, \varrho)\chi(\varrho, \kappa(\varrho))d\varrho.$$

It is clear that if κ^* is a fixed point of the mapping Γ, then it a solution of the problem (26). Suppose that $\kappa, \mu \in \Omega$, we can get $\varpi(\Gamma\kappa(t), \Gamma\mu(t)) > 0$,

$$\begin{aligned}
& s^2 \left(|\Gamma\kappa(t)| + |\Gamma\mu(t)|\right)^m \\
&= s^2 \left(\left|\int_0^t \Lambda(t,\varrho)\chi(\varrho,\kappa(\varrho))d\varrho\right| + \left|\int_0^t \Lambda(t,\varrho)\chi(\varrho,\mu(\varrho))d\varrho\right|\right)^m \\
&\leq s^2 \left(\int_0^t |\Lambda(t,\varrho)\chi(\varrho,\kappa(\varrho))|\, d\varrho + \int_0^t |\Lambda(t,\varrho)\chi(\varrho,\mu(\varrho))d\varrho|\right)^m \\
&= s^2 \left(\int_0^t \Lambda(t,\varrho)\left(|\chi(\varrho,\kappa(\varrho))| + |\chi(\varrho,\mu(\varrho))|\right)d\varrho\right)^m \\
&\leq s^2 \left(\int_0^t \Lambda(t,\varrho)\tau^2 \left(\frac{e^{-\tau}}{s^2}\right)^{\frac{1}{m}} B^{\frac{1}{m}}(\kappa,\mu)d\varrho\right)^m \\
&= s^2 \left(\int_0^t \Lambda(t,\varrho)\tau^2 e^{2\tau\varrho} e^{-2\tau\varrho} \left(\frac{e^{-\tau}}{s^2}\right)^{\frac{1}{m}} B^{\frac{1}{m}}(\kappa,\mu)d\varrho\right)^m \\
&\leq e^{-\tau} \|B(\kappa,\mu)\|_\infty \left(\tau^2 \int_0^t e^{2\tau\varrho}\Lambda(t,\varrho)d\varrho\right)^m \\
&\leq e^{-\tau} \|B(\kappa,\mu)\|_\infty \left(e^{2\tau t}\left[1 - 2\tau t + \tau t e^{-\tau t} - e^{-\tau t}\right]\right)^m,
\end{aligned}$$

so we have

$$s^2 \left(|\Gamma\kappa(t)| + |\Gamma\mu(t)|\right)^m . e^{-2t\tau m} \leq e^{-\tau} \|B(\kappa,\mu)\|_\infty \left[1 - 2\tau t + \tau t e^{-\tau t} - e^{-\tau t}\right]^m,$$

which leads to,

$$s^2 \left(\|\Gamma\kappa(t)\|_\infty + \|\Gamma\mu(t)\|_\infty\right)^m \leq e^{-\tau} \|B(\kappa,\mu)\|_\infty \left[1 - 2\tau t + \tau t e^{-\tau t} - e^{-\tau t}\right]^m,$$

since $1 - 2\tau t + \tau t e^{-\tau t} - e^{-\tau t} \leq 1$, we obtain that

$$s^2 \varpi(\Gamma\kappa(t), \Gamma\mu(t)) \leq e^{-\tau} \|B(\kappa,\mu)\|_\infty.$$

Taking $F(\theta) = \ln(\theta)$, for all $\theta > 0$, which is $F \in \Sigma$, we obtain

$$\ln(s^2\varpi(\Gamma\kappa(t),\Gamma\mu(t))) \leq \ln(e^{-\tau}\|B(\kappa,\mu)\|_\infty),$$

or

$$\tau + \ln(s^2\varpi(\Gamma\kappa(t),\Gamma\mu(t))) \leq \ln(\|B(\kappa,\mu)\|_\infty).$$

Equivalently

$$\tau + F(s^q\varpi(\Gamma\kappa(\rho),\Gamma\mu(\rho))) \leq F(\alpha\frac{\varpi(\kappa,\Gamma\kappa).\varpi(\mu,\Gamma\mu)}{\varpi(\kappa,\mu)} + \beta\varpi(\kappa,\mu) + \gamma\varpi(\mu,\Gamma\kappa)).$$

By Theorem 3 and taking the coefficient $q = 2$, we deduce that Γ has a fixed point, which is a solution of the differential equation arising in the electric circuit equation. This finished the proof. □

The following example satisfy all required hypotheses of Theorem 5.

Example 7. *Consider the following nonlinear integral equation*

$$\kappa(t) = \frac{\tau^2 e^{-4t\tau}}{50s^3}\int_0^1 \varrho^3\kappa(\varrho)d\varrho, \ \varrho \in [0,1]. \tag{27}$$

Then it has a solution in Ω.

Proof. Let $\Gamma : \Omega \to \Omega$ be defined by $\Gamma\kappa(t) = \frac{\tau^2 e^{-4t\tau}}{50s^3}\int_0^1 \varrho^3\kappa(\varrho)d\varrho$. By specifying $\Lambda(t,\varrho) = \frac{\varrho^3}{5}, \chi(t,\kappa) = \frac{\tau^2 e^{-4t\tau}\kappa(\varrho)}{10s^3}$ in Theorem 5, it follows that:
(i) the function $\Lambda(t,\varrho)$ is continuous on $[0,1]\times[0,1]$,
(ii) $\chi(\varrho,\kappa(\varrho))$ is monotone increasing on $[0,1]\times\mathbb{R}$ for all $\varrho \in [0,1]$,
(iii) By taking $m = 2$ and $\tau = \ln(4)$, hence $s = 2$, so, for all $(t,\varrho) \in [0,1]\times[0,1]$ and $\kappa,\mu \in \mathbb{R}^+$, we obtain that

$$\begin{aligned}
|\chi(t,\kappa) + \chi(t,\mu)| &= \frac{\tau^2 e^{-4t\tau}}{10s^3}|\kappa(\varrho) + \mu(\varrho)| \leq \frac{\tau^2 e^{-4t\tau}}{20s^2}(|\kappa(\varrho)| + |\mu(\varrho)|) \\
&\leq \frac{\tau^2 e^{-4t\tau}}{4s^2}(|\kappa(\varrho)| + |\mu(\varrho)|)^2 \\
&= \left(\frac{\tau}{s}\right)^2 \left(\frac{1}{e^{\ln 4}.2^2}\right)^{\frac{1}{2}}\left([|\kappa(\varrho)| + |\mu(\varrho)|]e^{-4t\tau}\right)^2 \\
&= \left(\frac{\tau}{s}\right)^2 \left(\frac{1}{e^\tau.s^2}\right)^{\frac{1}{m}}\left([|\kappa(\varrho)| + |\mu(\varrho)|]e^{-2t\tau m}\right)^m \\
&= \left(\frac{\tau}{s}\right)^2 \left(\frac{1}{e^\tau.s^2}\right)^{\frac{1}{m}}(\|\kappa(\varrho)\|_\infty + \|\mu(\varrho)\|_\infty)^m \\
&= \left(\frac{\tau}{s}\right)^2 \left(\frac{1}{e^\tau.s^2}\right)^{\frac{1}{m}}\varpi(\kappa,\mu) \\
&\leq \tau^2\left(\frac{e^{-\tau}}{s^2}\right)^{\frac{1}{m}}B^{\frac{1}{m}}(\kappa,\mu).
\end{aligned}$$

Finally

$$\sup_{t,\varrho\in[0,1]}\int_0^1 \Lambda(t,\varrho)d\varrho = \sup_{t\in[0,1]}\int_0^1 \frac{\varrho^3}{5}d\varrho \leq \sup_{\rho\in[0,1]}\frac{1}{20} = \frac{1}{20} < 1.$$

Therefore, all conditions of Theorem 5 are satisfied, therefore a mapping Γ has a fixed point in Ω, which is a solution to the problem (27). □

5. Solution of Second-Order Differential Equations

In this part, we shall apply the previous theoretical results of Theorem 3 to study the existence and uniqueness of solutions for the following second-order differential equation:

$$\begin{cases} \kappa''(t) = -\chi(t, \kappa(t)), & t \in [0,1] \\ \kappa(0) = \kappa(1) = 0 \end{cases}, \tag{28}$$

where $\chi : [0,1] \times [0,1] \to \mathbb{R}$ is a continuous functions.

The problem (28) is equivalent to the following integral equation:

$$\kappa(t) = \int_0^1 \Lambda(t, \varrho)\chi(t, \kappa(\varrho))d\varrho, \quad \forall t \in [0,1] \tag{29}$$

where Λ is the Green function defined by

$$\Lambda(\rho, \varrho) = \begin{cases} t(1-\varrho) & \text{if } 0 \le t \le \varrho \le 1 \\ \varrho(1-t) & \text{if } 0 \le \varrho \le t \le 1 \end{cases},$$

and χ be a function as in Theorem 5. Hence if $\kappa \in C^2([0,1], \mathbb{R})$, then κ is a solution of (28) if and only if κ is a solution of (29).

Let $\Omega = C([0,1], \mathbb{R})$ be the set of all continuous functions defined on $[0,1]$, endowed with the same distance of the above section. Then (Ω, ϖ) is a complete $b-$ metric-like space with parameter $s = 2^{m-1}$.

Now, we introduce the main theorem of this part.

Theorem 6. *Let Γ be a nonlinear self mapping on Ω of a $b-$metric-like space (Ω, ϖ), such that there exists monotone nondecreasing mapping $\chi : [0,1] \times \mathbb{R} \to \mathbb{R}$ such that*

$$|\chi(\varrho, \kappa) + \chi(\varrho, \mu)| \le \left(\frac{7\beta e^{-\tau}}{s^2}\right)^{\frac{1}{m}} (|\kappa| + |\mu|),$$

for all $\varrho \in [0,1]$, $\kappa, \mu \in \mathbb{R}$, $0 \le \beta < \frac{1}{8}$ and for $m > 1$.

Then the problem (28) has a unique solution $\kappa \in C([0,1], \mathbb{R})$, provided that the conditions (i) and (ii) of Theorem 5 are satisfied.

Proof. Let us define a nonlinear self-mapping Γ on a set Ω by

$$\Gamma \kappa(t) = \int_0^1 \Lambda(t, \varrho)\chi(t, \kappa(\varrho))d\varrho,$$

for all $t \in [0,1]$ and $\kappa \in \Omega$. The solution of the problem (28) is equivalent to find a fixed point κ of Γ on Ω. Suppose that $\kappa, \mu \in \Omega$, we have $\varpi(\Gamma \kappa, \Gamma \mu) > 0$,

$$s^2 \left(|\Gamma\kappa(t)| + |\Gamma\mu(t)|\right)^m = s^2 \left(\left|\int_0^1 \Lambda(t,\varrho)\chi(t,\kappa(\varrho))d\varrho\right| + \left|\int_0^1 \Lambda(t,\varrho)\chi(t,\mu(\varrho))d\varrho\right|\right)^m$$

$$\leq s^2 \left(\int_0^1 |\Lambda(t,\varrho)\chi(t,\kappa(\varrho))|\,d\varrho + \int_0^1 |\Lambda(t,\varrho)\chi(t,\mu(\varrho))|\,d\varrho\right)^m$$

$$= s^2 \left(\int_0^1 \Lambda(t,\varrho)\,|\chi(t,\kappa(\varrho)) + \chi(t,\mu(\varrho))|\,d\varrho\right)^m$$

$$\leq s^2 \left(\int_0^1 \Lambda(t,\varrho) \left(\frac{7\beta e^{-\tau}}{s^2}\right)^{\frac{1}{m}} (|\kappa| + |\mu|)d\varrho\right)^m$$

$$= 7\beta e^{-\tau} \left(\int_0^1 \Lambda(t,\varrho)e^{2t\tau}e^{-2t\tau}(|\kappa| + |\mu|)\,d\varrho\right)^m$$

$$\leq 7\beta e^{-\tau}(\|\kappa\|_\infty + \|\mu\|_\infty)^m e^{2t\tau m} \left(\int_0^1 \Lambda(t,\varrho)d\varrho\right)^m$$

$$= \frac{7}{7^m}\beta e^{-\tau}e^{2t\tau m}(\|\kappa\|_\infty + \|\mu\|_\infty)^m$$

$$\leq \beta e^{-\tau}e^{2t\tau m}(\|\kappa\|_\infty + \|\mu\|_\infty)^m.$$

For instance above, for all $t \in [0,1]$, we can get $\int_0^1 \Lambda(t,\varrho)d\varrho = \frac{1}{2}(1-t)$ and thus, we choose $\sup_{\varrho \in [0,1]} \int_0^1 \Lambda(t,\varrho)d\varrho = \frac{1}{7}$. Hence

$$s^2 (\Gamma\kappa(t) + |\Gamma\mu(t)|)^m \cdot e^{-2t\tau m} \leq e^{-\tau}\beta(\|\kappa\|_\infty + \|\mu\|_\infty)^m,$$

or

$$s^2 (\|\Gamma\kappa\|_\infty + \|\Gamma\mu\|_\infty)^m \leq e^{-\tau}\beta(\|\kappa\|_\infty + \|\mu\|_\infty)^m, \qquad (30)$$

Let $\alpha = \gamma = 0$, hence $\alpha + \beta + 2s\gamma < \frac{1}{8}$. It follows from (30) that

$$s^2\omega(\Gamma\kappa, \Gamma\mu) \leq (e^{-\tau})[\beta\omega(\kappa,\mu)]. \qquad (31)$$

Taking the function $F(\theta) = \ln(\theta)$ in (31), such that $F \in \Sigma$, we can obtain

$$\tau + F(s^2\omega(\Gamma\kappa, \Gamma\mu)) \leq F(\beta\omega(\kappa,\mu))$$
$$= F(\alpha\frac{\omega(\kappa,\Gamma\kappa).\omega(\mu,\Gamma\mu)}{\omega(\kappa,\mu)} + \beta\omega(\kappa,\mu) + \gamma\omega(\mu,\Gamma\kappa)).$$

Hence all requirements of Theorem 3 are holds by taking the coefficient $q = 2$, therefore Γ has a fixed point $\kappa \in \Omega$, that is, (28) has a unique solution $\kappa \in C^2([0,1], \mathbb{R})$. □

6. Question

It was proved in [22] that if $F(\theta) = \frac{-1}{\sqrt[p]{\theta}}$, where $p > 1$ and $\theta > 0$, then $F \in \Sigma$. The question that arises here, what are the properties of the contraction mapping under this function?

7. Conclusions

The paper generalizes known contraction conditions and the obtained fixed point results, generalized several results known before such as Banach contraction [1], Jaggi-contraction [28], [29], and Ciric almost contraction [30]. Furthermore, as it has been observed in studies, fixed point results in $b-$metric-like spaces can be derived from the results of ordinary and $b-$metric spaces under some suitable conditions. We have applied our results to get the existence of a solution for electric circuit equation and second-order differential equation.

Author Contributions: H.A.H. contributed in conceptualization, investigation,methodology, validation and writing the original draft; M.D.l.S. contributed in funding acquisition, methodology, project administration, supervision, validation, visualization, writing and editing. Both Authors agree and approve the final version of this manuscript. All authors have read and agreed to the published version of the manuscript.

Funding: This work was supported in part by the Basque Government under Grant IT1207-19.

Acknowledgments: The authors are grateful to the Spanish Government and the European Commission for Grant IT1207-19.

Conflicts of Interest: The authors declare that they have no competing interests concerning the publication of this article.

References

1. Banach, S. Sur les opérations dans les ensembles abstraits et leur application aux équations intégrales. *Fund. Math.* **1922**, *3*, 133–181. [CrossRef]
2. Ran, A.C.M.; Reuring, M.C.B. A fixed point theorem in partially ordered sets and some applications to matrix equations. *Proc. Am. Math. Soc.* **2004**, *132*, 1435–1443. [CrossRef]
3. Kirk, W.A.; Srinavasan, P.S.; Veeramani, P. Fixed points for mapping satisfying cyclical contractive conditions. *Fixed Point Theory* **2003**, *4*, 79–89.
4. Shatanawi, W.; Postolache, M. Common fixed point results for mappings under nonlinear contraction of cyclic form in ordered metric spaces. *Fixed Point Theory Appl.* **2013**. [CrossRef]
5. Bakhtin, I.A. The contraction mapping principle in quasi-metric spaces. *Funct. Anal. Ulianowsk Gos. Ped. Inst.* **1989**, *298*, 26–37.
6. Czerwik, S. Contraction mappings in $b-$metric spaces. *Acta Math. Inform. Univ. Ostrav.* **1993**, *1*, 5–11.
7. Yamaod, O.; Sintunavarat, W.; Je Cho, Y. Existence of a common solution for a system of nonlinear integral equations via fixed point methods in $b-$metric spaces. *Open Math.* **2017**, *14*, 128–145. [CrossRef]
8. Aydi, H. $\alpha-$implicit contractive pair of mappings on quasi $b-$metric spaces and an application to integral equations. *J. Nonlinear Convex Anal.* **2016**, *17*, 2417–2433.
9. Amini-Harandi, A. Metric-like spaces, partial metric spaces and fixed points. *Fixed Point Theory Appl.* **2012**, *204*. [CrossRef]
10. Sarwar, M.; Rahman, M. Fixed point theorems for Ciric's and generalized contractions in $b-$metric spaces. *Int. J. Anal. Appl.* **2015**, *7*, 70–78.
11. Păcurar, M. Sequences of almost contractions and fixed points in $b-$metric spaces. *Anal. Univ. Vest Timis. Ser. Mat-Inform.* **2010**, *48*, 125–137.
12. Roshan, J.R.; Parvaneh, V.; Altun, I. Some coincidence point results in ordered $b-$metric spaces and applications in a system of integral equations. *Appl. Math. Comput.* **2014**, *262*, 725–737.
13. Roshan, J.R.; Parvaneh, V.; Sedghi, S.; Shobkolaei, N.; Shatanawi, W. Common fixed points of almost generalized $(\psi, \phi)-$contractive mappings in ordered $b-$metric spaces. *Fixed Point Theory Appl.* **2013**, *2013*, 159.
14. Lukács, A.; Kajántó, S. Fixed point theorems for various types of $F-$contractions in complete $b-$metric spaces. *Fixed Point Theory Appl.* **2018**, *19*, 321–334. [CrossRef]
15. Aydi, H.; Felhi, A.; Sahmim, S. Common fixed points via implicit contractions on $b-$metric-like spaces. *J. Nonlinear Sci. Appl.* **2017**, *10*, 1524–1537. [CrossRef]
16. Nashine, H.K.; Kadelburg, Z. Existence of solutions of Cantilever Beam Problem via $\alpha - \beta - FG-$contractions in $b-$metric-like spaces. *Filomat* **2017**, *31*, 3057–3074. [CrossRef]

17. Aydi, H.; Felhi, A.; Sahmim, S. Ciric-Berinde fixed point theorems for multi-valued mappings on α complete metric-like spaces. *Filomat* **2017**, *31*, 3727–3740. [CrossRef]
18. Aydi H.; Karapinar, E. Fixed point results for generalized $\alpha - \psi$-contractions in metric-like spaces and applications. *Electron. J. Diff. Eq.* **2015**, *2015*, 1–15.
19. Alsulami, H.; Gulyaz, S.; Karapinar, E.; Erha, I.M. An Ulam stability result on quasi−b−metric-like spaces. *Open Math.* **2016** *14*, 1087–1103. [CrossRef]
20. Joshi, V.; Singh, D.; Petrusel, A. Existence results for integral equations and boundary value problems via fixed point theorems for generalized F−contractions in b−metric-like spaces. *J. Function Spaces* **2017**, *2017*, 1649864. [CrossRef]
21. Wardowski, D. Fixed points of a new type of contractive mappings in complete metric spaces. *Fixed Point Theory Appl.* **2012**, *2012*, 94. [CrossRef]
22. Hammad, H.A.; De la Sen, M. A coupled fixed point technique for solving coupled systems of functional and nonlinear integral equations. *Mathematics* **2019**, *7*, 634. [CrossRef]
23. Abbas, M.; Ali, B.; Romaguera, S. Fixed and periodic points of generalized contractions in metric spaces. *Fixed Point Theory Appl.* **2013**, *2013*, 243. [CrossRef]
24. Batra, R.; Vashistha, S. Fixed points of an F−contraction on metric spaces with a graph. *Int. J. Comput. Math.* **2014**, *91*, 2483–2490. [CrossRef]
25. Batra, R.; Vashistha, S.; Kumar, R. A coincidence point theorem for F−contractions on metric spaces equipped with an altered distance. *J. Math. Comput. Sci.* **2014**, *4*, 826–833.
26. Cosentino, M.; Vetro, P. Fixed point results for F−contractive mappings of Hardy-Rogers-type. *Filomat* **2014**, *28*, 715–722. [CrossRef]
27. Piri, H.; Kuman, P. Some fixed point theorems concerning F−contraction in complete metric spaces. *Fixed Point Theory Appl.* **2014**, *2014*, 210. [CrossRef]
28. Jaggi, D.S. Some unique fixed point theorems. *Indian J. Pure Appl. Math.* **1977**, *8*, 223–230.
29. Zoto, K.; Rhoades, B.E.; Radenović, S. Common fixed point theorems for a class of (s,q)−contractive mappings in b−metric-like spaces and application to integral equations. *Math. Slovaca* **2019**, *69*, 233–247. [CrossRef]
30. Berinde, V. Approximating fixed points of weak contractions using the Picard iteration. *Nonlinear Anal. Forum.* **2004**, *9*, 43–53.
31. Berinde, V. General constructive fixed point theorems for Ciric-type almost contractions in metric spaces. *Carpathian J. Math.* **2008**, *24*, 10–19.
32. Alghmandi, M.A.; Hussain, N.; Salimi, P. Fixed point and coupled fixed point theorems on b−metric-like spaces. *J. Inequal. Appl.* **2013**, *2013*, 402.
33. Hammad, H.A.; De la Sen, M. Solution of nonlinear integral equation via fixed point of cyclic α_L^ψ−rational contraction mappings in metric-like spaces. *Bull. Braz. Math. Soc. New Ser.* **2019**, 347. [CrossRef]
34. Hammad, H.A.; De la Sen, M. Generalized contractive mappings and related results in b−metric like spaces with an application. *Symmetry* **2019**, *11*, 667. [CrossRef]
35. Anitar, T.; Ritu, S. Some coincidence and common fixed point theorems concerning F−contraction and applications. *J. Int. Math. Virtual Inst.* **2018**, *8*, 181–198.
36. Saipara, P.; Khammahawong, K.; Kumam, P. Fixed-point theorem for a generalized almost Hardy-Rogers-type F−contraction on metric-like spaces. *Math. Meth. Appl. Sci.* **2019**. [CrossRef]

© 2020 by the authors. Licensee MDPI, Basel, Switzerland. This article is an open access article distributed under the terms and conditions of the Creative Commons Attribution (CC BY) license (http://creativecommons.org/licenses/by/4.0/).

Article

Direct and Inverse Fractional Abstract Cauchy Problems

Mohammed AL Horani [1], Angelo Favini [2,*] and Hiroki Tanabe [3]

[1] Department of Mathematics, The University of Jordan, Amman 11942, Jordan; horani@ju.edu.jo
[2] Dipartimento di Matematica, Universita di Bologna, Piazza di Porta S. Donato 5, 40126 Bologna, Italy
[3] Takarazuka, Hirai Sanso 12-13, Osaka 665-0817, Japan; bacbx403@jttk.zaq.ne.jp
* Correspondence: angelo.favini@unibo.it

Received: 7 September 2019; Accepted: 21 October 2019; Published: 25 October 2019

Abstract: We are concerned with a fractional abstract Cauchy problem for possibly degenerate equations in Banach spaces. This form of degeneration may be strong and some convenient assumptions about the involved operators are required to handle the direct problem. Moreover, we succeeded in handling related inverse problems, extending the treatment given by Alfredo Lorenzi. Some basic assumptions on the involved operators are also introduced allowing application of the real interpolation theory of Lions and Peetre. Our abstract approach improves previous results given by Favini–Yagi by using more general real interpolation spaces with indices $\theta, p, p \in (0, \infty]$ instead of the indices θ, ∞. As a possible application of the abstract theorems, some examples of partial differential equations are given.

Keywords: fractional derivative; abstract Cauchy problem; C_0−semigroup; inverse problem

1. Introduction

Consider the abstract equation
$$BMu - Lu = f \tag{1}$$
where B, M, L are closed linear operators on the complex Banach space E, the domain of L is contained in domain of M, i.e., $D(L) \subseteq D(M)$, $0 \in \rho(L)$, the resolvent set of L, $f \in E$ and u is the unknown. The first approach to handle existence and uniqueness of the solution u to (1) was given by Favini–Yagi [1], see in particular the monograph [2]. By using real interpolation spaces, see [3,4], suitable assumptions on the operators B, M, L guarantee that (1) has a unique solution. Such a result was improved by Favini, Lorenzi and Tanabe in [5], see also [6–8]. In order to describe the results, we list the basic assumptions:

(H_1) Operator B has a resolvent $(z - B)^{-1}$ for any $z \in \mathbb{C}$, $\operatorname{Re} z < a$, $a > 0$ satisfying
$$\|(z - B)^{-1}\|_{\mathcal{L}(E)} \leq \frac{c}{|\operatorname{Re} z| + 1}, \quad \operatorname{Re} z < a, \tag{2}$$
where $\mathcal{L}(E)$ denotes the space of all continuous linear operators from E into E.

(H_2) Operators L, M satisfy
$$\|M(\lambda M - L)^{-1}\|_{\mathcal{L}(E)} \leq \frac{c}{(|\lambda| + 1)^\beta} \tag{3}$$
for any $\lambda \in \Sigma_\alpha := \{z \in \mathbb{C} : \operatorname{Re} z \geq -c(1 + |\operatorname{Im} z|)^\alpha, \ c > 0, \ 0 < \beta \leq \alpha \leq 1\}$.

(H₃) Let A be the possibly multivalued linear operator $A = LM^{-1}$, $D(A) = M(D(L))$. Then A and B commute in the resolvent sense:
$$B^{-1}A^{-1} = A^{-1}B^{-1}.$$

Let $(E, D(B))_{\theta,\infty}$, $0 < \theta < 1$, denote the real interpolation space between E and $D(B)$. The main result holds

Theorem 1. *Let $\alpha + \beta > 1$, $2 - \alpha - \beta < \theta < 1$. Then under hypotheses (H₁)–(H₃), Equation (1) admits a unique strict solution u such that Lu, $BMu \in (E, D(B))_{\omega,\infty}$, $\omega = \theta - 2 + \alpha + \beta$, provided that $f \in (E, D(B))_{\theta,\infty}$.*

It is straightforward to verify that if B generates a bounded c_0–group in E, then assumption (H₁) holds for B. Analogously, if $-B$ generates a bounded c_0–semigroup in E, then assumption (H₁) holds for B. It was also shown, in a previous paper, that Theorem 1 works well for solving degenerate equations on the real axis, too, see [9].

The first aim of this paper is to extend Theorem 1 to the interpolation spaces $(E, D(B))_{\theta,p}$, $1 < p < \infty$. This affirmation is not immediate. Section 2 is devoted to this proof. In Section 3, we apply the abstract results to solve concrete differential equations. In Section 4, we handle related inverse problems. In Section 5, we study abstract equations generalizing second-order equations in time. In Section 6, we present our conclusions and remarks. For some related results, we refer to Guidetti [10] and Bazhlekova [11].

2. Fundamental Results

To begin with, we recall, from Favini–Yagi [2], p. 16, that if E_0, E_1 are two Banach spaces such that (E_0, E_1) is an interpolation couple, i.e., there exists a locally convex topological space X such that $E_i \subset X$, $i = 0, 1$, continuously, then the following injections
$$E_0 \cap E_1 \subset_d (E_0, E_1)_{\xi,q} \subset_d (E_0, E_1)_{\eta,1} \subset_d (E_0, E_1)_{\eta,\infty} \subset_d (E_0, E_1)_{\zeta,q} \subset E_0 + E_1$$
are true for $1 \le q < \infty$, $0 < \xi < \eta < \zeta < 1$, where $\subset_d$ denotes continuous and dense embedding. Moreover,
$$(E_0, E_1)_{\theta,q} \subset_d (E_0, E_1)_{\theta,r} \text{ for } 1 \le q < r < \infty, \quad 0 < \theta < 1.$$

Taking into account the previous embedding and Theorem 1, we easily deduce that if ϵ, ϵ_1 are suitable small positive numbers, since $(E, D(B))_{\theta+\epsilon,q} \subset (E, D(B))_{\theta,\infty}$, then Equation (1) admits a unique solution u with Lu, $BMu \in (E, D(B))_{\theta-2+\alpha+\beta,\infty}$ and Lu, $BMu \in (E, D(B))_{\theta-2+\alpha+\beta-\epsilon_1,q}$, that is a weaker result than case $q = \infty$.

Our aim is to extend Theorem 1 to $1 < p < \infty$. In order to establish the corresponding result, we need the following lemma concerning multiplicative convolution. We recall that $L^p_*(\mathbb{R}^+) = L^p(\mathbb{R}^+; t^{-1}dt)$ and that the multiplicative convolution of two (measurable) functions $f, g : \mathbb{R}^+ \to \mathbb{C}$ is defined by
$$(f * g)(x) = \int_0^\infty f(xt^{-1})g(t)t^{-1}\,dt$$
where the integral exists a.e. for $x \in \mathbb{R}_+$.

Lemma 1. *For any $f_1 \in L^p_*(\mathbb{R}^+)$ and $g \in L^1_*(\mathbb{R}^+)$, the multiplicative convolution $f_1 * g \in L^p_*(\mathbb{R}^+)$ and satisfies*
$$\|f_1 * g\|_{L^p_*} \le \|f_1\|_{L^p_*} \|g\|_{L^1_*}$$

Consider now the chain of estimates

$$t^{\theta+\alpha+\beta-2}\|B(B+t)^{-1}v\| \leq t^{\theta+\alpha+\beta-2} \int_0^\infty \frac{(1+y)^{3-\alpha-\beta-\theta}}{(1+y+t)} (1+y)^\theta \|B(B+1+y)^{-1}f\| \frac{dy}{1+y}$$

$$\leq t^{\theta+\alpha+\beta-2} \int_0^\infty \frac{y^{3-\alpha-\beta-\theta}}{y+t} y^\theta \|B(B+y)^{-1}f\| \frac{dy}{y}$$

$$= t^{\theta+\alpha+\beta-2} \int_0^\infty \frac{y^{3-\alpha-\beta-\theta}}{y(1+ty^{-1})} y^\theta \|B(B+y)^{-1}f\| \frac{dy}{y}$$

$$= \int_0^\infty \frac{(ty^{-1})^{\theta+\alpha+\beta-2}}{1+ty^{-1}} y^\theta \|B(B+y)^{-1}f\| \frac{dy}{y}$$

where

$$v = (2\pi i)^{-1} \int_\Gamma z^{-1}(zT-1)^{-1}B(B-z)^{-1}f\, dz, \quad T = ML^{-1},$$

$\Gamma = \Gamma_\alpha$ being the oriented contour

$$\Gamma = \{z = a - c(1+|y|)^\alpha + iy, \quad -\infty < y < \infty\},$$

with $a \in (c, c + a_0)$. Such a function v is the unique solution to $BTv - v = f$, that is, u with $v = Lu$ satisfies (1).

Let $f_1(y) = y^\theta \|B(B+y)^{-1}f\|$, $g(y) = \dfrac{y^{\theta+\alpha+\beta-2}}{1+y}$, $y \in \mathbb{R}^+$, and note that $f \in (E, D(B))_{\theta, p}$ if and only if $f_1 \in L_*^p(\mathbb{R}^+)$. Moreover, $g \in L_*^p(\mathbb{R}^+)$ since $\theta > 2 - \alpha - \beta$ and obviously $\theta < 3 - \alpha - \beta$. Therefore, from Lemma 1, we deduce that $f \in (E, D(B))_{\omega, p}$, where $\omega = \theta + \alpha + \beta - 2$. Thus, we can establish the fundamental result concerning Equation (1).

Theorem 2. *Let B, M, L be three closed linear operators on the Banach space E satisfying (H_1)–(H_3), $0 < \beta \leq \alpha \leq 1$. Then for all $f \in (E, D(B))_{\theta, p}$, $2 - \alpha - \beta < \theta < 1$, $1 < p < \infty$, Equation (1) admits a unique solution u. Moreover, $Lu, BMu \in (E, D(B))_{\omega, p}$, $\omega = \theta + \alpha + \beta - 2$.*

3. Fractional Derivative

Let $\tilde{\alpha} > 0$, $m = \lceil \tilde{\alpha} \rceil$ is the smallest integer greater or equal to $\tilde{\alpha}$, $I = (0, T)$ for some $T > 0$. Define

$$g_\beta(t) = \begin{cases} \frac{1}{\Gamma(\beta)} t^{\beta-1} & t > 0, \\ 0 & t \leq 0, \end{cases} \quad \beta \geq 0,$$

where $\Gamma(\beta)$ is the Gamma function. Note that $g_0(t) = 0$ because $\Gamma(0)^{-1} = 0$. The Riemann–Liouville fractional derivative of order $\tilde{\alpha}$, or, more precisely, the so-called left handed Riemann–Liouville fractional derivative of order $\tilde{\alpha}$, is defined for all $f \in L^1(I)$, $g_{m-\tilde{\alpha}} * f \in W^{m,1}(I)$ by

$$D_t^{\tilde{\alpha}} f(t) = D_t^m(g_{m-\tilde{\alpha}} * f)(t) = D_t^m J_t^{m-\tilde{\alpha}} f(t)$$

where $D_t^m := \dfrac{d^m}{dt^m}$, $m \in \mathbb{N}$. $D_t^{\tilde{\alpha}}$ is a left inverse of $J_t^{\tilde{\alpha}}$, but in general it is not a right inverse. The Riemann–Liouville fractional integral of order $\tilde{\alpha} > 0$ is defined as:

$$J_t^{\tilde{\alpha}} f(t) := (g_{\tilde{\alpha}} * f)(t), \quad f \in L^1(I), \ t > 0, \quad J_t^0 f(t) := f(t).$$

If X is a complex Banach space, $\tilde{\alpha} > 0$, then we define the operator $\mathcal{J}_{\tilde{\alpha}}$ as:

$$D(\mathcal{J}_{\tilde{\alpha}}) := L^p(I; X), \quad \mathcal{J}_{\tilde{\alpha}} u = g_{\tilde{\alpha}} * u, \quad p \in [1, \infty).$$

Define the spaces $R^{\tilde{\alpha},p}(I;X)$ and $R_0^{\tilde{\alpha},p}(I;X)$ as follows. If $\tilde{\alpha} \notin \mathbb{N}$, set

$$R^{\tilde{\alpha},p}(I;X) := \{u \in L^p(I;X) \: : \: g_{m-\tilde{\alpha}} * u \in W^{m,p}(I;X)\},$$

$$R_0^{\tilde{\alpha},p}(I;X) := \{u \in L^p(I;X) \: : \: g_{m-\tilde{\alpha}} * u \in W_0^{m,p}(I;X)\},$$

where

$$W_0^{m,p}(I;X) = \{y \in W^{m,p}(I;X), \: y^{(k)}(0) = 0, \: k = 0, 1, ..., m-1\}.$$

For the Sobolev space $W^{\beta,p}(I;X)$ of fractional order $\beta > 0$, we define

$$W_0^{\beta,p}(I;X) = \{y \in W^{\beta,p}(I;X), \: y^{(k)}(0) = 0, \: k = 0, 1, ..., \lfloor \beta - 1/p \rfloor\},$$

$\beta - 1/p \notin \mathbb{N}_0 = \mathbb{N} \cup \{0\}$, and $\lfloor \beta - 1/p \rfloor$ is the greatest integer less or equal to $\beta - 1/p$. If $\tilde{\alpha} \in \mathbb{N}$, we take

$$R^{\tilde{\alpha},p}(I;X) := W^{\tilde{\alpha},p}(I;X), \quad R_0^{\tilde{\alpha},p}(I;X) := W_0^{\tilde{\alpha},p}(I;X).$$

Denote the extensions of the operators of fractional differentiation in $L^p(I;X)$ by $\mathcal{L}_{\tilde{\alpha}}$, i.e.,

$$D(\mathcal{L}_{\tilde{\alpha}}) := R_0^{\tilde{\alpha},p}(I;X), \quad \mathcal{L}_{\tilde{\alpha}} u := D_t^{\tilde{\alpha}} u,$$

where $D_t^{\tilde{\alpha}}$ is the Riemann–Liouville fractional derivative. Notice that if $\tilde{\alpha} \in (0,1)$, $u \in D(\mathcal{L}_{\tilde{\alpha}})$, then $(g_{1-\tilde{\alpha}} * u)(0) = 0$.

We illustrate the previous abstract concepts in the following example

Example 1. *For $u \in L^p(I,X) = L^p(0,T;X)$ set $u(t) = 0$ for $t < 0$. Then, if $u \in W_0^{1,p}(I,X)$, we have $u \in W^{1,p}(-\infty,T;X)$. Let $U(\tau)$, $\tau \geq 0$, be the semigroup in $L^p(0,T;X)$ defined by*

$$(U(\tau)u)(t) = u(t-\tau), \quad t \in I.$$

Clearly $U(\tau) = 0$ if $\tau > T$. For $\mathrm{Re}\lambda > 0, t > 0$

$$\left(\int_0^\infty e^{-\lambda\tau} U(\tau) u \, d\tau\right)(t) = \int_0^\infty e^{-\lambda\tau}(U(\tau)u)(t)d\tau = \int_0^\infty e^{-\lambda\tau} u(t-\tau)d\tau$$
$$= \int_0^t e^{-\lambda\tau} u(t-\tau)d\tau = \int_0^t e^{-\lambda(t-s)} u(s)ds. \tag{4}$$

Since $D(D_t) = W_0^{1,p}(I,X)$, $D_t = d/dt$, equation $(\lambda + D_t)u = f$ is

$$\begin{cases} \lambda u(t) + u'(t) = f(t), \: 0 < t < T, \\ u(0) = 0. \end{cases}$$

The solution is

$$u(t) = \int_0^t e^{-\lambda(t-s)} f(s)ds,$$

i.e.,

$$((\lambda + D_t)^{-1} f)(t) = \int_0^t e^{-\lambda(t-s)} f(s)ds, \quad \lambda > 0. \tag{5}$$

From (4) and (5) it follows that

$$(\lambda + D_t)^{-1} = \int_0^\infty e^{-\lambda\tau} U(\tau)d\tau, \quad \lambda > 0.$$

Therefore $-D_t$ is the infinitesimal generator of the semigroup $U(\tau)$, $\tau \geq 0$. Let $\tilde{\alpha} > 0$. Then for $f \in L^p(I,X)$

$$(D_t^{-\tilde{\alpha}}f)(t) = \frac{1}{\Gamma(\tilde{\alpha})} \int_0^\infty \tau^{\tilde{\alpha}-1}(U(\tau)f)(t)d\tau = \frac{1}{\Gamma(\tilde{\alpha})} \int_0^t \tau^{\tilde{\alpha}-1} f(t-\tau)d\tau$$

$$= \frac{1}{\Gamma(\tilde{\alpha})} \int_0^t (t-s)^{\tilde{\alpha}-1} f(s)ds, \ 0 \leq t \leq T.$$

If $m \in \mathbb{N}$, $m-1 < \tilde{\alpha} < m$,

$$(D_t^{-\tilde{\alpha}}f)(t) = 0 \ \forall t \in [0,T] \iff \int_0^t (t-s)^{\tilde{\alpha}-1} f(s)ds = 0 \ \forall t \in [0,T]$$

$$\implies \int_0^\tau (\tau-t)^{m-\tilde{\alpha}-1} \int_0^t (t-s)^{\tilde{\alpha}-1} f(s)dsdt = 0 \ \forall \tau \in [0,T]$$

$$\implies \int_0^\tau \int_s^\tau (\tau-t)^{m-\tilde{\alpha}-1}(t-s)^{\tilde{\alpha}-1}dt f(s)ds = 0 \ \forall \tau \in [0,T]$$

$$\implies \frac{\Gamma(m-\tilde{\alpha})\Gamma(\tilde{\alpha})}{\Gamma(m)} \int_0^\tau (\tau-s)^{m-1} f(s)ds = 0 \ \forall \tau \in [0,T] \implies f(\tau) = 0 \ \forall \tau \in [0,T].$$

If $\tilde{\alpha} = m \in \mathbb{N}$ and

$$(D_t^{-\tilde{\alpha}}f)(t) = (D_t^{-m}f)(t) = \frac{1}{(m-1)!} \int_0^t (t-s)^{m-1} f(s)ds = 0 \ \forall t \in [0,T] \implies f(t) = 0 \ \forall t \in [0,T].$$

Therefore $D_t^{-\tilde{\alpha}}$ has an inverse which is denoted by $D_t^{\tilde{\alpha}}$. We have

$$D_t^{\tilde{\alpha}+\beta} = D_t^{\tilde{\alpha}} D_t^{\beta} \quad \forall \tilde{\alpha}, \beta \in \mathbb{R}.$$

Therefore, if $m \in \mathbb{N}$, $m-1 < \tilde{\alpha} < m$,

$$D_t^{\tilde{\alpha}} = D_t^m D_t^{\tilde{\alpha}-m} = D_t^m g_{m-\tilde{\alpha}}*, \quad D(D_t^{\tilde{\alpha}}) = \{u; g_{m-\tilde{\alpha}} * u \in D(D_t^m) = W_0^{m,p}(I;X)\} = R_0^{\tilde{\alpha},p}(I;X).$$

Let us now list the main properties of $\mathcal{L}_{\tilde{\alpha}}$, see [11], Lemma 1.8, p. 15.

Lemma 2. *Let $\tilde{\alpha} > 0$, $1 < p < \infty$, X a complex Banach space, and $\mathcal{L}_{\tilde{\alpha}}$ be the operator introduced above. Then*
(a) $\mathcal{L}_{\tilde{\alpha}}$ is closed, linear and densely defined
(b) $\mathcal{L}_{\tilde{\alpha}} = \mathcal{J}_{\tilde{\alpha}}^{-1}$
(c) $\mathcal{L}_{\tilde{\alpha}} = \mathcal{L}_1^{\tilde{\alpha}}$, the $\tilde{\alpha}$-th power of the operator $\mathcal{L}_1$
(d) if $\tilde{\alpha} \in (0,2)$, operator $\mathcal{L}_{\tilde{\alpha}}$ is positive with spectral angle $\omega_{\mathcal{L}_{\tilde{\alpha}}} = \tilde{\alpha}\pi/2$
(e) if $\tilde{\alpha} \in (0,1]$, then $\mathcal{L}_{\tilde{\alpha}}$ is m-accretive
(f) $R_0^{\tilde{\alpha},p}(I;X) \hookrightarrow C^{\tilde{\alpha}-1/p}(I;X)$, $\tilde{\alpha} > 1/p$, $\tilde{\alpha} - 1/p \notin \mathbb{N}$, see [11], Theorem 1.10, p. 17
(g) if $\tilde{\alpha}\gamma - 1/p \notin \mathbb{N}_0$,

$$\left(L^p(I;X), R_0^{\tilde{\alpha},p}(I;X)\right)_{\gamma,p} = W_0^{\tilde{\alpha}\gamma,p}(I;X),$$

see [11], Proposition 11, p. 18.

Statement (e) implies that if $\tilde{\alpha} \in (0,1]$,

$$\|\lambda(\lambda+\mathcal{L}_{\tilde{\alpha}})^{-1}\|_{L^p(I;X)} \leq C, \ |\arg \lambda| < \pi(1-\frac{\tilde{\alpha}}{2}).$$

However, this reads equivalently $\|(\lambda - \mathcal{L}_{\tilde{\alpha}})^{-1}\| \le C/|\lambda|$ provided that λ is in a sector of the complex plane containing $\operatorname{Re}\lambda \le 0$. Therefore, if $\tilde{\alpha} \le 1$, operator $\mathcal{L}_{\tilde{\alpha}} = \dfrac{d}{dt^{\tilde{\alpha}}} = D_t^{\tilde{\alpha}}$ satisfies assumption (H$_1$) in Theorem 1. Therefore, we can handle abstract equations of the type

$$D_t^{\tilde{\alpha}}(My(t)) = Ly(t) + f(t), \quad 0 \le t \le T,$$

in a Banach space X with an initial condition $(g_{1-\tilde{\alpha}} * u)(0) = 0$. Then the results follow easily from the abstract model.

Example 2. *Let M be the multiplication operator in $L^p(\Omega)$, Ω a bounded open set in $\mathbb{R}^n$ with a C^n boundary $\partial\Omega$, $1 < p < \infty$, by $m(x)$, m is continuous and bounded, and take $L = \Delta - c$, $D(L) = W^{2,p}(\Omega) \cap W_0^{1,p}(\Omega)$, $c > 0$. Then it is seen in Favini–Yagi [2], pp. 79–80,*

$$\|M(zM - L)^{-1}f\|_{L^p(\Omega)} \le \dfrac{c}{(1 + |z|)^{1/p}} \|f\|_{L^p(\Omega)}$$

for all z in a sector containing $\operatorname{Re} z \ge 0$.

In order to solve our problem, $0 < \tilde{\alpha} \le 1$,

$$D_t^{\tilde{\alpha}}(My(t)) = Ly(t) + f(t), \quad 0 \le t \le T,$$

we must recall, see (g) in Lemma 2, that if $\tilde{\alpha}\gamma - 1/p \notin \mathbb{N}$, the interpolation space

$$\left(L^p(I;X), R_0^{\tilde{\alpha},p}(I;X)\right)_{\gamma,p} = W_0^{\tilde{\alpha}\gamma,p}(I;X).$$

Therefore, using Theorem 2, for any $f \in W_0^{\tilde{\alpha}\theta,p}(I;X)$, $1 - \dfrac{1}{p} < \theta < 1$, $1 < p < \infty$, $\tilde{\alpha}\theta - \dfrac{1}{p} \notin \mathbb{N}_0$, the problem above admits a unique strict solution y such that

$$\Delta y, \; D_t^{\tilde{\alpha}} m(\cdot)y \in W_0^{\tilde{\alpha}(\theta + \frac{1}{p} - 1),p}(I;X).$$

Remark 1. *Since $\dfrac{-1}{p} < \tilde{\alpha}\theta - \dfrac{1}{p} < 1$, then the only integer that $\tilde{\alpha}\theta - \dfrac{1}{p}$ can take is the zero integer.*

We refer to to the monograph [2] for many further examples of concrete degenerate partial differential equations to which Theorem 2 applies.

4. Inverse Problems

Given the problem

$$D_t^{\tilde{\alpha}}(My(t)) = Ly(t) + f(t)z + h(t), \quad 0 \le t \le T, \tag{6}$$

then corresponding to an initial condition and following the strategy in various previous papers, see in particular Lorenzi [12], we could study existence and regularity of solutions (y, f) to the above problem such that $\Phi[My(t)] = g(t)$, where g is a complex-valued function on $[0, T]$. This is, of course, an inverse problem. Applying Φ to both sides of Equation (6) we get

$$D_t^{\tilde{\alpha}} g(t) = \Phi[Ly(t)] + \Phi[h(t)] + f(t)\Phi[z].$$

If $\Phi[z] \neq 0$, we obtain necessarily

$$f(t) = \dfrac{D_t^{\tilde{\alpha}} g(t) - \Phi[Ly(t)] - \Phi[h(t)]}{\Phi[z]}.$$

Therefore,
$$D_t^{\tilde{\alpha}}(My(t)) = Ly(t) + h(t) - \frac{\Phi[Ly(t)]}{\Phi[z]}z - \frac{\Phi[h(t)]}{\Phi[z]}z + \frac{D_t^{\tilde{\alpha}}g(t)}{\Phi[z]}z.$$

If L_1 is defined by
$$D(L_1) = D(L), \quad L_1 y = -\frac{\Phi[Ly(t)]}{\Phi[z]}z,$$
one can introduce assumptions on the given operators ensuring that the direct problem
$$D_t^{\tilde{\alpha}}(My(t)) = Ly(t) + L_1 y + h(t) - \frac{\Phi[h(t)]}{\Phi[z]}z + \frac{D_t^{\tilde{\alpha}}g(t)}{\Phi[z]}z$$
has a unique strict solution, see [13]. The main step is to verify that assumption (H$_2$) holds for the operators $L + L_1$ and M.

Introduce the multivalued linear operator $A := LM^{-1}$, $D(A) = M(D(L))$ such that (H$_2$) holds. This means that $\|(\lambda I - A)^{-1}\|_{\mathcal{L}(X)} \leq \frac{c}{(|\lambda|+1)^\beta}$, $\lambda \in \Sigma_\alpha$. Theorem 1 in [13], pp. 148–149, affirms that if L, L_1, M are closed linear operators on X, $D(L) \subseteq D(L_1) \subseteq D(M)$, $0 \in \rho(L)$, such that (H$_2$) holds and $L_1 \in \mathcal{L}\left(D(L), X_A^{\theta_1}\right)$, $1 - \beta < \theta_1 < 1$, where
$$X_A^{\theta_1} = \left\{u \in X, \sup_{t>0} t^{\theta_1}\|A^0(t-A)^{-1}u\|_X < \infty\right\},$$
with $A^0(t-A)^{-1} = -I + t(t-A)^{-1}$, then
$$\|M(\lambda M - L - L_1)^{-1}\|_{\mathcal{L}(X)} \leq c(1 + |\lambda|)^{-\beta}, \quad \forall \lambda \in \Sigma_\alpha, \ |\lambda| \text{ large}.$$

In order to apply this theorem in our case, we must suppose that z belongs to $X_A^{\theta_1}$ for some $\theta_1 \in (1 - \beta, 1)$. Then
$$D_t^{\tilde{\alpha}}(My(t)) = Ly(t) + L_1 y + h(t) - \frac{\Phi[h(t)]}{\Phi[z]}z + \frac{D_t^{\tilde{\alpha}}g(t)}{\Phi[z]}z$$
$$(g_{1-\tilde{\alpha}} * My)(0) = 0$$
will admit a unique strict solution y provided that
$$h(t) - \frac{\Phi[h(t)]}{\Phi[z]}z + \frac{D_t^{\tilde{\alpha}}g(t)}{\Phi[z]}z \in W_0^{\tilde{\alpha}\theta,p}(I;X)$$
with $\tilde{\alpha}\theta - 1/p \notin \mathbb{N}$ and then $D_t^{\tilde{\alpha}}(My(t))$ and $Ly(t) \in W_0^{\tilde{\alpha}(\theta+\alpha+\beta-2),p}(I;X)$, $\tilde{\alpha}(\theta + \alpha + \beta - 2) \notin \mathbb{N}$. Notice that if $\tilde{\alpha}\theta - 1/p \notin \mathbb{N}$, then $\tilde{\alpha}(\theta + \alpha + \beta - 2) - 1/p = \tilde{\alpha}\theta - \frac{1}{p} + \tilde{\alpha}(\alpha + \beta - 2) \notin \mathbb{N}$.

5. Application: Generalized Second-Order Abstract Equation

Let us consider the abstract equation, generalizing second-order equation in time,
$$B_2 C B_1 u + B B_1 u + A u = f$$
where A, B, C are some closed linear operators in the complex Banach space X, B_1, B_2 are suitable operators defined on suitable Banach spaces. The change of variables $B_1 u = v$ transforms the given equation to the system

$$B_1 u = v,$$
$$B_2 C v + B v + A u = f,$$

which can be written in the matrix form

$$\begin{bmatrix} B_1 & 0 \\ 0 & B_2 \end{bmatrix} \begin{bmatrix} I & 0 \\ 0 & C \end{bmatrix} \begin{bmatrix} u \\ v \end{bmatrix} + \begin{bmatrix} 0 & -I \\ A & B \end{bmatrix} \begin{bmatrix} u \\ v \end{bmatrix} = \begin{bmatrix} 0 \\ f \end{bmatrix}.$$

The basic idea is to use a convenient space and a domain of operator matrices. Noting

$$\mathbb{B} = \begin{bmatrix} B_1 & 0 \\ 0 & B_2 \end{bmatrix}, \quad M = \begin{bmatrix} I & 0 \\ 0 & C \end{bmatrix}, \quad L = \begin{bmatrix} 0 & I \\ -A & -B \end{bmatrix}, \quad F = \begin{bmatrix} 0 \\ f \end{bmatrix},$$

it assumes the form

$$(\mathbb{B} M - L) U = F, \quad U = (u, v)^T.$$

In order to simplify the argument, we take $D(B) \subseteq D(A) \cap D(C)$. Moreover, we assume that for all $z \in \Sigma_\alpha$, where

$$\Sigma_\alpha := \left\{ z \in \mathbb{C} : \operatorname{Re} z \geq -c(1 + |\operatorname{Im} z|)^\alpha, \ c > 0, \ 0 < \beta \leq \alpha \leq 1, \ \alpha + \beta > 1 \right\},$$

the involved operators satisfy

$$\| C(zC + B)^{-1} \|_{\mathcal{L}(X)} \leq \frac{c_1}{(1 + |z|)^\beta}, \tag{7}$$

which guarantees that the problem is of parabolic type. Take $Y = D(B) \times X$ with the usual product norm. Then it is shown in Favini–Yagi [2], page 184, that the resolvent estimate

$$\| M(zM - L)^{-1} \|_{\mathcal{L}(Y)} \leq \frac{c}{(1 + |z|)^\beta}, \quad \forall z \in \Sigma_\alpha$$

holds. Therefore assumption (H$_2$) is satisfied.

Take B_1 the Riemann–Liouville fractional derivative of order $\tilde{\alpha}$, $0 < \tilde{\alpha} \leq 1$, in $L^p(0, T; D(B))$, $1 < p < \infty$; similarly, take B_2 the Riemann–Liouville fractional derivative of order $\tilde{\beta}$, $0 < \tilde{\beta} \leq 1$, in $L^p(0, T; X)$, $1 < p < \infty$. Then assumptions (H$_1$) and (H$_2$) hold. Therefore, according to Theorem 2, see also Bazhlekova [11], problem

$$D_t^{\tilde{\alpha}} u = v,$$
$$D_t^{\tilde{\beta}} C v + B v + A u = f(t), \quad 0 \leq t \leq T,$$
$$(g_{1-\tilde{\alpha}} * u)(0) = 0, \quad (g_{1-\tilde{\beta}} * Cv)(0) = 0$$

admits a unique strict solution (u, v) in $L^p(0, T; D(B)) \times L^p(0, T; X)$, provided that $D(B) \subseteq D(A) \cap D(C)$, $f \in W_0^{\tilde{\beta}\theta, p}(I; X)$, $2 - \alpha - \beta < \theta < 1$, $0 < \beta \leq \alpha \leq 1$, $\alpha + \beta > 1$, $\tilde{\beta}\theta - 1/p \neq 0$. Moreover, $D_t^{\tilde{\alpha}} u = v \in W_0^{\tilde{\alpha}\omega, p}(I; D(B))$, $D_t^{\tilde{\beta}} C v \in W_0^{\tilde{\beta}\omega, p}(I; X)$, $Au + Bv \in W_0^{\tilde{\beta}\omega, p}(I; X)$, $\omega = \theta + \alpha + \beta - 2$, $\tilde{\alpha}\omega = \tilde{\alpha}\theta + \tilde{\alpha}(\alpha + \beta - 2) - \frac{1}{p} \notin \mathbb{N}_0$, $\tilde{\beta}\omega = \tilde{\beta}\theta + \tilde{\beta}(\alpha + \beta - 2) - \frac{1}{p} \notin \mathbb{N}_0$.

Example 3. *Consider the problem*

$$D_t^{\tilde{\beta}} \left(m(x) D_t^{\tilde{\alpha}} u \right) - \Delta D_t^{\tilde{\alpha}} u + A(x; D) u = f(x, t), \quad (x, t) \in \Omega \times [0, T],$$
$$u(x, 0) = 0, \quad x \in \Omega,$$
$$m(x) D_t^{\tilde{\alpha}} u(x, 0) = 0, \quad x \in \Omega,$$

where Ω is a bounded open set in $\mathbb{R}^n$, $n \geq 1$, with a smooth boundary $\partial\Omega$, $m \in L^\infty$, $m(x) \geq 0$ in Ω, $A(x;D)$ is a second order linear differential operator on Ω with continuous coefficients in $\overline{\Omega}$, $f(x,t)$ is a scalar valued continuous function on $\overline{\Omega} \times [0,T]$, then we take $B = -\Delta_1$, the Laplacian with respect to x, $D(B) = H_0^1(\Omega) \cap H^2(\Omega)$, $X = L^2(\Omega)$. Therefore, (H_2) holds with $\alpha = 1$, $\beta = 1/2$.

6. Conclusions

It was shown that the degenerate problem including Riemann–Liouville fractional derivative can be handled by means of a general abstract equation. Applications to degenerate fractional differential equations with some related inverse problems were studied. Moreover, generalized second-order abstract equations were well-treated.

Author Contributions: All authors have equally contributed to this work. All authors wrote, read, and approved the final manuscript.

Funding: There is no external fund.

Conflicts of Interest: The authors declare no conflict of interest.

References

1. Favini, A.; Yagi, A. Multivalued Linear Operators and Degenerate Evolution Equations. *Ann. Mat. Pura Appl.* **1993**, *163*, 353–384. [CrossRef]
2. Favini, A.; Yagi, A. *Degenerate Differential Equations in Banach Spaces*; Marcel Dekker. Inc.: New York, NY, USA, 1999.
3. Lions, J.L.; Peetre, J. Sur Une Classe d'Espaces d'Interpolation,Publications Mathmatiques de l'IHS. *Publ. Math. Inst. Hautes Etudes Sci.* **1964**, *19*, 5–68. [CrossRef]
4. Lions, J.L.; Magenes, E. *Non-Homogeneous Boundary Value Problems and Applications*; Springer: Berlin, Germany, 1972; Volume 1, p. 187.
5. Favini, A.; Lorenzi, A.; Tanabe, H. Degenerate Integrodifferential Equations of Parabolic Type With Robin Boundary Conditions: L^p–Theory. *J. Math. Anal. Appl.* **2017**, *447*, 579–665. [CrossRef]
6. Favini, A.; Lorenzi, A.; Tanabe, H. Direct and Inverse Degenerate Parabolic Differential Equations with Multi-Valued Operators. *Electron. J. Differ. Equ.* **2015**, *2015*, 1–22.
7. Favini, A.; Lorenzi, A.; Tanabe, H. Singular integro-differential equations of parabolic type. *Adv. Differ. Equ.* **2002**, *7*, 769–798.
8. Favini, A.; Tanabe, H. Degenerate Differential Equations and Inverse Problems. In Proceedings of the Partial Differential Equations, Osaka, Japan, 21–24 August 2013; pp. 89–100.
9. Al Horani, M.; Fabrizio, M.; Favini, A.; Tanabe, H. Direct and Inverse Problems for Degenerate Differential Equations. *Ann. Univ. Ferrara* **2018**, *64*, 227–241. [CrossRef]
10. Guidetti, D. On Maximal Regularity For The Cauchy-Dirichlet Mixed Parabolic Problem with Fractional Time Derivative. *arXiv* **2018**, arXiv:1807.05913.
11. Bazhlekova, E.G. *Fractional Evolution Equations in Banach Spaces*; Eindhoven University of Technology: Eindhoven, The Netherlands, 2001.
12. Lorenzi, A. *An Introduction to Identification Problems Via Functional Analysis*; VSP: Utrecht, The Netherland, 2001.
13. Favini, A.; Lorenzi, A.; Marinoschi, G.; Tanabe, H. Perturbation Methods and Identifcation Problems for Degenerate Evolution Systems. In *Advances in Mathematics, Contributions at the Seventh Congress of Romanian Mathematicians, Brasov, 2011*; Beznea, L., Brinzanescu, V., Iosifescu, M., Marinoschi, G., Purice, R., Timotin, D., Eds.; Publishing House of the Romanian Academy: Bucharest, Romania, 2013; pp. 145–156.

© 2019 by the authors. Licensee MDPI, Basel, Switzerland. This article is an open access article distributed under the terms and conditions of the Creative Commons Attribution (CC BY) license (http://creativecommons.org/licenses/by/4.0/).

Article

Fractional Cauchy Problems for Infinite Interval Case-II

Mohammed Al Horani [1], Mauro Fabrizio [2], Angelo Favini [2,*] and Hiroki Tanabe [3]

1. Department of Mathematics, The University of Jordan, Amman 11942, Jordan; horani@ju.edu.jo
2. Dipartimento di Matematica, Università di Bologna, Piazza di Porta S. Donato 5, 40126 Bologna, Italy; mauro.fabrizio@unibo.it
3. Takarazuka, Hirai Sanso 12-13, Osaka 665-0817, Japan; acbx403@jttk.zaq.ne.jp
* Correspondence: angelo.favini@unibo.it or favini@dm.unibo.it

Received: 29 October 2019; Accepted: 19 November 2019; Published: 2 December 2019

Abstract: We consider fractional abstract Cauchy problems on infinite intervals. A fractional abstract Cauchy problem for possibly degenerate equations in Banach spaces is considered. This form of degeneration may be strong and some convenient assumptions about the involved operators are required to handle the direct problem. Required conditions on spaces are also given, guaranteeing the existence and uniqueness of solutions. The fractional powers of the involved operator B_X have been investigated in the space which consists of continuous functions u on $[0, \infty)$ without assuming $u(0) = 0$. This enables us to refine some previous results and obtain the required abstract results when the operator B_X is not necessarily densely defined.

Keywords: fractional derivative; abstract Cauchy problem; evolution equations; degenerate equations

MSC: 26A33

---

1. Introduction

In recent years, many studies were devoted to the problem of recovering the solution u to

$$BMu - Lu = f \tag{1}$$

where B, M, and L are closed linear operators on the complex Banach space E with $D(L) \subseteq D(M)$, $0 \in \rho(L)$, $f \in E$ and u is unknown. The first approach to handle existence and uniqueness of the solution u to Equation (1) was given by Favini-Yagi [1] (see in particular the monograph [2]). By using the real interpolation space $(E, D(B))_{\theta,\infty}, 0 < \theta < 1$ (see [3,4]), suitable assumptions on the operators B, M, L guarantee that Equation (1) has a unique solution. This result was improved by Favini, Lorenzi, and Tanabe [5] (see also [6–8]).

In all cases, the basic assumptions read as follows:

(H$_1$) Operator B has a resolvent $(z - B)^{-1}$ for any $z \in \mathbb{C}$, $Re\, z < a$, $a > 0$ satisfying

$$\|(z - B)^{-1}\|_{\mathcal{L}(E)} \leq \frac{c}{|Re\, z| + 1}, \quad Re\, z < a. \tag{2}$$

(H$_2$) Operators L, M satisfy

$$\|M(zM - L)^{-1}\|_{\mathcal{L}(E)} \leq \frac{c}{(|z| + 1)^\beta} \tag{3}$$

for any $z \in \Sigma_\alpha := \{z \in \mathbb{C} : Re\, z \geq -c(1 + |Im\, z|)^\alpha, \ c > 0, \ 0 < \beta \leq \alpha \leq 1\}$.

(H₃) Let A be the possibly multivalued linear operator $A = LM^{-1}$, $D(A) = M(D(L))$. Then, A and B commute in the resolvent sense:

$$B^{-1}A^{-1} = A^{-1}B^{-1}.$$

Very recently, Al Horani et al. [9], see also [10], generalized the previous results to the interpolation space $(E, D(B))_{\theta,p}$, $1 < p \leq \infty$, i.e.,

Lemma 1. *Let B, M, L be three closed linear operators on the complex Banach space E satisfying (H₁)–(H₃), $0 < \beta \leq \alpha \leq 1$, $\alpha + \beta > 1$. Then, for all $f \in (E, D(B))_{\theta,p}$, $2 - \alpha - \beta < \theta < 1$, $1 < p \leq \infty$, Equation (1) admits a unique solution u such that $Lu, BMu \in (E, D(B))_{\theta,p}$.*

There are many choices of the operator B verifying Assumption (H₁). In [9], the authors handled the abstract equation of the form

$$D_t^{\tilde{\alpha}}(Mu(t)) - Lu(t) = f(t), \ 0 \leq t \leq T < \infty \tag{4}$$

in the Banach space X with initial condition $(g_{1-\tilde{\alpha}} * My)(0) = 0$, where

$$g_\beta(t) = \begin{cases} \frac{1}{\Gamma(\beta)} t^{\beta-1} & t > 0, \\ 0 & t \leq 0, \end{cases} \quad \beta \geq 0,$$

and $\Gamma(\beta)$ is the Gamma function.

For Riemann–Liouville derivative $D_t^{\tilde{\alpha}}$ of order $\tilde{\alpha}$, we address the monograph [11] (see also [12,13]). Very recent applications concerning Caputo fractional derivative operator are also discussed in [14] by the same authors using a completely different method than Sviridyuk's group (see [15,16]). Some related topics can be found in [17–19].

In [20], the authors extended the results of direct and inverse problems, given in [9], to degenerate differential equations on the half line $[0, \infty)$. Precisely, let X be a complex Banach space and

$$E = \{u \in C([0,\infty); X); u \text{ is uniformly continuous and bounded in } [0,\infty), u(0) = 0\}$$

endowed with the sup norm. If B_X is the operator defined by

$$D(B_X) = \{u \in C^1([0,\infty); X); u \text{ and } u' \text{ are uniformly continuous and bounded in } [0,\infty), u(0) = 0 = u'(0)\},$$
$$B_X u = u' \text{ for } u \in D(B_X),$$

and M, L are two closed linear operators in the complex Banach space E satisfying

$$\|M(\lambda M - L)^{-1}\| \leq \frac{\tilde{c}}{(|\lambda|+1)^\beta} \quad \forall \lambda \in \Sigma_\alpha = \{\lambda; \text{Re}\lambda \geq -c(1+|\text{Im}\lambda|)^\alpha\}, \ c > 0, \ \tilde{c} > 0,$$

$0 < \beta \leq \alpha \leq 1$, $0 < \tilde{\alpha} < 1$, then for all $f \in (E, D(B_X^{\tilde{\alpha}}))_{\theta,p}$, $2 - \alpha - \beta < \theta < 1$, $1 < p \leq \infty$, equation

$$B_X^{\tilde{\alpha}} Mu - Lu = f$$

admits a unique solution u. Moreover, $Lu, B_X^{\tilde{\alpha}} Mu \in (E, D(B_X^{\tilde{\alpha}}))_{\omega,p}$, $\omega = \theta + \alpha + \beta - 2$.

In this paper, we refine our results in [20] by investigating the fractional power of the operator B_X in the space of continuous functions u defined on $[0,\infty)$ without assuming $u(0) = 0$, i.e.,

$$E = \{u \in C([0,\infty); X); u \text{ is uniformly continuous and bounded in } [0,\infty)\},$$
$$\|u\|_E = \sup_{0 \le t < \infty} \|u(t)\|_X,$$

where X is a complex Banach space. In this case, B_X is not densely defined. In such a case, it is not known whether $B_X^\alpha B_X^\beta = B_X^{\alpha+\beta}$ is true or not, since in the proof of Lemma A2 of T. Kato [21] it seems it is essentially used that A is densely defined. To obtain our results on such a new space E, we should investigate the previous fractional power problem in case $A = B_X$.

The interpolation space $(E, D(B_X^{\tilde{\alpha}}))_{\theta,p}$, $0 < \theta < 1$, $p \in (1, \infty]$ could be characterized by using the famous results of P. Grisvard. Since the operator B_X is of type $(\pi/2, 1)$ and $\tilde{\alpha} \in (0,1)$, $-B_X^{\tilde{\alpha}}$ is the infinitesimal generator of an analytic semigroup $\{e^{-tB_X^{\tilde{\alpha}}}\}_{t>0}$ (see [2], Proposition 0.9, p. 19), the interpolation space $(E, D(B_X^{\tilde{\alpha}}))_{\theta,p}$ could be characterized by

$$(E, D(B_X^{\tilde{\alpha}}))_{\theta,p} = \left\{u \in E; \, \|t^{1-\theta} B_X^{\tilde{\alpha}} e^{-tB_X^{\tilde{\alpha}}} u\|_{L_p^*(E)} + \|u\|_E < \infty \right\}$$
$$= \left\{u \in E; \, \|\zeta^\theta B_X^{\tilde{\alpha}} (\zeta + B_X^{\tilde{\alpha}})^{-1} u\|_{L_p^*(E)} < \infty \right\},$$

where $L_p^*(E)$ denotes the space of all strongly measurable E−valued functions f on $(0,\infty)$ such that

$$\int_0^\infty \|f(t)\|_E^p \frac{dt}{t} < \infty, \quad 1 \le p < \infty,$$
$$\|f(t)\|_{L_\infty^*(E)} = \sup_{0 < t < \infty} \|f(t)\|_E, \quad p = \infty.$$

The following lemma is also needed:

Lemma 2. $\|f * g\|_{L_*^p} \le \|f\|_{L_*^p} \|g\|_{L_*^1} \quad \forall f \in L_*^p(\mathbb{R}_+), \, \forall g \in L_*^1(\mathbb{R}_+)$

Section 2 is devoted to our main results. In Section 3, we present our conclusions and remarks.

2. Main Results

Let X be a complex Banach space and

$$E = \{u \in C([0,\infty); X); u \text{ is uniformly continuous and bounded in } [0,\infty)\}, \quad (5)$$
$$\|u\|_E = \sup_{0 \le t < \infty} \|u(t)\|_X. \quad (6)$$

Let B_X be an operator defined by

$$D(B_X) = \{u \in C^1([0,\infty); X); u, u' \in E, u(0) = 0\}$$
$$= \{u \in C^1([0,\infty); X); u \text{ and } u' \text{ are uniformly continuous and bounded in } [0,\infty)$$
and u(0)=0\} \quad (7)
$$= \{u \in C^1([0,\infty); X); u \text{ and } u' \text{ are bounded and } u' \text{ is uniformly continuous in } [0,\infty)$$
and $u(0) = 0\}$,
$$B_X u = u' \text{ for } u \in D(B_X). \quad (8)$$

Let $f \in E$, $\text{Re}\lambda > 0$. Consider the problem

$$\frac{d}{dt}u(t) + \lambda u(t) = f(t), \quad 0 < t < \infty,$$
$$u(0) = 0.$$
(9)

The solution is
$$u(t) = \int_0^t e^{-\lambda(t-s)} f(s) ds,$$
(10)

and

$$\|u(t)\| = \left\|\int_0^t e^{-\lambda(t-s)} f(s) ds\right\| \leq \int_0^t e^{-\text{Re}\lambda(t-s)} ds \|f\|_E = \frac{1 - e^{-\text{Re}\lambda t}}{\text{Re}\lambda} \|f\|_E \leq \frac{1}{\text{Re}\lambda} \|f\|_E.$$

Hence, u is bounded in $[0, \infty)$, and so is $u' = f - \lambda u$. This implies that u is uniformly continuous in $[0, \infty)$. Furthermore, $u' = f - \lambda u$ is uniformly continuous. Therefore, $u \in D(B_X)$ and $(B_X + \lambda)u = f$. Since $f \in E$ is arbitrary, one concludes that $R(B_X + \lambda) = E$ and

$$((B_X + \lambda)^{-1} f)(t) = \int_0^t e^{-\lambda(t-s)} f(s) ds,$$
(11)

$$\|(B_X + \lambda)^{-1}\|_{\mathcal{L}(E)} \leq \frac{1}{\text{Re}\lambda} \quad \forall \lambda : \text{Re}\lambda > 0.$$
(12)

Here, we make some preparations. Suppose that A is a not necessarily densely defined closed linear operator in a Banach space X satisfying

(i) $\rho(-A) \supset \{\lambda; |\arg\lambda| < \pi - \omega\}, 0 < \omega < \pi$;
(ii) $\lambda(\lambda + A)^{-1}$ is uniformly bounded in each smaller sector $\{\lambda; |\arg\lambda| < \pi - \omega - \epsilon\}, 0 < \epsilon < \pi - \omega$; and
(iii) $\|\lambda(\lambda + A)^{-1}\|_{\mathcal{L}(X)} \leq M$ for $\lambda > 0$ with some $M > 0$.

The first Assumption (i) is equivalent to $\rho(A) \supset \{\lambda; \omega < |\arg\lambda| \leq \pi\}$.

CASE $0 \in \rho(A)$. Set for $\alpha > 0$

$$R_\alpha(\lambda) = -\frac{1}{2\pi i} \int_C (\lambda + z^\alpha)^{-1} (z - A)^{-1} dz, \quad \lambda \geq 0,$$
(13)

where C runs in the resolvent set of A from $+\infty e^{-i\theta}$ to $+\infty e^{i\theta}$, $\omega < \theta \leq \pi$, avoiding the negative real axis and 0, where $+\infty e^{\pm i\infty} = \lim_{r \to \infty} re^{\pm i\infty}$. Let $\lambda, \mu \geq 0$. Let C' be another contour which has the same property as C and is located to the right of C without intersecting C. Then,

$$R_\alpha(\lambda) R_\alpha(\mu) = \frac{1}{2\pi i} \int_C (\lambda + z^\alpha)^{-1}(z-A)^{-1} dz \frac{1}{2\pi i} \int_{C'} (\mu + \zeta^\alpha)^{-1}(\zeta - A)^{-1} d\zeta$$
$$= \left(\frac{1}{2\pi i}\right)^2 \int_{C'} \int_C (\lambda + z^\alpha)^{-1}(\mu + \zeta^\alpha)^{-1}(z-A)^{-1}(\zeta - A)^{-1} dz d\zeta.$$

$$R_\alpha(\lambda)R_\alpha(\mu) = \left(\frac{1}{2\pi i}\right)^2 \int_{C'}\int_C (\lambda+z^\alpha)^{-1}(\mu+\zeta^\alpha)^{-1}\frac{(z-A)^{-1}-(\zeta-A)^{-1}}{\zeta-z}dzd\zeta$$

$$= \left(\frac{1}{2\pi i}\right)^2 \int_{C'}\int_C (\lambda+z^\alpha)^{-1}(\mu+\zeta^\alpha)^{-1}\frac{(z-A)^{-1}}{\zeta-z}dzd\zeta$$

$$- \left(\frac{1}{2\pi i}\right)^2 \int_{C'}\int_C (\lambda+z^\alpha)^{-1}(\mu+\zeta^\alpha)^{-1}\frac{(\zeta-A)^{-1}}{\zeta-z}dzd\zeta$$

$$= \frac{1}{2\pi i}\int_C (\lambda+z^\alpha)^{-1}\frac{1}{2\pi i}\int_{C'}\frac{(\mu+\zeta^\alpha)^{-1}}{\zeta-z}d\zeta(z-A)^{-1}dz$$

$$- \frac{1}{2\pi i}\int_{C'}\frac{1}{2\pi i}\int_C\frac{(\lambda+z^\alpha)^{-1}}{\zeta-z}dz(\mu+\zeta^\alpha)^{-1}(\zeta-A)^{-1}d\zeta$$

$$= -\frac{1}{2\pi i}\int_{C'}(\lambda+\zeta^\alpha)^{-1}(\mu+\zeta^\alpha)^{-1}(\zeta-A)^{-1}d\zeta$$

$$= -\frac{1}{2\pi i}\int_{C'}\frac{[(\lambda+\zeta^\alpha)^{-1}-(\mu+\zeta^\alpha)^{-1}]}{\mu-\lambda}(\zeta-A)^{-1}d\zeta.$$

This yields

$$(\mu-\lambda)R_\alpha(\lambda)R_\alpha(\mu) = -\frac{1}{2\pi i}\int_{C'}(\lambda+\zeta^\alpha)^{-1}(\zeta-A)^{-1}d\zeta + \frac{1}{2\pi i}\int_{C'}(\mu+\zeta^\alpha)^{-1}(\zeta-A)^{-1}d\zeta$$
$$= R_\alpha(\lambda) - R_\alpha(\mu).$$

Hence, $\{R_\alpha(\lambda), \lambda \geq 0\}$ is a pseudo resolvent:

$$R_\alpha(\lambda) - R_\alpha(\mu) = (\mu-\lambda)R_\alpha(\lambda)R_\alpha(\mu), \tag{14}$$

and

$$R_\alpha(0) = -\frac{1}{2\pi i}\int_C z^{-\alpha}(z-A)^{-1}dz. \tag{15}$$

For $\alpha > 0$, $\beta > 0$,

$$R_\alpha(0)R_\beta(0) = \left(\frac{1}{2\pi i}\right)^2\int_{C'}\int_C z^{-\alpha}\zeta^{-\beta}(z-A)^{-1}(\zeta-A)^{-1}dzd\zeta$$

$$= \left(\frac{1}{2\pi i}\right)^2\int_{C'}\int_C z^{-\alpha}\zeta^{-\beta}\frac{(z-A)^{-1}-(\zeta-A)^{-1}}{\zeta-z}dzd\zeta$$

$$= \left(\frac{1}{2\pi i}\right)^2\int_{C'}\int_C z^{-\alpha}\zeta^{-\beta}\frac{(z-A)^{-1}}{\zeta-z}dzd\zeta - \left(\frac{1}{2\pi i}\right)^2\int_{C'}\int_C z^{-\alpha}\zeta^{-\beta}\frac{(\zeta-A)^{-1}}{\zeta-z}dzd\zeta.$$

The first term of the last side vanishes, and the second term is equal to

$$-\frac{1}{2\pi i}\int_{C'}\frac{1}{2\pi i}\int_C \frac{z^{-\alpha}}{\zeta-z}dz\,\zeta^{-\beta}(\zeta-A)^{-1}d\zeta = -\frac{1}{2\pi i}\int_{C'}\zeta^{-\alpha-\beta}(\zeta-A)^{-1}d\zeta = R_{\alpha+\beta}(0).$$

Therefore, the following formula is obtained:

$$R_\alpha(0)R_\beta(0) = R_{\alpha+\beta}(0), \quad \alpha > 0, \beta > 0. \tag{16}$$

By virtue of Cauchy's representation formula of holomorphic functions, one has

$$R_1(0) = -\frac{1}{2\pi i}\int_C z^{-1}(z-A)^{-1}dz = A^{-1}. \tag{17}$$

Let $0 < \alpha < 1$. Then,

$$R_{1-\alpha}(0)R_\alpha(0) = R_\alpha(0)R_{1-\alpha}(0) = R_1(0) = A^{-1}.$$

Therefore, if $R_\alpha(0)u = 0$, then $A^{-1}u = R_{1-\alpha}(0)R_\alpha(0)u = 0$. This implies $u = 0$. Hence, $R_\alpha(0)$ has an inverse. Since

$$R_\alpha(0)u - R_\alpha(\lambda)u = \lambda R_\alpha(0)R_\alpha(\lambda)u,$$

$R_\alpha(\lambda)u = 0$ implies $R_\alpha(0)u = 0$, and hence $u = 0$, $R_\alpha(\lambda)$ has an inverse $\forall \lambda \geq 0$. Since $\forall u \in X$

$$R_\alpha(\lambda)u - R_\alpha(\mu)u = (\mu - \lambda)R_\alpha(\lambda)R_\alpha(\mu)u,$$

one observes $R_\alpha(\mu)u \in R(R_\alpha(\lambda)) = D(R_\alpha(\lambda)^{-1})$, and

$$u - R_\alpha(\lambda)^{-1}R_\alpha(\mu)u = (\mu - \lambda)R_\alpha(\mu)u.$$

Let $v \in D(R_\alpha(\mu)^{-1}) = R(R_\alpha(\mu))$ and $v = R_\alpha(\mu)u$. Then, $u = R_\alpha(\mu)^{-1}v$, $v \in D(R_\alpha(\lambda)^{-1})$ and

$$R_\alpha(\mu)^{-1}v - R_\alpha(\lambda)^{-1}v = (\mu - \lambda)v.$$

This yields

$$R_\alpha(\mu)^{-1}v - \mu v = R_\alpha(\lambda)^{-1}v - \lambda v, \quad \forall v \in D(R_\alpha(\lambda)^{-1}) = D(R_\alpha(\mu)^{-1}).$$

Set $A^\alpha = R_\alpha(\lambda)^{-1} - \lambda$ for some $\lambda \geq 0$.
Then, $A^\alpha = R_\alpha(\lambda)^{-1} - \lambda \ \forall \lambda \geq 0$, and $R_\alpha(\lambda)^{-1} = A^\alpha + \lambda$.
This implies

$$R_\alpha(\lambda) = (\lambda + A^\alpha)^{-1}.$$

Furthermore, $A^\alpha = R_\alpha(0)^{-1}$ and

$$(A^\alpha)^{-1} = R_\alpha(0) = -\frac{1}{2\pi i}\int_C z^{-\alpha}(z - A)^{-1}dz.$$

Letting $\alpha = 1$, one gets $(A^1)^{-1} = R_1(0) = A^{-1}$. Hence, $A^1 = A$. Thus, writing $(A^\alpha)^{-1} = A^{-\alpha}$

$$A^{-\alpha} = R_\alpha(0) = -\frac{1}{2\pi i}\int_C z^{-\alpha}(z - A)^{-1}dz, \quad \alpha > 0,$$
$$A^{-\alpha}A^{-\beta} = R_\alpha(0)R_\beta(0) = R_{\alpha+\beta}(0) = A^{-\alpha-\beta}, \quad \alpha > 0, \ \beta > 0.$$

It is not difficult to show that the following relation holds if $0 < \alpha < 1$:

$$(\lambda + A^\alpha)^{-1} = \frac{\sin \pi\alpha}{\pi}\int_0^\infty \frac{\mu^\alpha}{\lambda^2 + 2\lambda\mu^\alpha \cos \pi\alpha + \mu^{2\alpha}}(\mu + A)^{-1}d\mu, \ \lambda \geq 0. \tag{18}$$

Proposition 1. *Let A be a not necessarily densely defined closed linear operator in a Banach space X satisfying (i)–(iii) and $0 \in \rho(A)$. Then, the fractional power A^α of A is defined for $\alpha > 0$, and the followings hold:*

$$A^\alpha A^\beta = A^{\alpha+\beta}, \alpha > 0, \beta > 0, \quad A^1 = A, \quad A^{-\alpha} = (A^\alpha)^{-1} \text{ is bounded}$$

and Equation (10) *holds.*

GENERAL CASE In what follows, we assume (i)–(iii). Let $0 < \alpha < 1$. Set for $\lambda > 0$

$$R_\alpha(\lambda) = \frac{\sin \pi\alpha}{\pi}\int_0^\infty \frac{\mu^\alpha}{\lambda^2 + 2\lambda\mu^\alpha \cos \pi\alpha + \mu^{2\alpha}}(\mu + A)^{-1}d\mu. \tag{19}$$

If $\epsilon > 0$, $A_\epsilon = A + \epsilon$ satisfies (i)–(iii), and $0 \in \rho(A_\epsilon)$. Hence, A_ϵ^α is defined, and

$$(\lambda + A_\epsilon^\alpha)^{-1} = \frac{\sin \pi\alpha}{\pi} \int_0^\infty \frac{\mu^\alpha}{\lambda^2 + 2\lambda\mu^\alpha \cos \pi\alpha + \mu^{2\alpha}} (\mu + A_\epsilon)^{-1} d\mu, \tag{20}$$

$$(\lambda + A_\epsilon^\alpha)^{-1} - (\mu + A_\epsilon^\alpha)^{-1} = (\mu - \lambda)(\lambda + A_\epsilon^\alpha)^{-1}(\mu + A_\epsilon^\alpha)^{-1}, \quad \lambda > 0, \mu > 0. \tag{21}$$

In view of (iii)

$$\|(\mu + A_\epsilon)^{-1} - (\mu + A)^{-1}\|_X = \|-\epsilon(\mu + \epsilon + A)^{-1}(\mu + A)^{-1}\|_X \leq \epsilon \frac{M}{\mu + \epsilon} \frac{M}{\mu}.$$

Therefore, with the aid of the dominated convergence theorem one obtains from Equations (17) and (18)

$$(\lambda + A_\epsilon^\alpha)^{-1} \to R_\alpha(\lambda), \quad \lambda > 0 \text{ as } \epsilon \to 0. \tag{22}$$

Thus, we have obtained:

Proposition 2. *Let A be a not necessarily densely defined closed linear operator in a Banach space X satisfying (i)–(iii). Then, the bounded operator valued function $\{R_\alpha(\lambda), \lambda > 0\}$ defined by Equation (19) is a pseudo resolvent:*

$$R_\alpha(\lambda) - R_\alpha(\mu) = (\mu - \lambda) R_\alpha(\lambda) R_\alpha(\mu), \quad \lambda > 0, \mu > 0. \tag{23}$$

CASE $A = B_X$. Let $\epsilon > 0$. Then, $B_X + \epsilon > 0$ has a bounded inverse and

$$\|(B_X + \epsilon + \lambda)^{-1}\|_{\mathcal{L}(E)} \leq \frac{1}{\text{Re}\lambda + \epsilon} < \frac{1}{\text{Re}\lambda} \quad \forall \lambda : \text{Re}\lambda \geq 0. \tag{24}$$

By virtue of Proposition 1, the fractional power $(B_X + \epsilon)^\alpha$ of $B_X + \epsilon$ is defined for $\alpha > 0$, and the followings hold:

$$(B_X + \epsilon)^\alpha (B_X + \epsilon)^\beta = (B_X + \epsilon)^{\alpha+\beta}, \alpha > 0, \beta > 0, \quad (B_X + \epsilon)^1 = B_X + \epsilon,$$
$$(B_X + \epsilon)^{-\alpha} = ((B_X + \epsilon)^\alpha)^{-1} \text{ is bounded}, \tag{25}$$

and for $0 < \alpha < 1$

$$(\lambda + (B_X + \epsilon)^\alpha)^{-1} = \frac{\sin \pi\alpha}{\pi} \int_0^\infty \frac{\mu^\alpha}{\lambda^2 + 2\lambda\mu^\alpha \cos \pi\alpha + \mu^{2\alpha}} (\mu + B_X + \epsilon)^{-1} d\mu, \quad \lambda \geq 0. \tag{26}$$

Especially,

$$(B_X + \epsilon)^{-\alpha} = \frac{\sin \pi\alpha}{\pi} \int_0^\infty \mu^{-\alpha} (\mu + B_X + \epsilon)^{-1} d\mu. \tag{27}$$

Therefore,

$$((\lambda + (B_X + \epsilon)^\alpha)^{-1} f)(t)$$
$$= \frac{\sin \pi\alpha}{\pi} \int_0^\infty \frac{\mu^\alpha}{\lambda^2 + 2\lambda\mu^\alpha \cos \pi\alpha + \mu^{2\alpha}} ((\mu + B_X + \epsilon)^{-1} f)(t) d\mu$$
$$= \frac{\sin \pi\alpha}{\pi} \int_0^\infty \frac{\mu^\alpha}{\lambda^2 + 2\lambda\mu^\alpha \cos \pi\alpha + \mu^{2\alpha}} \int_0^t e^{-(\mu+\epsilon)(t-s)} f(s) ds d\mu$$
$$= \frac{\sin \pi\alpha}{\pi} \int_0^t \int_0^\infty \frac{\mu^\alpha e^{-(\mu+\epsilon)(t-s)}}{\lambda^2 + 2\lambda\mu^\alpha \cos \pi\alpha + \mu^{2\alpha}} d\mu f(s) ds \tag{28}$$

and

$$((B_X + \epsilon)^{-\alpha} f)(t) = \frac{\sin \pi\alpha}{\pi} \int_0^t \int_0^\infty \mu^{-\alpha} e^{-(\mu+\epsilon)(t-s)} d\mu f(s) ds. \tag{29}$$

By the change of the independent variable $\mu(t-s) = \tau$,

$$\int_0^\infty \mu^{-\alpha} e^{-(\mu+\epsilon)(t-s)} d\mu = e^{-\epsilon(t-s)} \int_0^\infty \mu^{-\alpha} e^{-\mu(t-s)} d\mu$$
$$= e^{-\epsilon(t-s)} \int_0^\infty (t-s)^\alpha \tau^{-\alpha} e^{-\tau} (t-s)^{-1} d\tau = (t-s)^{\alpha-1} e^{-\epsilon(t-s)} \int_0^\infty \tau^{-\alpha} e^{-\tau} d\tau$$
$$= (t-s)^{\alpha-1} e^{-\epsilon(t-s)} \Gamma(1-\alpha).$$

Hence, using $\Gamma(\alpha)\Gamma(1-\alpha) = \pi/\sin\pi\alpha$, one observes

$$((B_X + \epsilon)^{-\alpha} f)(t) = \frac{1}{\Gamma(\alpha)} \int_0^t (t-s)^{\alpha-1} e^{-\epsilon(t-s)} f(s) ds. \tag{30}$$

Set for $\lambda > 0$,

$$R(\lambda) = \frac{\sin\pi\alpha}{\pi} \int_0^\infty \frac{\mu^\alpha}{\lambda^2 + 2\lambda\mu^\alpha \cos\pi\alpha + \mu^{2\alpha}} (\mu + B_X)^{-1} d\mu. \tag{31}$$

Then, in view of Equation (23),

$$R(\lambda) - R(\mu) = (\mu - \lambda) R(\lambda) R(\mu), \quad \lambda > 0, \ \mu > 0. \tag{32}$$

From Equations (12) and (31), it follows that

$$\|R(\lambda)\|_{\mathcal{L}(E)} \le \frac{\sin\pi\alpha}{\pi} \int_0^\infty \frac{\mu^{\alpha-1}}{(\lambda + \mu^\alpha \cos\pi\alpha)^2 + (\mu \sin\pi\alpha)^2} d\mu. \tag{33}$$

For $f \in E$, in view of Equations (11) and (31),

$$(R(\lambda)f)(t) = \frac{\sin\pi\alpha}{\pi} \int_0^\infty \frac{\mu^\alpha}{\lambda^2 + 2\lambda\mu^\alpha \cos\pi\alpha + \mu^{2\alpha}} ((\mu + B_X)^{-1} f)(t) d\mu$$
$$= \frac{\sin\pi\alpha}{\pi} \int_0^\infty \frac{\mu^\alpha}{\lambda^2 + 2\lambda\mu^\alpha \cos\pi\alpha + \mu^{2\alpha}} \int_0^t e^{-\mu(t-s)} f(s) ds d\mu$$
$$= \frac{\sin\pi\alpha}{\pi} \int_0^t \int_0^\infty \frac{\mu^\alpha e^{-\mu(t-s)}}{\lambda^2 + 2\lambda\mu^\alpha \cos\pi\alpha + \mu^{2\alpha}} d\mu f(s) ds. \tag{34}$$

Using

$$\Gamma(\alpha)\Gamma(1-\alpha) = \frac{\pi}{\sin\pi\alpha}, \quad \int_0^\infty \mu^{-\alpha} e^{-(t-s)\mu} d\mu = \Gamma(1-\alpha)(t-s)^{\alpha-1}, \tag{35}$$

one deduces from Equation (34)

$$(R(\lambda)f)(t) - \frac{1}{\Gamma(\alpha)} \int_0^t (t-s)^{\alpha-1} f(s) ds$$
$$= \frac{\sin\pi\alpha}{\pi} \int_0^t \int_0^\infty \frac{\mu^\alpha e^{-\mu(t-s)}}{\lambda^2 + 2\lambda\mu^\alpha \cos\pi\alpha + \mu^{2\alpha}} d\mu f(s) ds - \frac{\sin\pi\alpha}{\pi} \int_0^t \int_0^\infty \mu^{-\alpha} e^{-\mu(t-s)} d\mu f(s) ds$$
$$= \frac{\sin\pi\alpha}{\pi} \int_0^t \int_0^\infty \left(\frac{\mu^\alpha}{\lambda^2 + 2\lambda\mu^\alpha \cos\pi\alpha + \mu^{2\alpha}} - \mu^{-\alpha} \right) e^{-\mu(t-s)} d\mu f(s) ds.$$

One has

$$\left\|\int_0^t \int_0^\infty \left(\frac{\mu^\alpha}{\lambda^2 + 2\lambda\mu^\alpha \cos \pi\alpha + \mu^{2\alpha}} - \mu^{-\alpha}\right) e^{-\mu(t-s)} d\mu f(s) ds\right\|_X$$
$$\leq \int_0^t \left|\int_0^\infty \left(\frac{\mu^\alpha}{\lambda^2 + 2\lambda\mu^\alpha \cos \pi\alpha + \mu^{2\alpha}} - \mu^{-\alpha}\right) e^{-\mu(t-s)} d\mu\right| \|f(s)\|_X ds$$
$$\leq \int_0^t \int_0^\infty \left|\frac{\mu^\alpha}{(\lambda + \mu^\alpha \cos \pi\alpha)^2 + (\mu^\alpha \sin \pi\alpha)^2} - \mu^{-\alpha}\right| e^{-\mu(t-s)} d\mu ds \|f\|_E. \quad (36)$$

Since

$$\left|\frac{\mu^\alpha}{(\lambda + \mu^\alpha \cos \pi\alpha)^2 + (\mu^\alpha \sin \pi\alpha)^2} - \mu^{-\alpha}\right| e^{-\mu(t-s)}$$
$$\leq \left(\frac{\mu^\alpha}{(\lambda + \mu^\alpha \cos \pi\alpha)^2 + (\mu^\alpha \sin \pi\alpha)^2} + \mu^{-\alpha}\right) e^{-\mu(t-s)} \leq \left(\frac{1}{(\sin \pi\alpha)^2} + 1\right) \mu^{-\alpha} e^{-\mu(t-s)},$$
$$\int_0^t \int_0^\infty \mu^{-\alpha} e^{-\mu(t-s)} d\mu ds = \Gamma(1-\alpha) \int_0^t (t-s)^{\alpha-1} ds = \Gamma(1-\alpha) \frac{t^\alpha}{\alpha} < \infty$$

and

$$\left|\frac{\mu^\alpha}{(\lambda + \mu^\alpha \cos \pi\alpha)^2 + (\mu^\alpha \sin \pi\alpha)^2} - \mu^{-\alpha}\right| e^{-\mu(t-s)} \to 0 \text{ as } \lambda \to 0, \; \forall (\mu, s) \in (0, \infty) \times (0, t),$$

the last right hand-side of Inequality (36) tends to 0 as $\lambda \to 0$. Therefore,

$$(R(\lambda)f)(t) \to \frac{1}{\Gamma(\alpha)} \int_0^t (t-s)^{\alpha-1} f(s) ds \text{ as } \lambda \to 0, \; \forall t \in (0, \infty). \quad (37)$$

Suppose $R(\lambda)f = 0 \; \exists \lambda > 0$. Then, $R(\lambda)f = 0 \; \forall \lambda > 0$, i.e., $(R(\lambda)f)(t) \equiv 0 \; \forall \lambda > 0$. Hence, by virtue of Equation (37),

$$\frac{1}{\Gamma(\alpha)} \int_0^t (t-s)^{\alpha-1} f(s) ds \equiv 0.$$

This yields

$$0 \equiv \int_0^t (t-\tau)^{-\alpha} \int_0^\tau (\tau-s)^{\alpha-1} f(s) ds d\tau = \int_0^t \int_s^t (t-\tau)^{-\alpha} (\tau-s)^{\alpha-1} d\tau f(s) ds$$
$$= B(1-\alpha, \alpha) \int_0^t f(s) ds \implies f(t) \equiv 0.$$

Therefore, $R(\lambda)$ has an inverse $\forall \lambda > 0$. Set $B_X^\alpha = R(\lambda)^{-1} - \lambda$ for some $\lambda > 0$. Then, $B_X^\alpha = R(\lambda)^{-1} - \lambda, \forall \lambda > 0$, and $R(\lambda) = (\lambda + B_X^\alpha)^{-1}, \forall \lambda > 0$. Hence, in view of Equation (31),

$$(\lambda + B_X^\alpha)^{-1} = \frac{\sin \pi\alpha}{\pi} \int_0^\infty \frac{\mu^\alpha}{\lambda^2 + 2\lambda\mu^\alpha \cos \pi\alpha + \mu^{2\alpha}} (\mu + B_X)^{-1} d\mu, \; \lambda > 0. \quad (38)$$

By virtue of Equations (34) and (38), one observes that, for $\lambda > 0$ and $f \in E$,

$$((\lambda + B_X^\alpha)^{-1} f)(t) = \frac{\sin \pi\alpha}{\pi} \int_0^t \int_0^\infty \frac{\mu^\alpha e^{-\mu(t-s)}}{\lambda^2 + 2\lambda\mu^\alpha \cos \pi\alpha + \mu^{2\alpha}} d\mu f(s) ds, \; 0 < t < \infty. \quad (39)$$

Since

$$\left\|\int_0^t \int_0^\infty \frac{\mu^\alpha e^{-(\mu+\epsilon)(t-s)}}{\lambda^2 + 2\lambda\mu^\alpha \cos \pi\alpha + \mu^{2\alpha}} d\mu f(s) ds - \int_0^t \int_0^\infty \frac{\mu^\alpha e^{-\mu(t-s)}}{\lambda^2 + 2\lambda\mu^\alpha \cos \pi\alpha + \mu^{2\alpha}} d\mu f(s) ds\right\|_X$$

$$= \left\|\int_0^t \left(e^{-\epsilon(t-s)} - 1\right) \int_0^\infty \frac{\mu^\alpha e^{-\mu(t-s)}}{\lambda^2 + 2\lambda\mu^\alpha \cos \pi\alpha + \mu^{2\alpha}} d\mu f(s) ds\right\|_X$$

$$\leq \int_0^t \left(1 - e^{-\epsilon(t-s)}\right) \int_0^\infty \frac{\mu^\alpha e^{-\mu(t-s)}}{\lambda^2 + 2\lambda\mu^\alpha \cos \pi\alpha + \mu^{2\alpha}} d\mu \|f(s)\|_X ds$$

$$\leq \int_0^t \left(1 - e^{-\epsilon(t-s)}\right) \int_0^\infty \frac{\mu^\alpha e^{-\mu(t-s)}}{(\mu^\alpha \sin \pi\alpha)^2} d\mu ds \|f\|_E$$

$$= \frac{1}{(\sin \pi\alpha)^2} \int_0^t \left(1 - e^{-\epsilon(t-s)}\right) \int_0^\infty \mu^{-\alpha} e^{-\mu(t-s)} d\mu ds \|f\|_E$$

$$= \frac{\Gamma(1-\alpha)}{(\sin \pi\alpha)^2} \int_0^t \left(1 - e^{-\epsilon(t-s)}\right)(t-s)^{\alpha-1} ds \|f\|_E$$

$$= \frac{\Gamma(1-\alpha)}{(\sin \pi\alpha)^2} \int_0^t \left(1 - e^{-\epsilon s}\right) s^{\alpha-1} ds \|f\|_E \to 0 \text{ uniformly in } [0,T],\ 0 < T < \infty, \text{ as } \epsilon \to 0,$$

noting Equations (28) and (39), one observes that, if $\lambda > 0$,

$$((\lambda + (B_X + \epsilon)^\alpha)^{-1} f)(t) \to ((\lambda + B_X^\alpha)^{-1} f)(t)$$

uniformly in $[0, T]$, $0 < T < \infty$ as $\epsilon \to 0$. By virtue of Equations (12) and (38), one deduces

$$\|(\lambda + B_X^\alpha)^{-1}\|_{\mathcal{L}(E)} \leq \frac{\sin \pi\alpha}{\pi} \int_0^\infty \frac{\mu^{\alpha-1}}{\lambda^2 + 2\lambda\mu^\alpha \cos \pi\alpha + \mu^{2\alpha}} d\mu$$

$$= \frac{\sin \pi\alpha}{\pi} \int_0^\infty \frac{r^{\alpha-1}}{1 + 2r^\alpha \cos \pi\alpha + r^{2\alpha}} dr \frac{1}{\lambda} = \frac{1}{\lambda},\ \lambda > 0.$$

For an arbitrary λ with $\operatorname{Re}\lambda > 0$, let μ be so large that $\mu > |\lambda|^2/(2\operatorname{Re}\lambda)$. Then, $\mu > |\mu - \lambda|$. One has

$$(\lambda + B_X^\alpha)^{-1} = (\mu + B_X^\alpha + \lambda - \mu)^{-1} = \left(\left(I + (\lambda - \mu)(\mu + B_X^\alpha)^{-1}\right)(\mu + B_X^\alpha)\right)^{-1}$$

$$= (\mu + B_X^\alpha)^{-1} \left(I + (\lambda - \mu)(\mu + B_X^\alpha)^{-1}\right)^{-1} = (\mu + B_X^\alpha)^{-1} \sum_{n=0}^\infty (-1)^n (\lambda - \mu)^n (\mu + B_X^\alpha)^{-n}.$$

Hence,

$$\left\|(\lambda + B_X^\alpha)^{-1}\right\|_{\mathcal{L}(E)} \leq \left\|(\mu + B_X^\alpha)^{-1}\right\|_{\mathcal{L}(E)} \sum_{n=0}^\infty |\lambda - \mu|^n \left\|(\mu + B_X^\alpha)^{-1}\right\|_{\mathcal{L}(E)}^n$$

$$= \frac{1}{\mu} \sum_{n=0}^\infty \left(\frac{|\lambda - \mu|}{\mu}\right)^n = \frac{1}{\mu} \frac{1}{1 - |\lambda - \mu|/\mu} = \frac{1}{\mu - |\lambda - \mu|}.$$

Since

$$\frac{1}{\mu - |\lambda - \mu|} = \frac{\mu + |\lambda - \mu|}{\mu^2 - |\lambda - \mu|^2} = \frac{\mu + |\lambda - \mu|}{\mu^2 - (\mu^2 - 2\mu\operatorname{Re}\lambda + |\lambda|^2)} = \frac{\mu + |\lambda - \mu|}{2\mu\operatorname{Re}\lambda - |\lambda|^2}$$

$$= \frac{1 + |\lambda/\mu - 1|}{2\operatorname{Re}\lambda - |\lambda|^2/\mu} \longrightarrow \frac{1}{\operatorname{Re}\lambda} \text{ as } \mu \longrightarrow \infty,$$

one concludes

$$\rho(B_X^\alpha) \supset \{\lambda; \operatorname{Re}\lambda < 0\} \text{ and } \|(\lambda + B_X^\alpha)^{-1}\|_{\mathcal{L}(E)} \leq \frac{1}{\operatorname{Re}\lambda}, \quad \operatorname{Re}\lambda > 0. \tag{40}$$

One has $\forall \lambda > 0, \forall f \in D(B_X^\alpha)$

$$R(\lambda)^{-1} f = \lambda f + B_X^\alpha f \Longrightarrow \lambda R(\lambda) f + R(\lambda) B_X^\alpha f = f.$$

Hence,

$$(R(\lambda) B_X^\alpha f)(t) = f(t) - \lambda (R(\lambda) f)(t), \quad t > 0.$$

Since Equation (37) holds for any $f \in E$, one has

$$(R(\lambda) B_X^\alpha f)(t) \longrightarrow \frac{1}{\Gamma(\alpha)} \int_0^t (t-s)^{\alpha-1} (B_X^\alpha f)(s) ds, \quad \forall t \in (0, \infty),$$

$$\lambda (R(\lambda) f)(t) \longrightarrow 0, \quad \forall t \in (0, \infty).$$

Therefore, one obtains

$$\frac{1}{\Gamma(\alpha)} \int_0^t (t-s)^{\alpha-1} (B_X^\alpha f)(s) ds = f(t).$$

This implies

$$\frac{1}{\Gamma(\alpha)} \int_0^t (t-\tau)^{-\alpha} \int_0^\tau (\tau-s)^{\alpha-1} (B_X^\alpha f)(s) ds d\tau = \int_0^t (t-\tau)^{-\alpha} f(\tau) d\tau.$$

The left hand side is equal to

$$\frac{1}{\Gamma(\alpha)} \int_0^t \int_s^t (t-\tau)^{-\alpha} (\tau-s)^{\alpha-1} d\tau (B_X^\alpha f)(s) ds = \Gamma(1-\alpha) \int_0^t (B_X^\alpha f)(s) ds.$$

Hence,

$$\int_0^t (B_X^\alpha f)(s) ds = \frac{1}{\Gamma(1-\alpha)} \int_0^t (t-s)^{-\alpha} f(s) ds.$$

From this, it follows that

$$\int_0^t (t-\tau)^{\alpha-1} \int_0^\tau (B_X^\alpha f)(s) ds d\tau = \frac{1}{\Gamma(1-\alpha)} \int_0^t (t-\tau)^{\alpha-1} \int_0^\tau (\tau-s)^{-\alpha} f(s) ds d\tau.$$

By the change of the order of the integration

$$\frac{1}{\alpha} \int_0^t (t-s)^\alpha (B_X^\alpha f)(s) ds = \Gamma(\alpha) \int_0^t f(s) ds.$$

By the differentiation of both sides

$$\int_0^t (t-s)^{\alpha-1} (B_X^\alpha f)(s) ds = \Gamma(\alpha) f(t).$$

Therefore, B_X^α has an inverse $B_X^{-\alpha}$, and for $f \in D(B_X^{-\alpha})$

$$(B_X^{-\alpha} f)(t) = \frac{1}{\Gamma(\alpha)} \int_0^t (t-s)^{\alpha-1} f(s) ds, \quad 0 < t < \infty. \tag{41}$$

Consequently, the following proposition is established:

Proposition 3. *Let B_X be the operator defined by Equations (7) and (8).*

Then, B_X satisfies $\rho(B_X) \supset \{\lambda; \operatorname{Re}\lambda < 0\}$ and Equations (11) and (12) hold. The fractional power B_X^α, $0 < \alpha < 1$, of B_X is defined implicitly by Equation (38) or Equation (39). B_X^α has an inverse $B_X^{-\alpha}$ and for $f \in D(B_X^{-\alpha})$ Equation (41) holds.

Especially if $f \in D(B_X^{-\alpha})$, then the function $\int_0^t (t-s)^{\alpha-1} f(s) ds$ belongs to E. The converse is given in the next proposition.

Proposition 4. *Suppose that both functions f and $\frac{1}{\Gamma(\alpha)} \int_0^\cdot (\cdot - s)^{\alpha-1} f(s) ds$, $0 < \alpha < 1$, belong to E. Then, $f \in D(B_X^{-\alpha})$ and Equation (41) holds.*

Proof. As a preparation, we first consider the case of a finite interval. Let $0 < T < \infty$. Let

$$D(B_X) = \{u \in C^1([0,T];X); u(0) = 0\}, \quad B_X u = u'.$$

Then, $\forall \lambda \in \mathbb{C}$

$$((\lambda + B_X)^{-1} f)(t) = \int_0^t e^{-\lambda(t-s)} f(s) ds, \text{ especially } (B_X^{-1} f)(t) = \int_0^t f(s) ds, \ 0 \leq t \leq T. \quad (42)$$

Therefore, B_X satisfies the assumptions of Proposition 1 with $\omega = \pi/2$, and hence its fractional power B_X^α is defined for $0 < \alpha < 1$, and we have:

$$(\lambda + B_X^\alpha)^{-1} = \frac{\sin \pi \alpha}{\pi} \int_0^\infty \frac{\mu^\alpha}{\lambda^2 + 2\lambda \mu^\alpha \cos \pi \alpha + \mu^{2\alpha}} (\mu + B_X)^{-1} d\mu, \ \lambda \geq 0. \quad (43)$$

Analogously to Equation (40), the following statement is established:

$$\rho(B_X^\alpha) \supset \{\lambda; \operatorname{Re}\lambda < 0\} \text{ and } \|(\lambda + B_X^\alpha)^{-1}\|_{\mathcal{L}(C([0,T];X))} \leq \frac{1}{\operatorname{Re}\lambda}, \ \operatorname{Re}\lambda > 0. \quad (44)$$

It follows from Equations (42) and (43) that for $f \in C([0,T];X)$, $\lambda \geq 0$

$$((\lambda + B_X^\alpha)^{-1} f)(t) = \frac{\sin \pi \alpha}{\pi} \int_0^t \int_0^\infty \frac{\mu^\alpha e^{-\mu(t-s)}}{\lambda^2 + 2\lambda \mu^\alpha \cos \pi \alpha + \mu^{2\alpha}} d\mu f(s) ds, \ 0 \leq t \leq T. \quad (45)$$

Especially if $\lambda = 0$,

$$(B_X^{-\alpha} f)(t) = \frac{\sin \pi \alpha}{\pi} \int_0^t \int_0^\infty \mu^{-\alpha} e^{-\mu(t-s)} d\mu f(s) ds = \frac{1}{\Gamma(\alpha)} \int_0^t (t-s)^{\alpha-1} f(s) ds, \ 0 \leq t \leq T. \quad (46)$$

For $f \in C([0,T];X)$ and $\lambda > 0$,

$$(\lambda + B_X^\alpha)^{-1} (\lambda B_X^{-\alpha} + 1) f = (\lambda + B_X^\alpha)^{-1} (\lambda + B_X^\alpha) B_X^{-\alpha} f = B_X^{-\alpha} f. \quad (47)$$

From Equation (45) with f replaced by $(\lambda B_X^{-\alpha}+1)f$ and Equation (46), it follows that

$$\left((\lambda+B_X^\alpha)^{-1}(\lambda B_X^{-\alpha}+1)f\right)(t)$$
$$=\frac{\sin\pi\alpha}{\pi}\int_0^t\int_0^\infty\frac{\mu^\alpha e^{-\mu(t-s)}}{\lambda^2+2\lambda\mu^\alpha\cos\pi\alpha+\mu^{2\alpha}}d\mu\,((\lambda B_X^{-\alpha}+1)f)(s)ds$$
$$=\frac{\sin\pi\alpha}{\pi}\int_0^t\int_0^\infty\frac{\mu^\alpha e^{-\mu(t-s)}}{\lambda^2+2\lambda\mu^\alpha\cos\pi\alpha+\mu^{2\alpha}}d\mu\left(\frac{\lambda}{\Gamma(\alpha)}\int_0^s(s-\sigma)^{\alpha-1}f(\sigma)d\sigma+f(s)\right)ds$$
$$=\frac{\sin\pi\alpha}{\pi}\frac{\lambda}{\Gamma(\alpha)}\int_0^t\left(\int_0^\infty\frac{\mu^\alpha e^{-\mu(t-s)}}{\lambda^2+2\lambda\mu^\alpha\cos\pi\alpha+\mu^{2\alpha}}d\mu\right)\int_0^s(s-\sigma)^{\alpha-1}f(\sigma)d\sigma ds$$
$$+\frac{\sin\pi\alpha}{\pi}\int_0^t\int_0^\infty\frac{\mu^\alpha e^{-\mu(t-s)}}{\lambda^2+2\lambda\mu^\alpha\cos\pi\alpha+\mu^{2\alpha}}d\mu f(s)ds. \tag{48}$$

By the changes of the order of integration,

$$\int_0^t\left(\int_0^\infty\frac{\mu^\alpha e^{-\mu(t-s)}}{\lambda^2+2\lambda\mu^\alpha\cos\pi\alpha+\mu^{2\alpha}}d\mu\right)\int_0^s(s-\sigma)^{\alpha-1}f(\sigma)d\sigma ds$$
$$=\int_0^t\int_\sigma^t\left(\int_0^\infty\frac{\mu^\alpha e^{-\mu(t-s)}(s-\sigma)^{\alpha-1}}{\lambda^2+2\lambda\mu^\alpha\cos\pi\alpha+\mu^{2\alpha}}d\mu\right)dsf(\sigma)d\sigma$$
$$=\int_0^t\int_0^\infty\frac{\mu^\alpha}{\lambda^2+2\lambda\mu^\alpha\cos\pi\alpha+\mu^{2\alpha}}\int_\sigma^t e^{-\mu(t-s)}(s-\sigma)^{\alpha-1}dsd\mu f(\sigma)d\sigma. \tag{49}$$

Substituting Equation (49) (with s and σ interchanged) into Equation (48), one deduces

$$\left((\lambda+B_X^\alpha)^{-1}(\lambda B_X^{-\alpha}+1)f\right)(t)$$
$$=\frac{\sin\pi\alpha}{\pi}\frac{\lambda}{\Gamma(\alpha)}\int_0^t\int_0^\infty\frac{\mu^\alpha}{\lambda^2+2\lambda\mu^\alpha\cos\pi\alpha+\mu^{2\alpha}}\int_s^t e^{-\mu(t-\sigma)}(\sigma-s)^{\alpha-1}d\sigma d\mu f(s)ds *$$
$$+\frac{\sin\pi\alpha}{\pi}\int_0^t\int_0^\infty\frac{\mu^\alpha e^{-\mu(t-s)}}{\lambda^2+2\lambda\mu^\alpha\cos\pi\alpha+\mu^{2\alpha}}d\mu f(s)ds$$
$$=\frac{\sin\pi\alpha}{\pi}\int_0^t\int_0^\infty\frac{\mu^\alpha}{\lambda^2+2\lambda\mu^\alpha\cos\pi\alpha+\mu^{2\alpha}}$$
$$\times\left[\frac{\lambda}{\Gamma(\alpha)}\int_s^t e^{-\mu(t-\sigma)}(\sigma-s)^{\alpha-1}d\sigma+e^{-\mu(t-s)}\right]d\mu f(s)ds. \tag{50}$$

From Equations (46), (47), and (50), it follows that the following equality holds $\forall f\in C([0,T];X)$:

$$\frac{\sin\pi\alpha}{\pi}\int_0^t\int_0^\infty\frac{\mu^\alpha}{\lambda^2+2\lambda\mu^\alpha\cos\pi\alpha+\mu^{2\alpha}}\left[\frac{\lambda}{\Gamma(\alpha)}\int_s^t e^{-\mu(t-\sigma)}(\sigma-s)^{\alpha-1}d\sigma+e^{-\mu(t-s)}\right]d\mu f(s)ds$$
$$=\frac{1}{\Gamma(\alpha)}\int_0^t(t-s)^{\alpha-1}f(s)ds,\ 0\le t\le T.$$

This yields that, for $<t\le T$:

$$\frac{\sin\pi\alpha}{\pi}\int_0^\infty\frac{\mu^\alpha}{\lambda^2+2\lambda\mu^\alpha\cos\pi\alpha+\mu^{2\alpha}}\left[\frac{\lambda}{\Gamma(\alpha)}\int_0^t e^{-\mu(t-\sigma)}\sigma^{\alpha-1}d\sigma+e^{-\mu t}\right]d\mu=\frac{t^{\alpha-1}}{\Gamma(\alpha)}. \tag{51}$$

Since $T>0$ is arbitrary, one concludes that Equation (51) holds for $0<t<\infty$.

We return to the case of the infinite interval $(0, \infty)$. Suppose that both functions f and $u(t) = \frac{1}{\Gamma(\alpha)} \int_0^t (t-s)^{\alpha-1} f(s) ds$ belong to E. One has by virtue of Equation (39) with $\lambda = 1$

$$\left((1 + B_X^\alpha)^{-1} u\right)(t) = \frac{\sin \pi \alpha}{\pi} \int_0^t \int_0^\infty \frac{\mu^\alpha e^{-\mu(t-s)}}{1 + 2\mu^\alpha \cos \pi \alpha + \mu^{2\alpha}} d\mu u(s) ds$$

$$= \frac{\sin \pi \alpha}{\pi} \int_0^t \left(\int_0^\infty \frac{\mu^\alpha e^{-\mu(t-s)}}{1 + 2\mu^\alpha \cos \pi \alpha + \mu^{2\alpha}} d\mu\right) \frac{1}{\Gamma(\alpha)} \int_0^s (s-\sigma)^{\alpha-1} f(\sigma) d\sigma ds$$

$$= \frac{\sin \pi \alpha}{\pi \Gamma(\alpha)} \int_0^t \int_\sigma^t \left(\int_0^\infty \frac{\mu^\alpha e^{-\mu(t-s)}}{1 + 2\mu^\alpha \cos \pi \alpha + \mu^{2\alpha}} d\mu\right) (s-\sigma)^{\alpha-1} ds f(\sigma) d\sigma$$

$$= \frac{\sin \pi \alpha}{\pi} \int_0^t \int_0^\infty \frac{\mu^\alpha}{1 + 2\mu^\alpha \cos \pi \alpha + \mu^{2\alpha}} \frac{1}{\Gamma(\alpha)} \int_\sigma^t e^{-\mu(t-s)} (s-\sigma)^{\alpha-1} ds d\mu f(\sigma) d\sigma, \quad (52)$$

and

$$\left((1 + B_X^\alpha)^{-1} f\right)(t) = \frac{\sin \pi \alpha}{\pi} \int_0^t \int_0^\infty \frac{\mu^\alpha e^{-\mu(t-s)}}{1 + 2\mu^\alpha \cos \pi \alpha + \mu^{2\alpha}} d\mu f(s) ds. \quad (53)$$

Adding Equations (52) and (53), and using Equation (51) with $\lambda = 1$, one observes

$$\left((1 + B_X^\alpha)^{-1}(u + f)\right)(t)$$

$$= \frac{\sin \pi \alpha}{\pi} \int_0^t \int_0^\infty \frac{\mu^\alpha}{1 + 2\mu^\alpha \cos \pi \alpha + \mu^{2\alpha}} \left[\frac{1}{\Gamma(\alpha)} \int_s^t e^{-\mu(t-\sigma)} (\sigma - s)^{\alpha-1} d\sigma + e^{-\mu(t-s)}\right] d\mu f(s) ds$$

$$= \frac{1}{\Gamma(\alpha)} \int_0^t (t-s)^{\alpha-1} f(s) ds = u(t).$$

This yields that $u \in D(B_X^\alpha)$ and $u + f = (I + B_X^\alpha) u = u + B_X^\alpha u$. Consequently, $f \in D(B_X^{-\alpha})$ and $B_X^{-\alpha} f = u$. □

In view of Propositions 3 and 4, the following statement is obtained:

Corollary 1. *Let $f \in E$. Then, $f \in D(B_X^{-\alpha})$ if and only if $\int_0^{\cdot} (\cdot - s)^{\alpha-1} f(s) ds \in E$. For $f \in D(B_X^{-\alpha})$, Equation (41) holds.*

For $f \in E$, $\alpha > 0$, $\beta > 0$,

$$\frac{1}{\Gamma(\alpha)} \int_0^t (t-s)^{\alpha-1} \frac{1}{\Gamma(\beta)} \int_0^s (s-\sigma)^{\beta-1} f(\sigma) d\sigma ds$$

$$= \frac{1}{\Gamma(\alpha)\Gamma(\beta)} \int_0^t \int_\sigma^t (t-s)^{\alpha-1} (s-\sigma)^{\beta-1} ds f(\sigma) d\sigma = \frac{1}{\Gamma(\alpha + \beta)} \int_0^t (t-\sigma)^{\alpha+\beta-1} f(\sigma) d\sigma.$$

Suppose $f \in D(B_X^{-\beta})$, $0 < \beta < 1$. Then, in view of Corollary 1, $\int_0^{\cdot} (\cdot - s)^{\beta-1} f(s) ds \in E$. Hence,

$$\frac{1}{\Gamma(\alpha)} \int_0^t (t-s)^{\alpha-1} (B_X^{-\beta} f)(s) d\sigma ds = \frac{1}{\Gamma(\alpha + \beta)} \int_0^t (t-\sigma)^{\alpha+\beta-1} f(\sigma) d\sigma.$$

Therefore, under the assumption $f \in D(B_X^{-\beta})$ $B_X^{-\beta} f \in D(B_X^{-\alpha})$ if and only if $f \in D(B_X^{-\alpha-\beta})$, and in this case $B_X^{-\alpha} B_X^{-\beta} f = B_X^{-\alpha-\beta} f$ holds. In particular, it is obtained that

$$B_X^{-\alpha} B_X^{-\beta} \subset B_X^{-\alpha-\beta}. \quad (54)$$

PROBLEM $\beta > \alpha \implies D(B_X^{-\beta}) \subset D(B_X^{-\alpha})$?

Let $0 < \tilde{\alpha} < 1, 0 < \beta \leq \alpha \leq 1$. Let L and M be densely defined closed linear operators in X such that $0 \in \rho(L), D(L) \subset D(M)$ and

(H)
$$\|M(\lambda M - L)^{-1}\|_{\mathcal{L}(X)} \leq \frac{\tilde{c}}{(|\lambda| + 1)^\beta}, \ \tilde{c} > 0, \tag{55}$$
$$\forall \lambda \in \Sigma_\alpha = \{\lambda; \text{Re}\lambda \geq -c(1 + |\text{Im}\lambda|)^\alpha\}, \ c > 0.$$

Consider the equation
$$B_X^{\tilde{\alpha}} Mu - Lu = f. \tag{56}$$

Let a_0 and a be such that
$$0 < a_0 < \min\{c, 1\}, \quad c < a < c + a_0. \tag{57}$$

Equation (56) is equivalent to
$$(B_X^{\tilde{\alpha}} + a_0)Mu - (L + a_0 M)u = f. \tag{58}$$

Since $-a_0 > -c$, in view of (H) $M(L + a_0 M)^{-1} \in \mathcal{L}(X)$ and if $\text{Re}(\lambda - a_0) \geq -c(1 + |\text{Im}(\lambda - a_0)|)^\alpha$,
$$\|M((\lambda - a_0)M - L)^{-1}\|_{\mathcal{L}(X)} \leq \frac{\tilde{c}}{(|\lambda - a_0| + 1)^\beta},$$

i.e., if $\text{Re}\lambda \geq a_0 - c(1 + |\text{Im}\lambda|)^\alpha$,
$$\|M(\lambda M - (L + a_0 M))^{-1}\|_{\mathcal{L}(X)} \leq \frac{\tilde{c}}{(|\lambda - a_0| + 1)^\beta}. \tag{59}$$

Since $|\lambda - a_0| + 1 \geq |\lambda| - a_0 + 1 = |\lambda| + (1 - a_0) \geq (1 - a_0)(|\lambda| + 1)$,
$$\frac{\tilde{c}}{(|\lambda - a_0| + 1)^\beta} \leq \frac{\tilde{c}}{(1 - a_0)^\beta (|\lambda| + 1)^\beta} = \frac{c_1}{(|\lambda| + 1)^\beta}, \ c_1 = \tilde{c}(1 - a_0)^{-\beta}.$$

Therefore,
$$\|M(\lambda M - (L + a_0 M))^{-1}\|_{\mathcal{L}(X)} \leq \frac{c_1}{(|\lambda| + 1)^\beta} \ \forall \lambda : \text{Re}\lambda \geq a_0 - c(1 + |\text{Im}\lambda|)^\alpha. \tag{60}$$

The inequality in Equation (40) implies
$$\|(B_X^{\tilde{\alpha}} + a_0 - \lambda)^{-1}\|_{\mathcal{L}(E)} \leq \frac{1}{a_0 - \text{Re}\lambda} \ \forall \lambda : \text{Re}\lambda < a_0. \tag{61}$$

By virtue of Equation (38),
$$(B_X^{\tilde{\alpha}} + a_0)^{-1} = \frac{\sin \pi \tilde{\alpha}}{\pi} \int_0^\infty \frac{\mu^{\tilde{\alpha}}}{a_0^2 + 2a_0 \mu^{\tilde{\alpha}} \cos \pi \tilde{\alpha} + \mu^{2\tilde{\alpha}}} (B_X + \mu)^{-1} d\mu. \tag{62}$$

Let $f \in L^p(0, \infty; X)$ and $T \in \mathcal{L}(X)$. Since
$$(T(B_X + \mu)^{-1} f)(t) = T \cdot ((B_X + \mu)^{-1} f)(t) = T \int_0^t e^{-\mu(t-s)} f(s) ds = \int_0^t e^{-\mu(t-s)} T f(s) ds$$
$$= \int_0^t e^{-\mu(t-s)} (Tf)(s) ds = ((B_X + \mu)^{-1} Tf)(t),$$

one observes $T(B_X + \mu)^{-1}f = (B_X + \mu)^{-1}Tf$, i.e., $T(B_X + \mu)^{-1} = (B_X + \mu)^{-1}T$. Therefore, with the aid of Equation (62), one obtains

$$T(B_X^{\tilde{a}} + a_0)^{-1} = (B_X^{\tilde{a}} + a_0)^{-1}T. \tag{63}$$

Applying this to $T = M(L + a_0 M)^{-1}$, one obtains

$$(B_X^{\tilde{a}} + a_0)^{-1} M(L + a_0 M)^{-1} = M(L + a_0 M)^{-1}(B_X^{\tilde{a}} + a_0)^{-1}. \tag{64}$$

Let Γ be the curve
$$\Gamma = \{z = a - c(1 + |y|)^\alpha + iy,\ y \in \mathbb{R}\}.$$

In view of Equation (57), one has $a_0 < c < a$, $a - c < a_0$. Hence, if $\lambda = a - c(1 + |y|)^\alpha + iy \in \Gamma$ with $y \in \mathbb{R}$, one has

$$a_0 - c(1 + |\mathrm{Im}\lambda|)^\alpha = a_0 - c(1 + |y|)^\alpha < a - c(1 + |y|)^\alpha = \mathrm{Re}\lambda \leq a - c < a_0.$$

Therefore, Equations (60) and (61) hold on Γ.

We verify
$$\|(B_X^{\tilde{a}} + a_0 - \lambda)^{-1}\|_{\mathcal{L}(E)} \leq \frac{c_2}{1 + |\mathrm{Re}\lambda|} \quad \forall \lambda \in \Gamma,\ c_2 = \frac{a - c + 1}{a_0 - a + c}. \tag{65}$$

We show that, if $\lambda \in \Gamma$, the following inequality holds:

$$a_0 - \mathrm{Re}\lambda \geq \frac{a_0 - a + c}{a - c + 1}(1 + |\mathrm{Re}\lambda|). \tag{66}$$

Note here $a_0 - a + c > 0$ and $a - c + 1 > a - c > 0$ in view of Equation (57). Hence,

$$a_0 - \frac{a_0 - a + c}{a - c + 1} = \frac{a_0 a - a_0 c + a - c}{a - c + 1} = \frac{(a_0 + 1)(a - c)}{a - c + 1} > 0. \tag{67}$$

This yields that

$$a_0 - \mathrm{Re}\lambda - \frac{a_0 - a + c}{a - c + 1}(1 + \mathrm{Re}\lambda) = a_0 - \mathrm{Re}\lambda - \frac{a_0 - a + c}{a - c + 1} - \frac{a_0 - a + c}{a - c + 1}\mathrm{Re}\lambda$$
$$= \frac{(a_0 + 1)(a - c)}{a - c + 1} - \frac{a_0 + 1}{a - c + 1}\mathrm{Re}\lambda = \frac{a_0 + 1}{a - c + 1}(a - c - \mathrm{Re}\lambda) \geq 0$$

if $\lambda \in \Gamma$. Therefore, Equation (66) holds if $\lambda \in \Gamma$ and $\mathrm{Re}\lambda \geq 0$. Recalling that $a_0 < 1$ we see if $\mathrm{Re}\lambda \leq 0$,

$$a_0 - \mathrm{Re}\lambda = a_0 + |\mathrm{Re}\lambda| \geq a_0(1 + |\mathrm{Re}\lambda|).$$

Therefore, recalling Equation (67) one observes that Equation (66) holds also in the case $\mathrm{Re}\lambda \leq 0$. Since $\mathrm{Re}\lambda \leq a - c < a_0$ if $\lambda \in \Gamma$, it follows from Equations (61) and (66) that Equation (65) holds. Set $B = B_X^{\tilde{a}} + a_0$, $T = M(L + a_0 M)^{-1}$. Then, Equations (61) and (64) are expressed as

$$B^{-1}T = TB^{-1}, \tag{68}$$

$$\|(B - \lambda)^{-1}\|_{\mathcal{L}(E)} \leq \frac{1}{a_0 - \mathrm{Re}\lambda} \quad \forall \lambda : \mathrm{Re}\lambda < a_0, \tag{69}$$

respectively. Let $v = (L + a_0 M)u$ be the new unknown variable. Then, $Mu = M(L + a_0 M)^{-1}v = Tv$ and Equation (58) is expressed as

$$BTv - v = f. \tag{70}$$

Our candidate of the solution to Equation (70) is

$$v = (2\pi i)^{-1} \int_\Gamma z^{-1}(zT-1)^{-1} B(B-z)^{-1} f\, dz. \tag{71}$$

We have
$$zT - 1 = zM(L+a_0 M)^{-1} - 1 = [zM - (L+a_0 M)](L+a_0 M)^{-1}.$$

If $z = a - c(1+|y|)^\alpha + iy \in \Gamma$, then $\operatorname{Re} z = a - c(1+|\operatorname{Im} z|)^\alpha > a_0 - c(1+|\operatorname{Im} z|)^\alpha$. Hence, in view of Equation (60)
$$\|M(zM - (L+a_0 M))^{-1}\|_{\mathcal{L}(X)} \le \frac{c_1}{(|z|+1)^\beta}.$$

Therefore, if $z \in \Gamma$,
$$\|(zT-1)^{-1}\|_{\mathcal{L}(E)} = \|(zT-1)^{-1}\|_{\mathcal{L}(X)} = \|(L+a_0 M)(zM - (L+a_0 M))^{-1}\|_{\mathcal{L}(X)}$$
$$= \|\{zM - (zM - (L+a_0 M))\}(zM - (L+a_0 M))^{-1}\|_{\mathcal{L}(X)}$$
$$= \|zM(zM - (L+a_0 M))^{-1} - I\|_{\mathcal{L}(X)} \le \|zM(zM - (L+a_0 M))^{-1}\|_{\mathcal{L}(X)} + 1$$
$$\le \frac{c_1 |z|}{(|z|+1)^\beta} + 1 \le c_1(|z|+1)^{1-\beta} + 1 \le c_2(|z|+1)^{1-\beta}, \quad c_2 = c_1 + 1. \tag{72}$$

This yields
$$\|v\|_E \le \frac{c_2}{2\pi} \int_\Gamma |z|^{-(1+\theta)}(1+|z|)^{1-\beta}|z|^\theta \|B(B-z)^{-1} f\|_E |dz|. \tag{73}$$

Let $z = a - c(1+|y|)^\alpha + iy \in \Gamma$. From
$$B(B-z)^{-1} = B(B+1+|y|)^{-1} + (1+|y|+z)(B-z)^{-1} B(B+1+|y|)^{-1}$$
$$= [1 + (1+|y|+z)(B-z)^{-1}] B(B+1+|y|)^{-1},$$

it follows that
$$\|B(B-z)^{-1} f\|_E = \|[1+(1+|y|+z)(B-z)^{-1}] B(B+1+|y|)^{-1} f\|_E$$
$$\le \|1 + (1+|y|+z)(B-z)^{-1}\|_{\mathcal{L}(E)} \|B(B+1+|y|)^{-1} f\|_E$$
$$\le \left(1 + \frac{1+|y|+|z|}{a_0 - \operatorname{Re} z}\right) \|B(B+1+|y|)^{-1} f\|_E. \tag{74}$$

Since
$$1 + |y| + |z| = 1 + |y| + |a - c(1+|y|)^\alpha + iy| \le 1 + |y| + a + c(1+|y|)^\alpha + |y|$$
$$= 1 + 2|y| + a + c(1+|y|)^\alpha,$$

one has
$$1 + \frac{1+|y|+|z|}{a_0 - \operatorname{Re} z} \le 1 + \frac{1+2|y|+a+c(1+|y|)^\alpha}{a_0 - a + c(1+|y|)^\alpha} = \frac{a_0 + 1 + 2|y| + 2c(1+|y|)^\alpha}{a_0 - a + c(1+|y|)^\alpha},$$

and
$$a_0 + 1 + 2|y| + 2c(1+|y|)^\alpha \le (a_0+1)(1+|y|) + 2(1+|y|) + 2c(1+|y|)$$
$$= (a_0 + 3 + 2c)(1+|y|).$$

Since $a_0 < a$,
$$a_0 - a + c(1+|y|)^\alpha \ge c(1+|y|)^\alpha - (a - a_0)(1+|y|)^\alpha = (c - a + a_0)(1+|y|)^\alpha.$$

Note here that $c - a + a_0 > 0$ (cf. Equation (57)). Hence,

$$1 + \frac{1 + |y| + |z|}{a_0 - \operatorname{Re} z} \leq \frac{(a_0 + 3 + 2c)(1 + |y|)}{(c - a + a_0)(1 + |y|)^\alpha} = c_3(1 + |y|)^{1-\alpha}, \quad c_3 = \frac{a_0 + 3 + 2c}{c - a + a_0}. \tag{75}$$

From Equations (74) and (75), it follows that

$$\|B(B - z)^{-1} f\|_E \leq c_3 (1 + |y|)^{1-\alpha} \|B(B + 1 + |y|)^{-1} f\|_E. \tag{76}$$

For $z = a - c(1 + |y|)^\alpha + iy \in \Gamma$, $y \in \mathbb{R}$, one has

$$\begin{aligned}|z| &= |a - c(1 + |y|)^\alpha + iy| \leq a + c(1 + |y|)^\alpha + |y| \leq a + c(1 + |y|) + |y| \\ &= a + c + (c + 1)|y| \leq c_4(1 + |y|), \quad c_4 = \max\{a, 1\} + c,\end{aligned} \tag{77}$$

$$(1 + |z|)^{1-\beta} \leq (1 + c_4(1 + |y|))^{1-\beta} \leq (1 + |y| + c_4(1 + |y|))^{1-\beta} = c_5(1 + |y|)^{1-\beta}, \\ c_5 = (1 + c_4)^{1-\beta}, \tag{78}$$

$$|1 + |y| + z| \leq 1 + |y| + |z| \leq 1 + |y| + c_4(1 + |y|) = (1 + c_4)(1 + |y|) = c_6(1 + |y|), \\ c_6 = 1 + c_4 = 1 + \max\{a, 1\} + c. \tag{79}$$

Here, we show that $\exists c_7$ such that

$$|z| \geq c_7(1 + |y|) \quad \forall z \in \Gamma : z = a - c(1 + |y|)^\alpha + iy, \ y \in \mathbb{R}. \tag{80}$$

Proof. Let $0 < b < a - c (\iff c - b < c < a - b)$.
(i) Case $|a - c(1 + |y|)^\alpha| \leq b$. In this case,

$$\begin{aligned}|a - c(1 + |y|)^\alpha| &\leq b \iff -b \leq a - c(1 + |y|)^\alpha \leq b \implies a - b \leq c(1 + |y|)^\alpha \\ &\iff \frac{a - b}{c} \leq (1 + |y|)^\alpha \iff \left(\frac{a - b}{c}\right)^{1/\alpha} \leq 1 + |y| \implies |y| \geq \left(\frac{a - b}{c}\right)^{1/\alpha} - 1 \equiv \delta > 0.\end{aligned}$$

Hence,

$$\frac{\delta}{1 + \delta}(1 + |y|) = \frac{\delta}{1 + \delta} + \frac{\delta|y|}{1 + \delta} \leq \frac{|y|}{1 + \delta} + \frac{\delta|y|}{1 + \delta} = |y|.$$

Therefore,

$$|z| = |a - c(1 + |y|)^\alpha + iy| \geq |y| \geq \frac{\delta}{1 + \delta}(1 + |y|).$$

(ii) Case $a - c(1 + |y|)^\alpha > b$. In this case

$$|z| \geq |a - c(1 + |y|)^\alpha| = a - c(1 + |y|)^\alpha > b,$$

and

$$a - c(1 + |y|)^\alpha > b \iff (1 + |y|)^\alpha < \frac{a - b}{c} \iff 1 + |y| < \left(\frac{a - b}{c}\right)^{1/\alpha} \\ \iff \left(\frac{c}{a - b}\right)^{1/\alpha}(1 + |y|) < 1.$$

Therefore,

$$|z| > b > b\left(\frac{c}{a - b}\right)^{1/\alpha}(1 + |y|).$$

(iii) Case $c(1+|y|)^\alpha - a > b$. In this case

$$c(1+|y|)^\alpha > a+b \iff (1+|y|)^\alpha > \frac{a+b}{c} \iff |y| > \left(\frac{a+b}{c}\right)^{1/\alpha} - 1 \equiv \gamma > 0.$$

Hence,

$$\frac{\gamma}{\gamma+1}(1+|y|) = \frac{\gamma}{\gamma+1} + \frac{\gamma}{\gamma+1}|y| < \frac{|y|}{\gamma+1} + \frac{\gamma}{\gamma+1}|y| = |y| \leq |z|.$$

Thus Equation (80) holds with $c_7 = \min\left\{\frac{\delta}{1+\delta}, b\left(\frac{c}{a-b}\right)^{1/\alpha}, \frac{\gamma}{\gamma+1}\right\}$. □

Hence, from Equations (73) and (76)–(80), it follows that

$$\|v\|_E \leq c_8 \int_\Gamma (1+|y|)^{1-\alpha-\beta-\theta}(1+|y|)^\theta \|B(B+1+|y|)^{-1}f\|_E |dz|,$$

where $c_8 = c_2 c_7^{-(1+\theta)} c_5 c_4^\theta c_3 / 2\pi$. For $z = a - c(1+|y|)^\alpha + iy$, $y \geq 0$

$$|dz| = |-c\alpha(1+y)^{\alpha-1}dy + idy| = \{(c\alpha)^2(1+y)^{2(\alpha-1)} + 1\}^{1/2} dy \leq ((c\alpha)^2 + 1)^{1/2} dy.$$

Therefore,

$$\|v\|_E \leq 2((c\alpha)^2+1)^{1/2} c_8 \int_0^\infty (1+y)^{1-\alpha-\beta-\theta}(1+y)^\theta \|B(B+1+y)^{-1}f\|_E dy$$

$$= 2((c\alpha)^2+1)^{1/2} c_8 \int_1^\infty y^{2-\alpha-\beta-\theta} y^\theta \|B(B+y)^{-1}f\|_E \frac{dy}{y}$$

$$\leq 2((c\alpha)^2+1)^{1/2} c_8 \left(\frac{p-1}{(\theta+\alpha+\beta-2)p}\right)^{(p-1)/p} \left(\int_1^\infty y^{\theta p}\|B(B+y)^{-1}f\|_E^p \frac{dy}{y}\right)^{1/p}$$

$$\leq 2((c\alpha)^2+1)^{1/2} c_8 \left(\frac{p-1}{(\theta+\alpha+\beta-2)p}\right)^{(p-1)/p} \|f\|_{(E,D(B))_{\theta,p}}.$$

Thus, it has been shown that v is well defined by Equation (71) if $f \in (E,D(B))_{\theta,p}$.

Next, we show that v satisfies Equation (70). We show that the following inequality holds with some constant c_9:

$$1 + |a - c(1+|y|)^\alpha| \geq c_9 (1+|y|)^\alpha \quad \forall y \in \mathbb{R}. \tag{81}$$

Proof. (i) Case $|a - c(1+|y|)^\alpha| \leq 1/2$. Since

$$|a - c(1+|y|)^\alpha| \leq \frac{1}{2} \iff -\frac{1}{2} \leq a - c(1+|y|)^\alpha \leq \frac{1}{2} \implies -1 \leq 2a - 2c(1+|y|)^\alpha$$

$$\iff 2c(1+|y|)^\alpha \leq 2a+1 \iff \frac{2c}{2a+1}(1+|y|)^\alpha \leq 1,$$

one observes

$$1 + |a - c(1+|y|)^\alpha| \geq 1 \geq \frac{2c}{2a+1}(1+|y|)^\alpha. \tag{82}$$

(ii) Case $a - c(1+|y|)^\alpha > 1/2$ (this can occur only in case $a - c > 1/2$). Since

$$a - c(1+|y|)^\alpha > \frac{1}{2} \iff a - \frac{1}{2} > c(1+|y|)^\alpha$$

$$\left(\text{multiplying both sides by } \frac{1+a}{a-1/2}\right) \implies 1+a > \frac{c(1+a)(1+|y|)^\alpha}{a-1/2},$$

one has

$$1 + a - c(1+|y|)^\alpha > \frac{c(1+a)(1+|y|)^\alpha}{a - 1/2} - c(1+|y|)^\alpha$$
$$= \left(\frac{1+a}{a-1/2} - 1\right)c(1+|y|)^\alpha = \frac{3/2}{a-1/2}c(1+|y|)^\alpha = \frac{3c}{2a-1}(1+|y|)^\alpha.$$

Therefore,

$$1 + |a - c(1+|y|)^\alpha| = 1 + a - c(1+|y|)^\alpha \geq \frac{3c}{2a-1}(1+|y|)^\alpha. \tag{83}$$

(iii) Case $c(1+|y|)^\alpha - a > 1/2$. In this case,

$$1 + |a - c(1+|y|)^\alpha| = 1 + c(1+|y|)^\alpha - a.$$

If $a \leq 1$,

$$1 + c(1+|y|)^\alpha - a \geq c(1+|y|)^\alpha. \tag{84}$$

If $a > 1$,

$$1 + c(1+|y|)^\alpha - a = c(1+|y|)^\alpha - (a-1) \geq c(1+|y|)^\alpha - (a-1)(1+|y|)^\alpha$$
$$= (c - a + 1)(1+|y|)^\alpha. \quad \text{(note that } a < c + a_0 < c + 1\text{)} \tag{85}$$

From Equations (84) and (85), it follows that

$$1 + |a - c(1+|y|)^\alpha| = 1 + c(1+|y|)^\alpha - a \geq \min\{c, c-a+1\}(1+|y|)^\alpha. \tag{86}$$

Note that $c - a + 1 > 0$ since $a < c + a_0 < c + 1$.
Consequently, it has been proved that Equation (81) holds with

$$c_9 = \begin{cases} \min\left\{\dfrac{2c}{2a+1}, \dfrac{3c}{2a-1}, \min\{c, c-a+1\}\right\} & \text{if } a - c > 1/2, \\ \min\left\{\dfrac{2c}{2a+1}, \min\{c, c-a+1\}\right\} & \text{if } a - c \leq 1/2. \end{cases}$$

□

Next, we show that v satisfies Equation (70). Since

$$z^{-1}T(zT-1)^{-1}B(B-z)^{-1} = z^{-2}(zT-1+1)(zT-1)^{-1}(B-z+z)(B-z)^{-1}$$
$$= z^{-2}\{1 + (zT-1)^{-1}\}\{1 + z(B-z)^{-1}\}$$
$$= z^{-2} + z^{-1}(B-z)^{-1} + z^{-2}(zT-1)^{-1} + z^{-1}(zT-1)^{-1}(B-z)^{-1},$$

we have

$$Tv = (2\pi i)^{-1}\int_\Gamma z^{-1}T(zT-1)^{-1}B(B-z)^{-1}fdz$$
$$= (2\pi i)^{-1}\int_\Gamma z^{-2}fdz + (2\pi i)^{-1}\int_\Gamma z^{-1}(B-z)^{-1}fdz$$
$$+ (2\pi i)^{-1}\int_\Gamma z^{-2}(zT-1)^{-1}fdz + (2\pi i)^{-1}\int_\Gamma z^{-1}(zT-1)^{-1}(B-z)^{-1}fdz. \tag{87}$$

Clearly, $(2\pi i)^{-1}\int_\Gamma z^{-2}fdz = 0$. By assumption $M\{(z-a_0)M - L\}^{-1}$ is holomorphc in $\mathrm{Re}z \geq a_0 - c(1+|\mathrm{Im}z|)^\alpha$. Since

$$zT - 1 = zM(L + a_0M)^{-1} - 1 = \{zM - (L + a_0M)\}(L + a_0M)^{-1}$$
$$= (zM - L - a_0M)(L + a_0M)^{-1},$$
$$(zT - 1)^{-1} = (L + a_0M)(zM - L - a_0M)^{-1} = (L - (z - a_0)M + zM)\{(z - a_0)M - L\}^{-1}$$
$$= \{zM - ((z - a_0)M - L)\}\{(z - a_0)M - L\}^{-1} = zM\{(z - a_0)M - L\}^{-1} - 1,$$

$(zT - 1)^{-1}$ is also holomorphic in $\text{Re} z \geq a_0 - c(1 + |\text{Im} z|)^\alpha$. If $z \in \Gamma$, then

$$\text{Re} z = a - c(1 + |\text{Im} z|)^\alpha > a_0 - c(1 + |\text{Im} z|)^\alpha.$$

Hence, Γ lies in the region where $(zT - 1)^{-1}$ is holomorphc. If

$$\text{Re} z \geq a_0 - c(1 + |\text{Im} z|)^\alpha \,(\iff \text{Re}(z - a_0) \geq -c(1 + |\text{Im}(z - a_0)|)^\alpha),$$

then

$$\|M\{(z - a_0)M - L\}^{-1}\|_{\mathcal{L}(X)} \leq \frac{\tilde{c}}{(|z - a_0| + 1)^\beta}.$$

Hence,

$$\|(zT - 1)^{-1}\|_{\mathcal{L}(X)} = \|zM\{(z - a_0)M - L\}^{-1} - 1\|_{\mathcal{L}(X)} \leq |z|\|M\{(z - a_0)M - L\}^{-1}\|_{\mathcal{L}(X)} + 1$$
$$\leq \frac{\tilde{c}|z|}{(|z - a_0| + 1)^\beta} + 1 = O(|z|^{1-\beta}) \text{ as } |z| \to \infty \text{ in } \text{Re} z > a_0 - c(1 + |\text{Im} z|)^\alpha. \tag{88}$$

Therefore,

$$\|z^{-2}(zT - 1)^{-1}\|_{\mathcal{L}(X)} = O(|z|^{-1-\beta}) \text{ as } |z| \to \infty \text{ in } \text{Re} z > a_0 - c(1 + |\text{Im} z|)^\alpha.$$

Hence, one observes

$$\int_\Gamma z^{-2}(zT - 1)^{-1} dz = 0. \tag{89}$$

Let R be a large positive number. The set $\Gamma \cap \{|z| = R\}$ consists of two points z_1, z_2. Let Γ_R be the closed curve which consists of the part of Γ in the disk $|z| \leq R$ and the part of the circle $|z| = R$ in $\text{Re} z \leq \text{Re} z_1 = \text{Re} z_2$. Since $(B - z)^{-1}$ is holomorphic in the region $\text{Re} z < a_0$, which contains the closed set surrounded by Γ_R,

$$(2\pi i)^{-1} \int_{\Gamma_R} z^{-1}(B - z)^{-1} dz = B^{-1}. \tag{90}$$

Since $z_k = a - c(1 + |\text{Im} z_k|)^\alpha + i\text{Im} z_k$ and $|z_k| = R$, $k = 1, 2$, one has

$$(a - c(1 + |\text{Im} z_k|)^\alpha)^2 + (\text{Im} z_k)^2 = R^2, \; k = 1, 2.$$

This implies $|\text{Im} z_k| = O(R)$ as $R \to \infty$, and hence $|\text{Re} z_k| = O(R^\alpha)$ as $R \to \infty, k = 1, 2$. Therefore, by virtue of Equation (69) for $z \in \Gamma_R \cap \{|z| = R\}$

$$\|(B - z)^{-1}\|_{\mathcal{L}(E)} \leq \frac{1}{a_0 - \text{Re} z} = \frac{1}{a_0 + |\text{Re} z|} \leq \frac{1}{a_0 + |\text{Re} z_k|} = O(R^{-\alpha}) \; (k = 1, 2),$$

as $R \to \infty$. Letting $R \to \infty$ in Equation (90), one observes

$$(2\pi i)^{-1} \int_\Gamma z^{-1}(B - z)^{-1} dz = B^{-1}. \tag{91}$$

From Equations (87), (89), and (91), one obtains

$$Tv = B^{-1}f + (2\pi i)^{-1} \int_\Gamma z^{-1}(zT-1)^{-1}(B-z)^{-1}f dz,$$

and hence

$$BTv = f + (2\pi i)^{-1} \int_\Gamma z^{-1}(zT-1)^{-1}B(B-z)^{-1}f dz = f + v. \tag{92}$$

Thus, we have established Equation (70).

Our next step is to establish the maximal regularity of solutions to Equation (56) or, equivalently, to Equation (58). By observing the resolvent identity

$$(B+t)^{-1}(B-z)^{-1} = -(t+z)^{-1}\{(B+t)^{-1} - (B-z)^{-1}\}$$
$$= -(t+z)^{-1}(B+t)^{-1} + (t+z)^{-1}(B-z)^{-1},$$

we get, for $t > 0$,

$$B(B+t)^{-1}(B-z)^{-1} = [I - t(B+t)^{-1}](B-z)^{-1} = (B-z)^{-1} - t(B+t)^{-1}(B-z)^{-1}$$
$$= (B-z)^{-1} - t[-(t+z)^{-1}(B+t)^{-1} + (t+z)^{-1}(B-z)^{-1}]$$
$$= (B-z)^{-1} + t(t+z)^{-1}(B+t)^{-1} - t(t+z)^{-1}(B-z)^{-1}.$$

Hence,

$$(B+t)^{-1}v = (2\pi i)^{-1} \int_\Gamma z^{-1}(zT-1)^{-1}B(B+t)^{-1}(B-z)^{-1}f dz$$
$$= (2\pi i)^{-1} \int_\Gamma z^{-1}(zT-1)^{-1}[(B-z)^{-1} + t(t+z)^{-1}(B+t)^{-1} - t(t+z)^{-1}(B-z)^{-1}]f dz$$
$$= (2\pi i)^{-1} \int_\Gamma z^{-1}(zT-1)^{-1}(B-z)^{-1}f dz + (2\pi i)^{-1} \int_\Gamma z^{-1}(zT-1)^{-1}t(t+z)^{-1}(B+t)^{-1}f dz$$
$$- (2\pi i)^{-1} \int_\Gamma z^{-1}(zT-1)^{-1}t(t+z)^{-1}(B-z)^{-1}f dz.$$

Therefore, we deduce

$$B(B+t)^{-1}v = v + (2\pi i)^{-1} \int_\Gamma z^{-1}(t+z)^{-1}(zT-1)^{-1}dz \, tB(B+t)^{-1}f$$
$$- (2\pi i)^{-1} \int_\Gamma z^{-1}t(t+z)^{-1}(zT-1)^{-1}B(B-z)^{-1}f dz$$
$$= v + J_1(f,t) + J_2(f,t), \, t \in \mathbb{R}_+.$$

One observes that $(zT-1)^{-1} = zM((z-a_0)M-L)^{-1} - I$ is holomorphic in $\{z; \text{Re} z \geq a_0 - c(1+|\text{Im} z|^\alpha)\}$. Hence, the integrand of $J_1(f,t)$ is holomorphic in $\{z; \text{Re} z \geq a - c(1+|\text{Im} z|^\alpha)\}$ and its norm is $O(|z|^{-1-\beta})$ as $|z| \to \infty$ in view of Equation (88). Therefore, $J_1(f,t) = 0$ for any $(t,f) \in (\mathbb{R}_+, E)$.
Moreover, $J_2(f,t)$ satisfies

$$J_2(f,t) = -(2\pi i)^{-1} \int_\Gamma z^{-1}(t+z-z)(t+z)^{-1}(zT-1)^{-1}B(B-z)^{-1}f dz$$
$$= -(2\pi i)^{-1} \int_\Gamma z^{-1}(zT-1)^{-1}B(B-z)^{-1}f dz + (2\pi i)^{-1} \int_\Gamma (t+z)^{-1}(zT-1)^{-1}B(B-z)^{-1}f dz$$
$$= -v + (2\pi i)^{-1} \int_\Gamma (t+z)^{-1}(zT-1)^{-1}B(B-z)^{-1}f dz.$$

Thus, we have obtained

$$B(B+t)^{-1}v = (2\pi i)^{-1} \int_\Gamma (t+z)^{-1}(zT-1)^{-1}B(B-z)^{-1}f dz. \tag{93}$$

Setting $\omega = \theta + \alpha + \beta - 2$, we can estimate $t^\omega \|B(B+t)^{-1}v\|$ taking into account the identity

$$B(B-z)^{-1} = B(B+1+|y|)^{-1} + (1+|y|+z)(B-z)^{-1}B(B+1+|y|)^{-1}$$
$$= [1 + (1+|y|+z)(B-z)^{-1}]B(B+1+|y|)^{-1},$$
$$z \in \Gamma: z = a - c(1+|y|)^\alpha + iy, \ y \in \mathbb{R}.$$

Here, we show that the following inequality holds for $t > 0, y \in \mathbb{R}$:

$$|t + a - c(1+|y|)^\alpha + iy| \geq c_{10}(t+1+|y|), \quad c_{10} = \frac{\min\{1, a-c\}}{\max\{2\sqrt{2}, 2c+1\}}. \tag{94}$$

Proof. (i) Case $|t + a - c(1+|y|)^\alpha| < (t+a-c)/2$. Recalling $a > c, 0 < \alpha \leq 1$, we deduce

$$|t + a - c(1+|y|)^\alpha| < (t+a-c)/2 \Longrightarrow t + a - c(1+|y|)^\alpha < (t+a-c)/2$$
$$\Longrightarrow (t+a+c)/2 < c(1+|y|)^\alpha \leq c(1+|y|) \Longrightarrow (t+a-c)/2 < c|y| \Longrightarrow t+a-c < 2c|y|.$$

Hence,
$$\min\{1, a-c\}(t+1) \leq t + a - c < 2c|y|.$$

This implies
$$\min\{1, a-c\}(t+1+|y|) < 2c|y| + |y| = (2c+1)|y|.$$

Therefore,
$$|t + a - c(1+|y|)^\alpha + iy| \geq |y| > \frac{\min\{1, a-c\}}{2c+1}(t+1+|y|).$$

(ii) Case $|t + a - c(1+|y|)^\alpha| \geq (t+a-c)/2$. From

$$|t + a - c(1+|y|)^\alpha + iy|^2 = (t + a - c(1+|y|)^\alpha)^2 + y^2 \geq (t+a-c)^2/4 + y^2$$
$$> \frac{1}{8}(2(t+a-c)^2 + 2y^2) \geq \frac{1}{8}(t+a-c+|y|)^2 \geq \frac{1}{8}\left(\min\{1, a-c\}(t+1+|y|)\right)^2$$

it follows that
$$|t + a - c(1+|y|)^\alpha + iy| \geq \frac{\min\{1, a-c\}}{2\sqrt{2}}(t+1+|y|).$$

□

From Equations (72), (76), (78), (93), and (94), it follows that

$$t^\omega \|B(B+t)^{-1}v\|_E = t^\omega \left\|(2\pi i)^{-1}\int_\Gamma (t+z)^{-1}(zT-1)^{-1}B(B-z)^{-1}f dz\right\|_E$$
$$\leq \frac{t^\omega}{2\pi}\int_\Gamma \frac{c_2 c_5(1+|y|)^{1-\beta}c_3(1+|y|)^{1-\alpha}}{c_{10}(t+1+|y|)} \|B(B+1+|y|)^{-1}f\|_E |dz|$$
$$= \frac{t^\omega}{2\pi}\frac{c_2 c_5 c_3}{c_{10}}\int_\Gamma \frac{(1+|y|)^{2-\alpha-\beta}}{t+1+|y|} \|B(B+1+|y|)^{-1}f\|_E |dz|$$
$$\leq c_{11}t^\omega \int_0^\infty \frac{(1+y)^{2-\alpha-\beta}}{t+1+y}\|B(B+1+y)^{-1}f\|_E dy, \tag{95}$$

where
$$c_{11} = \frac{1}{\pi}\frac{c_2 c_5 c_3}{c_{10}}((c\alpha)^2 + 1)^{1/2}.$$

One has

$$t^\omega \int_0^\infty \frac{(1+y)^{2-\alpha-\beta}}{t+1+y}\|B(B+1+y)^{-1}f\|_E dy = t^\omega \int_1^\infty \frac{y^{3-\alpha-\beta-\theta}}{t+y} y^\theta \|B(B+y)^{-1}f\|_E \frac{dy}{y}$$
$$= \int_1^\infty \frac{(ty^{-1})^{\alpha+\beta+\theta-2}}{ty^{-1}+1} y^\theta \|B(B+y)^{-1}f\|_E \frac{dy}{y} = \int_1^\infty g(ty^{-1}) f_1(y) \frac{dy}{y}, \tag{96}$$

where $f_1(y) = y^\theta \|B(B+y)^{-1}f\|_E$, $g(y) = \dfrac{y^{\theta+\alpha+\beta-2}}{1+y}$. Applying Lemma 2 and using Equation (96) one obtains

$$\left(\int_0^\infty \left(t^\omega \int_0^\infty \frac{(1+y)^{2-\alpha-\beta}}{t+1+y}\|B(B+1+y)^{-1}f\|_E dy\right)^p \frac{dt}{t}\right)^{1/p}$$
$$\leq \left(\int_0^\infty \left(\int_0^\infty g(ty^{-1}) f_1(y) \frac{dy}{y}\right)^p \frac{dt}{t}\right)^{1/p} \leq \int_0^\infty g(t) \frac{dt}{t} \left(\int_0^\infty f_1(y)^p \frac{dy}{y}\right)^{1/p}. \tag{97}$$

With the aid of the change of the independent variable $s = (1+t)^{-1}$, one observes

$$\int_0^\infty g(t) \frac{dt}{t} = \int_0^\infty \frac{t^{\theta+\alpha+\beta-2}}{1+t} \frac{dt}{t} = \int_0^1 (1-s)^{\theta+\alpha+\beta-3} s^{-\theta-\alpha-\beta+2} ds$$
$$= \Gamma(\theta+\alpha+\beta-2)\Gamma(3-\theta-\alpha-\beta) = \Gamma(\omega)\Gamma(1-\omega),$$

and

$$\int_0^\infty f_1(y)^p \frac{dy}{y} = \int_0^\infty y^{\theta p} \|B(B+y)^{-1}f\|_E^p \frac{dy}{y}.$$

Hence, one obtains from Equation (97)

$$\left(\int_0^\infty \left(t^\omega \int_0^\infty \frac{(1+y)^{2-\alpha-\beta}}{t+1+y}\|B(B+1+y)^{-1}f\|_E dy\right)^p \frac{dt}{t}\right)^{1/p}$$
$$\leq \Gamma(\omega)\Gamma(1-\omega) \left(\int_0^\infty y^{\theta p} \|B(B+y)^{-1}f\|_E^p \frac{dy}{y}\right)^{1/p}. \tag{98}$$

It follows from Equations (95) and (98) that

$$\left(\int_0^\infty \left(t^\omega \|B(B+t)^{-1}v\|_E\right)^p \frac{dt}{t}\right)^{1/p}$$
$$\leq \left(\int_0^\infty \left(c_{11} t^\omega \int_0^\infty \frac{(1+y)^{2-\alpha-\beta}}{t+1+y}\|B(B+1+y)^{-1}f\|_E dy\right)^p \frac{dt}{t}\right)^{1/p}$$
$$= c_{11} \left(\int_0^\infty \left(t^\omega \int_0^\infty \frac{(1+y)^{2-\alpha-\beta}}{t+1+y}\|B(B+1+y)^{-1}f\|_E dy\right)^p \frac{dt}{t}\right)^{1/p}$$
$$\leq c_{11}\Gamma(\omega)\Gamma(1-\omega) \left(\int_0^\infty y^{\theta p} \|B(B+y)^{-1}f\|_E^p \frac{dy}{y}\right)^{1/p}.$$

Hence, $v = (L+a_0 M)u \in (E, D(B))_{\omega,p}$. This implies $BTv = f + v \in (E, D(B))_{\omega,p}$. Since $BTv = BM(L+a_0 M)^{-1}v = BMu$, one has $BMu \in (E, D(B))_{\omega,p}$:

$$\int_0^\infty t^{\omega p} \|B(B+t)^{-1} BMu\|^p \frac{dt}{t} < \infty.$$

Hence,

$$\int_0^\infty t^{\omega p}\|B(B+t)^{-1}Mu\|^p \frac{dt}{t} = \int_0^\infty t^{\omega p}\|B^{-1}B(B+t)^{-1}BMu\|^p \frac{dt}{t}$$
$$\leq \|B^{-1}\|^p \int_0^\infty t^{\omega p}\|B(B+t)^{-1}BMu\|^p \frac{dt}{t} < \infty,$$

i.e., $Mu \in (E, D(B))_{\omega,p}$. In view of $v = (L + a_0 M)u \in (E, D(B))_{\omega,p}$ it follows that $Lu \in (E, D(B))_{\omega,p}$. Therefore, $B_X^{\tilde{\alpha}} Mu = Lu + f \in (E, D(B))_{\omega,p}$. Thus, the following result is established:

Theorem 1. *Let M, L be two closed linear operators in the complex Banach space E satisfying* (H), *and let $0 < \beta \leq \alpha \leq 1$. Let B_X be the operator defined by Equations* (7) *and* (8) *and $0 < \tilde{\alpha} < 1$. Then, for all $f \in (E, D(B_X^{\tilde{\alpha}}))_{\theta,p}$, $2 - \alpha - \beta < \theta < 1$, $1 < p \leq \infty$ equation $B_X^{\tilde{\alpha}} Mu + Lu = f$ admits a unique solution u. Moreover, $Lu, B_X^{\tilde{\alpha}} Mu \in (E, D(B_X^{\tilde{\alpha}}))_{\omega,p}$, $\omega = \theta + \alpha + \beta - 2$.*

3. Conclusions

The fractional powers of the involved operator B_X are investigated in the space of continuous functions which do not necessarily vanish at the origin. This enables us to prove some previous results in the case where the involved operator B_X is not necessarily densely defined. Precisely, a fractional abstract Cauchy problem for possibly degenerate equations in Banach spaces is considered and refined.

Author Contributions: All authors have equally contributed to this work. All authors wrote, read, and approved the final manuscript.

Funding: This research received no external funding.

Conflicts of Interest: The authors declare no conflict of interest.

References

1. Favini, A.; Yagi, A. Multivalued Linear Operators and Degenerate Evolution Equations. *Annali Di Matematica Pura ed Applicata* **1993**, *163*, 353–384. [CrossRef]
2. Favini, A.; Yagi, A. *Degenerate Differential Equations in Banach Spaces*; Marcel Dekker Inc.: New York, NY, USA, 1999.
3. Lions, J.L.; Magenes, E. *Non-Homogeneous Boundary Value Problems and Applications*; Springer: Berlin, Germany 1972; Volume 1, p. 187.
4. Lions, J.L.; Peetre, J. Sur Une Classe d'Espaces d'Interpolation. *Publ. Math. L'HIS* **1964**, *19*, 5–68. [CrossRef]
5. Favini, A.; Lorenzi, A.; Tanabe, H. Singular integro-differential equations of parabolic type. *Adv. Differ. Equ.* **2002**, *7*, 769–798.
6. Favini, A.; Lorenzi, A.; Tanabe, H. Degenerate Integrodifferential Equations of Parabolic Type with Robin Boundary Conditions: L^p-Theory. *J. Math. Anal. Appl.* **2017**, *447*, 579–665. [CrossRef]
7. Favini, A.; Lorenzi, A.; Tanabe, H. Direct and Inverse Degenerate Parabolic Differential Equations with Multi-Valued Operators. *Electron. J. Differ. Equ.* **2015**, *2015*, 1–22.
8. Favini, A.; Tanabe, H. Degenerate Differential Equations and Inverse Problems. In Proceedings of the Partial Differential Equations, Osaka, Japan, 21–24 August 2012; pp. 89–100.
9. Al Horani, M.; Fabrizio, M.; Favini, A.; Tanabe, H. Fractional Cauchy Problems and Applications. Discrete & Continuous Dynamical Systems-Series S. in press.
10. Al Horani, M.; Fabrizio, M.; Favini, A.; Tanabe, H. Direct and Inverse Problems for Degenerate Differential Equations. *Ann. Univ. Ferrara* **2018**, *64*, 227–241. [CrossRef]
11. Bazhlekova, E.G. *Fractional Evolution Equations in Banach Spaces*; Eindhoven University of Technology: Eindhoven, The Netherlands, 2001.
12. Al Horani, M.; Favini, A.; Tanabe, H. Direct and Inverse Fractional Abstract Cauchy Problems. *Mathematics* **2019**, *7*, 1016. [CrossRef]
13. Guidetti, D. On Maximal Regularity for The Cauchy-Dirichlet Mixed Parabolic Problem with Fractional Time Derivative. *arXiv* **2018**, arXiv:1807.05913.

14. Al Horani, M.; Fabrizio, M.; Favini, A.; Tanabe, H. Fractional Cauchy Problems for Degenerate Differential Equations. *Prog. Fract. Differ. Appl.* **2019**, *5*, 1–11.
15. Fedorov, V.E.; Ivanova, N.D. Identification Problem for Degenerate Evolution Equations of Fractional Order. *Fract. Calc. Appl. Anal.* **2017**, *20*, 706–721 [CrossRef]
16. Sviridyuk, G.A.; Fedorov, V.E. *Linear Sobolev Type Equations and Degenerate Semigroups of Operators*; VSP: Utrecht, The Netherlands; Boston, MA, USA, 2003.
17. Garrappa, R. Numerical Solution of Fractional Differential Equations: A Survey and a Software Tutorial. *Mathematics* **2018**, *6*, 16. [CrossRef]
18. Mainardi, F. Fractional Calculus: Theory and Applications. *Mathematics* **2018**, *6*, 145. [CrossRef]
19. Peride, N.; Carabineanu, A.; Craciun, E.M. Mathematical modelling of the interface crack propagation in a pre-stressed fiber reinforced elastic composite. *Comput. Mater. Sci.* **2009**, *45*, 684–692. [CrossRef]
20. Al Horani, M.; Fabrizio, M.; Favini, A.; Tanabe, H. Fractional Cauchy Problems for Infinite Interval Case. Discrete & Continuous Dynamical Systems-Series S. in press.
21. Kato, T. Fractional Powers of Dissipative Operators. *J. Math. Soc. Jpn.* **1961**, *13*, 246–274. [CrossRef]

© 2019 by the authors. Licensee MDPI, Basel, Switzerland. This article is an open access article distributed under the terms and conditions of the Creative Commons Attribution (CC BY) license (http://creativecommons.org/licenses/by/4.0/).

Article

Exact Solutions for a Modified Schrödinger Equation

Yassine Benia [1], Marianna Ruggieri [2] and Andrea Scapellato [3,*]

1. Department of Mathematics and Informatics, University of Benyoucef Benkhedda (Alger 1), Algiers 16000, Algeria; benia.yacine@yahoo.fr
2. Faculty of Engineering and Architecture, University of Enna "Kore", 94100 Enna, Italy; marianna.ruggieri@unikore.it
3. Department of Mathematics and Computer Science, University of Catania, 95125 Catania, Italy
* Correspondence: scapellato@dmi.unict.it

Received: 24 August 2019; Accepted: 27 September 2019; Published: 29 September 2019

Abstract: The aim of this paper was to propose a systematic study of a $(1+1)$-dimensional higher order nonlinear Schrödinger equation, arising in two different contexts regarding the biological science and the nonlinear optics. We performed a Lie symmetry analysis and here present exact solutions of the equation.

Keywords: Schrödinger equation; Davydov's model; partial differential equations; exact solutions

---

1. Introduction

In the framework of biological phenomena, the Davydov model has been a theme of intensive studies and has attracted the attention of many researchers [1–8]. The nonlinear dynamics of DNA molecule has been investigated by several authors [9,10]; an interesting study concerning the Davydov's model of α-helical proteins can be found in [11,12]. Motivated by all these applications, the purpose of this paper is to make a complete study of a generalized Davydov model arising in literature in two different contexts regarding the biological science and the nonlinear optics. Working with the systematic analysis of Lie's theory and without the restriction of the solitary wave ansatz, we showed that it admits some beautiful and most interesting reductions to ordinary differential equations; our results confirm that the equation can, in general, support periodic wave solutions, soliton solutions and interesting solutions expressed in terms of Bessel functions.

We consider the following higher order nonlinear Schrödinger equation:

$$i\frac{\partial w}{\partial t} + \frac{\partial^2 w}{\partial x^2} + |w|^2 w + i\varepsilon \left(\frac{\partial^3 w}{\partial x^3} + k_1 |w|^2 \frac{\partial w}{\partial x} + k_2 w^2 \frac{\partial w^*}{\partial x} \right) = 0, \qquad (1)$$

where $w(t,x)$ is a complex functions of x and t, $w^*(t,x)$ is the complex conjugate of $w(t,x)$. In what follows, subscripts denote partial derivatives with respect to space and time coordinates.

When $\varepsilon = 0$, the Equation (1) reduces to the completely integrable cubic nonlinear Schrödinger (NLS) equation, well studied in [13] and therefore we will not consider it further. Several interesting studies concerning NLS can be found in [14–17]; when $k_2 = 0$ it becomes the Hirota equation [16], while for $k_1 = 3k_2$ the equation is referred to as the Sasa–Satsuma equation (see, e.g., [18] and references therein). The hyperbolic secant and hyperbolic tangent solutions of (1) have been studied in [19], while in [20] it is also shown that a higher-order nonlinear Schrödinger equation is solvable by means of the inverse scattering transform.

In the context of biological sciences, the Equation (1) was proposed in [9] as a generalized Davydov model. In this interpretation, the variable w represents the vibrational coordinate of the amide-I vibrations, the coefficient k_1 is a contribution related to self-steepening (also known as Kerr dispersion) and k_2 is linked to stimulated Raman scattering effects, while ε represents the lattice parameter. In addition, the second term represents a dispersion term linked to dipole-dipole coupling, the third term appears as the nonlinear coupling to hydrogen bonds, while the last term represents a term related to a global interaction due to molecular excitations.

In the framework of nonlinear optics, with a interchange of independent variables ($t \leftrightarrow x$), the equation characterizes the propagation of femtosecond pulses in nonlinear fibres, where w is the slowly-varying envelope of the electromagnetic field, the parameters ε, k_1, k_2 are real coefficients linked to group velocity dispersion, self phase modulation, third-order dispersion, self steepening, and self-frequency shift due to stimulated Raman scattering, respectively.

We would like to point out that, today, a lot of problems related to regularity of solutions to Schrödinger equation in the framework of Morrey-type spaces (see, for instance, [21,22]) are studied. For an overview on Morrey spaces related on nonnegative potential, we refer the reader to [23] (and the references given there), where Guliyev collects some recently obtained results on Morrey classes and integral operators associated to Schrödinger operators and examines various versions of these spaces and the boundedness of some Schrödinger type operators on these spaces related to certain nonnegative potentials belonging to the reverse Hölder class. Furthermore, under the assumption that $\Omega \subseteq \mathbb{R}^n$ is an open set, in [23] are discussed some qualitative properties of solutions to the following Schrödinger equation:

$$(-\Delta + V)u = f(x), \quad \text{a.e. } x \in \Omega,$$

where

- V is a nonnegative potential belonging to the reverse Hölder class B_n, that is, the class of all nonnegative locally L^n integrable functions $V(x)$ on $\mathbb{R}^n$, for which there exists $C > 0$ such that the reverse Hölder inequality;

$$\left(\frac{1}{|B(x,r)|} \int_{B(x,r)} V^n(y) \, dy \right)^{1/n} \leq \frac{C}{|B(x,r)|} \int_{B(x,r)} V(y) \, dy$$

holds for every $x \in \mathbb{R}^n$, $r \in (0, \infty)$, $B(x,r)$ denotes the ball centered at x with radius r and $|B(x,r)|$ stands for the Lebesgue measure of the ball $B(x,r)$;
- the source term f belongs to some suitable Morrey spaces associated to nonnegative potential V.

We mention [24], where the authors recently studied fractional integrals associated with Schrödinger operators in the framework of vanishing generalized Morrey spaces and [25], where the authors investigated the action of commutator of fractional integral with Lipschitz functions with Schrödinger operator on local generalized Morrey space.

Despite to the general euclidean context in which the interesting regularity theory is currently studied, up to now, at our knowledge, in literature there are not regularity results, in the context of Morrey spaces, for higher order Schrödinger equation. For this reason, this paper seems to provide some basis for a potential subsequent study of qualitative properties of solutions to higher order Schrödinger equation in the context of Morrey spaces.

2. Main Results

It is convenient to rewrite the equation and after introducing $w(t,x) = u(t,x) + i\,v(t,x)$, being $u(t,x)$ and $v(t,x)$ real smooth functions, we substitute the complex-valued function $w(t,x)$ into (1) and then decomposing into real and imaginary parts yields two relations. The real and the imaginary parts gives the following system of real partial differential equations (PDEs):

$$u_t + v_{xx} + (u^2 + v^2)v + \varepsilon\left\{u_{xxx} + k_1(u^2 + v^2)u_x + k_2\left[(u^2 - v^2)u_x + 2uvv_x\right]\right\} = 0, \quad (2)$$

$$v_t - u_{xx} - (u^2 + v^2)u + \varepsilon\left\{v_{xxx} + k_1(u^2 + v^2)v_x + k_2\left[(v^2 - u^2)v_x + 2uvu_x\right]\right\} = 0. \quad (3)$$

The required theory, as well as the description of the method, can be found in the papers [26,27]; the symmetries of the system (2) and (3) will be generated by applying the third prolongation [26,27] of the infinitesimal operator, or *generator* Ξ

$$\Xi = \xi_1(t, x, u, v)\partial_t + \xi_2(t, x, u, v)\partial_x + \eta_1(t, x, u, v)\partial_u + \eta_2(t, x, u, v)\partial_v \quad (4)$$

to Equations (2) and (3), by which we obtain the following results:

$$\xi_1 = a_1 + 9\varepsilon a_2 t, \quad (5)$$

$$\xi_2 = a_3 + a_2(2t + 3\varepsilon x), \quad (6)$$

$$\eta_1 = -3\varepsilon a_2 u - (a_2 x - a_4) v, \quad (7)$$

$$\eta_2 = (a_2 x - a_4) u - 3\varepsilon a_2 v, \quad (8)$$

$$a_2(k_2 - k_1 + 3) = 0, \quad (9)$$

where a_i ($i = 1, 2, 3, 4$) are constants; from relation (9), we have the following two cases:

Case I: $a_2 = 0$

In this case, the Lie algebra is three-dimensional:

$$\Xi_1 = \partial_t, \quad \Xi_2 = \partial_x, \quad \Xi_3 = v\partial_u - u\partial_v. \quad (10)$$

Case II: $k_2 = k_1 - 3$.

It is worthwhile noticing that the assumption $k_2 = k_1 - 3$ is a *structural condition* of the Equation (1). In this case the Lie algebra is four-dimensional and is spanned by the three operators (10) plus the following fourth operator:

$$\Xi_4 = 9\varepsilon t\partial_t + (2t + 3\varepsilon x)\partial_x - (3\varepsilon u + xv)\partial_u + (xu - 3\varepsilon v)\partial_v. \quad (11)$$

3. Exact Solutions

The study of exact solutions to nonlinear PDEs is an important point in the applied science. Many interesting methods have been elaborated, such as ansatz method, Hirota bilinear method, inverse scattering method, Darboux method, multiple-exp function method, etc. [28–31]. Lie theory gives an important contribution in almost all the scientific fields; Sophus Lie proved that a differential equation can be invariant with respect to the so-called "continuous transformation groups". For the details, we suggest to the reader to see [26,27,32–34].

In this section, we use the symmetry group obtained in the Cases *I* and *II* of Section 2 in order to determine the reductions of (2) and (3) from which we are able to construct exact solutions which carry important physical meanings. In the study we performed, we consider the case $\varepsilon \neq 0$ since, as already stated in the introduction of the paper, the case $\varepsilon = 0$ has been well studied in the literature [13].

Reduction I: $a_2 = 0$

In this case we consider the operator Ξ as a linear combination of the operators (10), namely:

$$\Xi = \partial_t + c_1 \partial_x + c_2 (v \partial_u - u \partial_v), \tag{12}$$

where c_1 and c_2 are real constants, which give the following similarity variable:

$$z = x - c_1 t \tag{13}$$

and similarity solutions:

$$u = \phi(z) \sin(c_2 t) + \psi(z) \cos(c_2 t) \tag{14}$$
$$v = \phi(z) \cos(c_2 t) - \psi(z) \sin(c_2 t), \tag{15}$$

where ϕ and ψ satisfy the ODEs to which (2) and (3) are reduced by means of the operator (12):

$$c_1 \phi' + c_2 \psi + \psi'' + \left(\phi^2 + \psi^2\right) \psi$$
$$- \varepsilon \left\{ \phi''' + \left[(k_1 + k_2) \phi^2 + (k_1 - k_2) \psi^2\right] \phi' + 2 k_2 \phi \psi \psi' \right\} = 0, \tag{16}$$

$$c_1 \psi' - c_2 \phi - \phi'' - \left(\phi^2 + \psi^2\right) \phi$$
$$- \varepsilon \left\{ \psi''' + \left[(k_1 - k_2) \phi^2 + (k_1 + k_2) \psi^2\right] \psi' + 2 k_2 \phi \psi \phi' \right\} = 0. \tag{17}$$

A solution to (16) and (17) is

$$\phi = \cos z, \qquad \psi = \sin z, \tag{18}$$

under the condition $c_2 = c_1 + \varepsilon [(k_2 - k_1) + 1]$, with the constant of integration normalized to one. When we return to the original variables, we obtain that the system (2) and (3) of PDEs admits a periodic wave solution:

$$u = \cos(x - c_1 t) \sin(c_2 t) + \sin(x - c_1 t) \cos(c_2 t), \tag{19}$$
$$v = \cos(x - c_1 t) \cos(c_2 t) - \sin(x - c_1 t) \sin(c_2 t). \tag{20}$$

If, in particular, in (16) and (17), we have also the validity of the structural condition $k_2 = k_1 - 3$, the solution reads as:

$$\phi = \text{sech}(z), \qquad \psi = \text{sech}(z). \tag{21}$$

When $c_2 = -1$ and $c_1 = \varepsilon$, the system admits a soliton solution of the form:

$$u = \text{sech}(x - \varepsilon t) \cos(t) - \text{sech}(x - \varepsilon t) \sin(t), \tag{22}$$
$$v = \text{sech}(x - \varepsilon t) \sin(t) + \text{sech}(x - \varepsilon t) \cos(t). \tag{23}$$

Reduction II: $k_2 = k_1 - 3$

Proceeding as in the previous case, through the operator (11)

$$\Xi_4 = 9 \varepsilon t \partial_t + (2t + 3 \varepsilon x) \partial_x - (3 \varepsilon u + x v) \partial_u + (x u - 3 \varepsilon v) \partial_v$$

we obtain the similarity variable and solutions, respectively:

$$z = \left(x - \frac{1}{3\varepsilon}t\right) t^{-\frac{1}{3}},$$

$$u = \left[\phi(z) \cos\left(\frac{x}{3\varepsilon} - \frac{2t}{27\varepsilon^2}\right) - \psi(z) \sin\left(\frac{x}{3\varepsilon} - \frac{2t}{27\varepsilon^2}\right)\right] t^{-\frac{1}{3}},$$

$$v = \left[\phi(z) \sin\left(\frac{x}{3\varepsilon} - \frac{2t}{27\varepsilon^2}\right) + \psi(z) \cos\left(\frac{x}{3\varepsilon} - \frac{2t}{27\varepsilon^2}\right)\right] t^{-\frac{1}{3}}. \quad (24)$$

Additionally, here, ϕ and ψ satisfy the ODEs to which (2) and (3) are reduced by means of the operator (11), i.e.,

$$3\varepsilon \left\{\phi''' + \left[(2k_1 - 3)\phi^2 + 3\psi^2\right] \phi' + 2(k_1 - 3)\phi \psi \psi'\right\} - z\phi' - \phi = 0, \quad (25)$$

$$3\varepsilon \left\{\psi''' + \left[(2k_1 - 3)\psi^2 + 3\phi^2\right] \psi' + 2(k_1 - 3)\phi \psi \phi'\right\} - z\psi' - \psi = 0. \quad (26)$$

In this case, we obtain a rational solution of (25) and (26) which reads as:

$$\phi = \psi = H z^{-1}, \quad H = \pm \sqrt{\frac{3}{3 - 2k_1}}$$

under the condition $k_1 \neq \frac{3}{2}$.

If, on the contrary, $k_1 = \frac{3}{2}$, a solution of (25) and (26) can be expressed in terms of Bessel functions:

$$\phi = \psi = \sqrt{z} \left\{ h_1 \left[J_{\frac{1}{3}} \int \frac{\sqrt{z} J_{-\frac{1}{3}}}{i z \sqrt{3\varepsilon z} \left(J_{-\frac{4}{3}} J_{\frac{1}{3}} - J_{-\frac{2}{3}} J_{-\frac{1}{3}} \right) + 3\varepsilon J_{-\frac{1}{3}} J_{\frac{1}{3}}} dz \right. \right.$$

$$\left. \left. - J_{-\frac{1}{3}} \int \frac{\sqrt{z} J_{\frac{1}{3}}}{i z \sqrt{3\varepsilon z} \left(J_{-\frac{4}{3}} J_{\frac{1}{3}} - J_{-\frac{2}{3}} J_{-\frac{1}{3}} \right) + 3\varepsilon J_{-\frac{1}{3}} J_{\frac{1}{3}}} dz \right] + h_2 J_{\frac{1}{3}} + h_3 J_{-\frac{1}{3}} \right\}, \quad (27)$$

where we have set the Bessel functions as: $J_\alpha = J_\alpha \left(\frac{2 i z^{\frac{3}{2}}}{3 \sqrt{3\varepsilon}}\right)$, $\left(\alpha = \pm \frac{1}{3}, -\frac{2}{3}, -\frac{4}{3}\right)$, and h_1, h_2, h_3 are arbitrary constants.

4. Discussion and Conclusions

Usually, the nonlinear Schrödinger equation characterizes physical processes in which nonlinearity and dispersion cancel producing to solitons. The literature shows that the higher order nonlinear Schrödinger equation can, in general, support both soliton and periodic wave solutions. In this paper, we considered higher order nonlinear Schrödinger equation by using the Lie symmetry analysis in order to reduce it to ODEs; for each reduction exact solutions are obtained.

It is worth pointing out some details regarding the dimensionality of the parameters ε, k_1 and k_2, which appear in the model. It is interesting to observe the meaning of parameter ε in two different contexts; in the framework of biological sciences, ε represents the lattice parameter and in the context of nonlinear optics can be interpreted as a dispersion parameter. Usually, in the applications, it expresses the size of atoms and molecules, the length of chemical bonds and the arrangement of atoms in crystals; commonly it is expressed in Ångström (Å). The coefficient k_1 is related to the Kerr effect that is a change in the refractive index of a given material in response to an applied electric field. Precisely, the induced index change by the Kerr effect is directly proportional to the square of the electric field. Finally, the parameter k_2 is related to the Raman scattering. Several optical processes involve the simultaneous (instantaneous) absorption of an incident photon and emission of another photon. These processes are usually called *scattering processes* and the emitted photon is called the *scattered photon*. Scattering process can be classified into two main classes: *elastic scattering* and *inelastic scattering*. In the case of elastic scattering, the incident and scattered photons have the same energy, but they could have different direction and/or polarization. In the case of inelastic scattering, the scattered photon is at a different energy from that of the incident photon. The Raman scattering is an inelastic scattering which involves transitions between the vibrational/rotational levels. Then, when we regard to the solution (22) and (23) without losing the generality, we have posed in Figure 1 the value $\varepsilon = 1$.

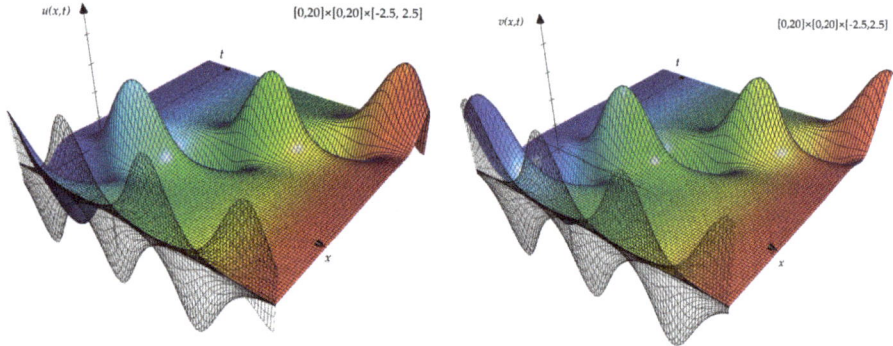

Figure 1. Snapshots of 3D view of the solution (22) and (23), with a interchange of independent variables $(t \leftrightarrow x)$ and $\varepsilon = 1$.

Remark 1. *In biological interpretation, as confirmed by Daniel [9], it turns out that the Equation (1), as well as variants of it, characterizes a protein chain where the nonlinear coupling, being liable to the development of solitons, derives from lightly different interaction. Futhermore, in two recent papers [5,6] the effects of anisotropy of left and right coupling between neighbouring peptide groups in proteins are studied in great details, and also moving solitons were illustrated.*

On the other hand, looking solutions (22) and (23) from the point of view of nonlinear optics, as predicted by Trippenbach et al. [35], the higher order nonlinearities definitively influences the dynamics.

Looking at the class of solutions (24), we observe that it is obtained for $\varepsilon \neq 0$ then it is no longer valid in the limit of cubic nonlinear Schrödinger equation. This fact is perfectly justified by the dimensionality of the constant ε that we have clarified at the beginning of this section. From Figure 2, we see a plot of the solution (24) under the structural condition for the equation $k_2 = k_1 - 3$, when $k_1 \neq \frac{3}{2}$.

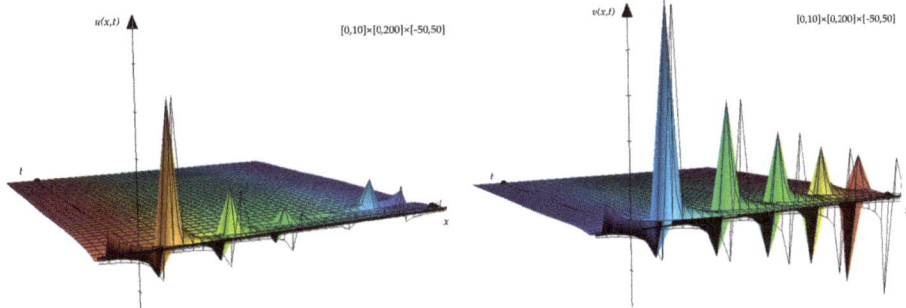

Figure 2. Snapshot of 3D view of the solution (24), when $\phi = \psi = Hz^{-1}$, $H = \sqrt{\frac{3}{3-2k_1}}$ with $k_1 = 1$ and $\epsilon = 1$.

The novelty of this work lies in the fact that the application of Lie symmetry analysis is being made for the first time, to the best of our knowledge, in such systems that lend themselves to a double interpretation in two different contexts of the applied sciences. Moreover, our future aim is to study qualitative properties of solutions to differential problems driven by nonlinear Schrödinger equations as (1) in the context of suitable Morrey-type spaces. The results contained in this paper provide a satisfactory background because they give us a precise shape of the solutions. On the other hand, an a priori analysis in some new function spaces allows us to overcome the computational difficulties that arise directly using the representation formulae for the solutions.

Author Contributions: Y.B., M.R., and A.S. worked together in the derivation of the mathematical results. All the authors provided critical feedback and helped shape the research, analysis, and manuscript.

Funding: This research received no external funding.

Acknowledgments: The document was written during the stay of the first author Y.B. in Catania as part of the project DIRE-MED (Dialogue Interculturel, REseaux et Mobilité en MEDiterranée). M.R. is a member of the INdAM Research group GNFM (National Group for the Mathematical Physics). A.S. is a member of UMI (Italian Mathematical Union) and EMS (European Mathematical Society).

Conflicts of Interest: The authors declare no conflict of interest.

Abbreviations

The following abbreviations are used in this manuscript:

NLS   Nonlinear Schrödinger
PDEs Partial Differential Equations
ODEs Ordinary Differential Equations

References

1. Davydov, A.S. The theory of contraction of proteins under their excitation. *J. Theor. Biol.* **1973**, *38*, 559–569. [CrossRef]
2. Davydov A.S.; Kislukha, N.I. Solitons in one-dimensional molecular chains. *Phys. Status Solidi B* **1976**, *75*, 735–742. [CrossRef]
3. Davydov, A.S. Solitons in molecular systems. *Phys. Scr.* **1979**, *20*, 387–394. [CrossRef]
4. Cisneros-Ake, L.A.; Brizhik, L.S. Charge and energy transport by Holstein solitons in anharmonic one-dimensional systems. *Chaos Solitons Fractals* **2019**, *119*, 343–354. [CrossRef]

5. Georgiev, D.D.; Glazebrook, J.F. On the quantum dynamics of Davydov solitons in protein α-helices. *Phys. A Stat. Mech. Appl.* **2019**, *517*, 257–269. [CrossRef]
6. Georgiev, D.D.; Glazebrook, J.F. Quantum tunneling of Davydov solitons through massive barriers. *Chaos Solitons Fractals* **2019**, *123*, 275–293. [CrossRef]
7. Luo, J.; Piette, B.M.A.G. A generalized Davydov-Scott model for polarons in linear peptide chains. *Eur. Phys. J. B* **2017**, *90*, 1–21. [CrossRef]
8. Brizhik, L.S.; Eremko, A.A.; Piette, B.; Zakrzewski, W.J.M. Solitons in α-helical proteins. *Phys. Rev. E* **2004**, *70*, 031914. [CrossRef] [PubMed]
9. Daniel, M.; Latha, M.M. A generalized Davydov soliton model for energy transfer in alpha helical proteins. *Phys. A* **2001**, *298*, 351–370. [CrossRef]
10. Komineas, S.; Kalosakas, G.; Bishop, A.R. Effects of intrinsic base-pair fluctuations on charge transport in DNA. *Phys. Rev. E* **2002**, *65*, 061905. [CrossRef]
11. Mvogo, A.; Ben-Bolie, G.H.; Kofane, T.C. Energy transport in the three coupled α-polypeptide chains of collagen molecule with long-range interactions effect. *Chaos* **2015**, *25*, 063115. [CrossRef] [PubMed]
12. Mvogo, A.; Ben-Bolie, G.H.; Kofane, T.C. Solitary waves in an inhomogeneous chain of α-helical proteins. *Int. J. Mod. Phys. B* **2014**, *28*, 1450109. [CrossRef]
13. Kivshar Y.S.; Malomed B.A. Dynamics of solitons in nearly integrable systems. *Rev. Mod. Phys.* **1989**, *61*, 763–915. [CrossRef]
14. Zakharov, V.E.; Takhtajan, L.A. Equivalence of a nonlinear Schrödinger equation and a Heisenberg ferromagnetic equation. *Theor. Math. Phys.* **1979**, *37*, 17–23. [CrossRef]
15. Zakharov, V.E.; Shabat, A.B. Exact theory of two-dimensional self-focussing and one-dimensional self-modulating waves in nonlinear media. *Sov. Phys. JETP* **1972**, *34*, 62–69.
16. Hirota, R. *Direct Methods in Soliton Theory*; Springer: Berlin, Germany, 1980.
17. Ablowitz, M.J.; Clarkson, P.A. *Solitons, Nonlinear Evolution Equations and Inverse Scattering*; Cambridge University Press: Cambridge, UK, 1992.
18. Liu, N.; Guo, B. Long-time asymptotics for the Sasa–Satsuma equation via nonlinear steepest descent method. *J. Math. Phys.* **2019**, *60*, 011504.
19. Potasek, M.J.; Tabor, M. Exact solutions for an extended nonlinear Schrödinger equation. *Phys. Lett. A* **1991**, *154*, 449–452. [CrossRef]
20. Kodama, Y. Optical solitons in a monomode fiber. *J. Stat. Phys.* **1985**, *39*, 597–614. [CrossRef]
21. Di Fazio, G.; Ragusa, M.A. Commutators and Morrey Spaces. *Boll. Unione Mat. Ital.* **1991**, *7*, 321–332.
22. Ragusa, M.A. Commutators of fractional integral operators on Vanishing-Morrey spaces. *J. Glob. Optim.* **2008**, *40*, 361–368. [CrossRef]
23. Guliyev, V.S. Function spaces and integral operators associated with Schrödinger operators: An overview. *Proc. Inst. Math. Mech.* **2014**, *40*, 178–202.
24. Akbulut, A.; Guliyev, R.V.; Celik, S.; Omarova, M.N. Fractional integral associated with Schrödinger operator on vanishing generalized Morrey spaces. *J. Math. Inequal.* **2018**, *12*, 789–805. [CrossRef]
25. Guliyev, V.S.; Akbulut, A. Commutator of fractional integral with Lipschitz functions associated with Schrödinger operator on local generalized Morrey spaces. *Bound. Value Probl.* **2018**, *80*. [CrossRef]
26. Ovsiannikov, L.V. *Group Analysis of Differential Equations*; Academic Press: New York, NY, USA, 1982.
27. Olver, P.J. *Applications of Lie Groups to Differential Equations*; Springer: New York, NY, USA, 1986.
28. Biswas, B.; Mirzazadeh, M.; Savescu, M.; Milovic, D.; Khan, K.R.; Mahmood, M.F.; Belic, M. Singular solitons in optical metamaterials by ansatz method and simplest equation approach. *J. Mod. Opt.* **2014**, *61*, 1550–1555. [CrossRef]
29. Hirota, R. *The Direct Method in Soliton Theory*; Cambridge University Press: Cambridge, UK, 2004.
30. Ma, W.X.; Huang, T.; Zhang, Y. A multiple exp-function method for nonlinear differential equations and its application. *Phys. Scr.* **2010**, *82*, 065003. [CrossRef]
31. Rogers, C.; Schief, W.K. *Backlund and Darboux Transformations: Geometry and Modern Applications in Soliton Theory*; Cambridge University Press: Cambridge, UK, 2002.

32. Ibragimov, N.H. *CRC Hanbook of Lie Group Analysis of Differential Equations*; CRC Press: Boca Raton, FL, USA, 1994.
33. Ames, W.F.; Rogers, C. *Nonlinear Equations in the Applied Sciences*; Academic Press: Boston, MA, USA, 1992.
34. Fushchych, W.I.; Shtelen, W.M. *Symmetry Analysis and Exact Solutions of Nonlinear Equations of Mathematical Physics*; Kluwer: Dordrecht, The Netherlands, 1993.
35. Trippenbach, M.; Band, Y.B. Effects of self-steepening and self-frequency shifting on short-pulse splitting in dispersive nonlinear media. *Phys. Rev. A* **1998**, *57*, 4791–4803. [CrossRef]

© 2019 by the authors. Licensee MDPI, Basel, Switzerland. This article is an open access article distributed under the terms and conditions of the Creative Commons Attribution (CC BY) license (http://creativecommons.org/licenses/by/4.0/).

Article

Mathematical Analysis of an Autoimmune Diseases Model: Kinetic Approach

Mikhail Kolev

Faculty of Mathematics and Computer Science, University of Warmia and Mazury, Słoneczna 54, 10-710 Olsztyn, Poland; kolev@matman.uwm.edu.pl

Received: 09 July 2019; Accepted: 25 October 2019; Published: 30 October 2019

Abstract: A new mathematical model of a general autoimmune disease is presented. Basic information about autoimmune diseases is given and illustrated with examples. The model is developed by using ideas from the kinetic theory describing individuals expressing certain functions. The modeled problem is formulated by ordinary and partial equations involving a variable for a functional state. Numerical results are presented and discussed from a medical view point.

Keywords: kinetic theory; active particles; autoimmune disease

1. Introduction

Among the tasks of the human immune system is the defence and protection of the organism from the invasion of foreign pathogens (viruses, bacteria, helminths, fungi, protozoa, etc.), the neutralization of extracellular invaders, the destruction of infected and degenerated own cells and so forth. In order to function properly, the immunological system should discriminate between own cells and foreign agents. The defend system must not attack healthy own cells, tissues and organs. This feature is known as "immune tolerance" [1,2].

However sometimes the mechanisms of immunological tolerance fail. Then the defence system attacks and destroys some of its own healthy cells and other structures. This can result in a disorder called autoimmune disease. Depending on the organs attacked by the immune system, more than 80 autoimmune diseases are known. Among them are insulin-dependent diabetes, rheumatoid arthritis (RA) and psoriatic arthritis, disseminated sclerosis (known also as multiple sclerosis (MS)), myocarditis, autoimmune hemolytic anemia, psoriasis, pancreatitis, lupus (known also as systemic lupus erythematosus (SLE)), Crohn's disease, ulcerative colitis, celiac sprue (gluten-sensitive enteropathy), autoimmune thyroid diseases—Graves' disease (thyroid overactivity), Hashimoto's thyroiditis (thyroid underactivity) and many others [2,3].

The opinion of the contemporary medicine concerning the causes of the autoimmune diseases is that they are multifactorial. It is believed that the autoimmune diseases are determined by multifaceted factors. They include environmental triggers (chemicals, heavy metals, bacterial and viral infections, mold etc.), genetic predispositions, sedentary habits, socioeconomic and emotional stress, drugs and so forth [3–5]. Many authors mention also the possible role of diet (the food quality, gluten consumption, vitamin D deficiency etc.) and gut dysbiosis for the development of autoimmune disorders [3,6–9].

Numerous immunological studies have been devoted to clarification of the role of infections in processes related to the failure of the self-tolerance of the immune system. Various pathogenic agents such as viruses, bacteria, fungi, parasites and so forth have been shown to participate in the development of autoimmunity [3,10]. For example, Streptococcus pyogenes bacteria has been associated with rheumatic fever [11,12], enteroviruses—with insulin-dependent diabetes [11,13], the hepatitis C virus has been shown to be able to trigger chronic liver disease and mixed cryoglobulinemia [14–16], the Epstein–Barr virus—to induce SLE, RA, juvenile idiopathic arthritis, type 1 diabetes, MS, celiac disease and inflammatory bowel disease [14,17–19].

Over the last decades a rapid growth of autoimmune diseases was observed in the developed countries. This led D. Strachan to a controversial idea called the "hygiene hypothesis" [11,14,20,21]. The idea consists in associating the rise of autoimmune disease incidence with the following from the better personal cleanliness lower rates of childhood infections. The hypothesis states that the immune system needs to be trained by multiple interactions with pathogenic agents in order to function correctly.

There is evidence that various natural and acquired immune components and mechanisms are involved in the processes of autoimmune diseases. For example, the infiltration of non-specific monocytes and macrophages in diseased tissues is often observed in the course of autoimmune diseases as well as the participation of specific antibodies, B and T cells and so forth [22–24].

This paper presents a new model of a general autoimmune disease for studying some mechanisms of autoimmune diseases, particularly the pathogenic role of viruses. The proposed model is a nonlinear system of Boltzmann-type equations.

During the last decades many mathematical models have been proposed for studying the mechanisms of the interactions between immune system and infectious agents. Some of them are related to autoimmune disorders, see among others the models in References [25–30] and references therein. The model presented in Reference [29] is developed within the framework of idiotypic networks. Most of the remaining deterministic models use ordinary differential equations describing populations which are assumed to be homogeneous. Taking into account immunological evidence that often the populations involved in interactions during the autoimmune reactions are heterogeneous [11,21,23,31], in the presented model the activity of interacting individuals is also considered.

The organisation of the paper is as follows. Section 2 is devoted to the the mathematical models of kinetic type. Further, the new model of autoimmune disease is described. Results of numerical experiments are given in Section 3. In the concluding Section 4 some remarks are presented.

2. Mathematical Model

2.1. Kinetic-Type Models

2.1.1. General Description

The presented model of a general autoimmune disease is developed by the use of ideas from the kinetic theory of active particles (in short KTAP)). It uses mathematical structures similar to statistical mechanics and phenomenological theory of gases, in particular Boltzmann-type equations. Within this framework a system of one or several large populations of interacting individuals characterized by an additional continuous or discrete variable describing their functional state, is considered. The introduction of this inner characteristic of the interacting entities makes this approach very suitable for describing the non-homogeneity often observed in complex living systems, see for example References [32–36] and references cited therein. In many cases the space and velocity variables of the modeled systems are homogeneously distributed or have no relevant meaning for the interactions [32,33,36].

Often it is possible to divide the system under consideration into several subsystems where the individuals have the same functional state [37,38].

2.1.2. Examples

The KTAP has been actively used for modeling phenomena not only in physics (see e.g., Reference [39,40]) but also in many other areas of applied and life sciences, for example in the political [41], economic and social [42–47] systems, psychological interactions [48], the dynamics of crowds [49–54] and swarms [55–61], traffic flow [62–68] and so forth.

Examples of applications of KTAP in biology are modeling of wound healing and disease [69–74], cancer dynamics [75–79] and multi-cellular dynamics [80,81].

Some of these models are formulated within the so-called thermostatted KTAP developed recently in a series of papers, see for example, References [36,82–90] and the references therein.

2.1.3. Functional State of Active Particles

The variable describing the inner functional states of the interacting individuals represents the function, strategy, type of behaviour, purpose or activity expressed by the interacting entities (particles). Often it is named "state of activity" (or "activation state"), which explains the term KTAP. The introduction of the microscopic functional states allows to account for heterogeneity of the corresponding subsystems. The functional states describe specific characteristics of the system (or its subsystems) under study.

For example, in models describing pedestrian/crowd dynamics, the functional (activation) state can be related to walking ability or walking strategy of pedestrians [50,51,53,67] depending on such characteristics of the environment as the quality of the road, light, interactions with other pedestrians, possible danger and so forth and their own emotional and physical state (including stress, panic, etc).

One of the first applications of KTAP to biology was the model by Jäger and Segel [91] describing the interactions between a general population of insects. The functional state there is the dominance governing their social dynamics and depending on the results of their encounters.

As an example of a decomposition of a complex system into several interacting functional subsystems can be considered can be considered. The model [69] describing fibrosis disease and in particular the process of keloid formation represents an example how a decomposition of a system into several interacting functional subsystems can be performed. In this model, the activity of normal-fibroblast cells and keloid fibroblast cells describes their ability to proliferate. The functional state of activated viruses refers to the value of their aggressiveness in regard to their proliferative capability. The functional state of immune cells describes their degree of activity and of responsiveness to pathogens. The activity of malignant cells refers to the magnitude of their progression capability.

Another example of a decomposition of a complex system into several interacting functional subsystems is the model [28] describing the development of autoimmune diseases. The model describes the competition between host cells, foreign cells, antigen presenting cells (APC) that do not expose any antigen on their surface, APC that expose a certain antigen on their surface, naive and active T cells. It is assumed that the population of APC that are not exposing any antigen are homogeneous with respect to their activation state. The functional state of the host cells and of the foreign cells refers to their antigenic expression. The activity of the structured population of APC denotes the antigen exposed by them. The functional state of the non-active T lymphocytes identifies their cognate antigen. The activity of the activated T cells stands for their target antigens.

2.2. Kinetic-Type Model of Autoimmune Disease

2.2.1. The Main Biological Assumptions

The main goal of the proposed mathematical model is to investigate the pathogenic impact of viruses and damaged cells for the autoimmune disorders.

One of the main mechanisms leading to such autoimmune diseases as type 1 diabetes, MS, primary biliary cirrhosis, SLE and so forth is the so-called molecular (or antigenic or epitopic) mimicry due to viral infection [12,25,26].

This phenomenon may occur when the peptides of foreign pathogens are similar to those of the host peptides. Such similarity can trigger an immune response not only to the pathogens but also to the host constituents with similar protein structure or shape. Thus we observe a cross-reaction of the immune system with self-antigens which can lead to damaging of self cells. The damaged cells can release other sequestered antigens that can be attacked by immune cells. Thus the autoimmune response can be enhanced leading to a prolonged disorder due to the continuous production of self-antigens becoming targets for the immune cells.

In order to analyse some pathogenic properties of viral infections for autoimmunity, the presented model describes the interactions between the following three populations:

1. host cells, denoted by the subscript h;
2. immune cells, denoted by the subscript i;
3. viral particles with molecular mimicry, denoted by the subscript v.

In the model it is supposed that certain viral infection leads to damage of some healthy host cells. This triggers a cross-reactive immune response. Specific immune cells are produced and activated, which can attack and destroy healthy cells possessing an antigen similar to the antigen presented by the viruses. In the model it is assumed that the healthy host cells can be produced from sources within the organism, for example in the thymus as well as due to the proliferation of existing healthy cells. It is supposed that the proliferation of existing healthy host cells is described by a logistic-like gain term. Immune cells are able to damage healthy cells, which also decay due to their natural death.

The immune cells can be produced due to self-antigens presented by damaged host cells as well as due to the presence of viral agents. It is assumed that it is more probably that the newly produced immune cells are with low activation states and they need some time in order to be activated. Immune cells are supposed also to be able to destroy virus particles.

2.2.2. Description of the Kinetic Model of Autoimmune Disease

It is assumed, that the general process of autoimmunity is modeled by a system consisting of three functional subsystems interacting with each other. They correspond to the populations described in the previous subsection. Due to the observations that the particular viral populations are homogeneous with respect to their ability to trigger an immune response, in the model it is supposed that the functional subsystem of the viruses is unstructured.

The remaining two functional subsystems of the host cells and of the immune cells are assumed to be structured, that is, to be able to express a specific biological function. The functional state of the population of the host cells is represented by the discrete variable u_h, whose value is equal to 1 if the corresponding host cell is healthy and is equal to 0 if the corresponding host cell is damaged.

The functional state of the population of immune cells is represented by the continuous variable u_i that spans the interval $[0;1]$. The activation states of the immune cells refer to their capability to damage healthy host cells. Increasing values of the activity denote the capability to damage more healthy cells. The assumption the activities of the immune cells to be represented by a continuous variable is due to the large number of active individuals participated in the functional subsystem of immune cells. The macroscopic state of the unstructured population of the viral particles is characterized, at time t, by their concentration $n_v(t)$.

The states of the structured populations are modeled, at time t, by their probability density functions $f_h(t,u)$ and $f_i(t,u)$.

The concentrations of the structured populations can be determined from the relations

$$\frac{f_h(t,0) + f_h(t,1)}{2}, \quad n_i(t) = \int_0^1 f_i(t,u_i) du_i. \qquad (1)$$

The present model is a modification of the kinetic model of autoimmune diseases proposed in Reference [92]. Here, the populations of target cells and damaged cells are described as only one population of host cells. The functional state of host cells may either be healthy or damaged.

Following the general suggestions for creating models within the KTAP [33], the dynamics of the system under consideration is modeled through derivation of evolution equations for the probability density functions of the structured subsystems as well as for the density of the unstructured subsystem.

To model the interactions between active particles belonging to the subsystems one can equate the variation rates of the numbers of active particles with the sums of inlet flux rates due to the proliferative interactions and outlet flow rates due to the destructive interactions described in the previous subsection. In addition, conservative interactions allowing the time change of the variable u_i should be taken into account.

As a result, the following system of ordinary and partial integro-differential equations is obtained for the present model:

$$\frac{\partial f_h}{\partial t}(t, u_h) = u_h \left[S_h(t) - d_{ht} f_h(t,1) + f_h(t,1)\left(p_h - \frac{p_h}{H_{Max}} f_h(t,1)\right) \right. $$
$$\left. - d_{hi} f_h(t,1) \int_0^1 u_i f_i(t, u_i) du_i \right] $$
$$+ (1 - u_h) \left[d_{hi} f_h(t,1) \int_0^1 u_i f_i(t, u_i) du_i - d_{hd} f_h(t,0) \right], \quad (2)$$

$$\frac{\partial f_i}{\partial t}(t, u_i) = (1 - u_i) \left[p_{id} f_h(t,0) + p_{iv} n_v(t) \right] - d_i f_i(t, u_i) $$
$$+ c_i \left[2 \int_0^{u_i} (u_i - v) f_i(t, v) dv - (1 - u_i)^2 f_i(t, u_i) \right], \quad (3)$$

$$\frac{d}{dt} n_v(t) = p_v f_h(t,1) n_v(t) - d_{vv} n_v(t) - d_{vi} n_v(t) \int_0^1 f_i(t, u_i) du_i, \quad (4)$$

with non-negative initial conditions

$$f_h(0, u_h) = f_h^{(0)}(u_h), \quad f_i(0, u_i) = f_i^{(0)}(u_i), \quad n_v(0) = n_v^{(0)}.$$

Naturally, the parameters in the system (2)–(4) should be non-negative.

Equation (2) describes the time change of the distribution density $f_h(t, u_h)$ of the host cells. The participating parameters have the following meaning:

- $S_h(t) - d_{ht} f_h(t,1)$ models the birth rate of healthy host cells from sources within the organism: it is assumed that their production is limited by the amount $f_h(t,1)$ of healthy cells present in the organism;
- p_h characterizes the proliferation rate of the healthy host cells;
- H_{Max} refers to the concentration of the healthy cells at which their proliferation turns off;
- d_{ht} describes the natural mortality rate of the healthy cells;
- d_{hi} describes the rate of damaging of the healthy cells by the immune cells;
- d_{hd} is the natural mortality rate of the damaged host cells.

Equation (3) describes the evolution of the distribution density $f_i(t, u_i)$ of the immune cells. The participating parameters have the following meaning:

- p_{id} is the birth rate of immune cells due to the presence of self-antigens presented by damaged host cells;
- p_{iv} is the birth rate of immune cells due to the presence of viruses;
- d_i is the natural mortality rate of the immune cells;

- the factor $1 - u_i$ is related to the assumption that the state of activity of the newly produced immune cells is low and they need time for activation;
- Due to the viral infection and certain cytokines and chemokines released by the damaged cells the activity of the immune cells can increase. The raising of the functional state of the immune cells is described by the conservative term $c_i \left[2 \int_0^{u_i} (u_i - v) f_i(t, v) dv - (1 - u_i)^2 f_i(t, u_i) \right]$, which do not change the concentration of the immune cells.

Equation (4) describes the evolution of the the concentration $n_v(t)$ of the viral particles. The participating parameters have the following meaning:

- p_v is the rate of replication of viruses;
- d_{vi} is the rate of destruction of the viruses due to the immune response;
- d_{vv} is the natural mortality rate of the viruses.

Consider the following spaces:

$$X = \{f = (f_h, f_i, n_v) : |f_h| < \infty, |n_v| < \infty, f_i \in L_1(0,1)\},$$

$$X^+ = \{f = (f_h, f_i, n_v) \in X : f_h \geq 0, f_i \geq 0, n_v \geq 0, \text{ a.e.}\}.$$

By the use of standard arguments it is easy to verify the validity of the following statement:

Theorem 1. *Let $S_1 \in C^0([0, \infty); R_+^1)$. For each $T > 0$, the system (2)–(4) with initial datum $f^{(0)} = (f_h^{(0)}, f_i^{(0)}, n_v^{(0)})$, $f^{(0)} \in X^+$, possesses a unique solution*

$$f \in C^0([0,T]; X) \cap C^1((0,T); X).$$

The solution possesses the property:
$f(t) \in X^+, \forall t \in [0, T]$.

3. Results of Simulations

The computational algorithm for solving the Cauchy problem corresponding to (2)–(4) includes the discretization of Equation (3) regarding the activation state $u_i \in [0, 1]$ by uniform mesh. The included integrals are calculated by using the Simpson's rule.

The corresponding system of ordinary differential equations has been solved for various parameter sets by the use of the Matlab ODE suite [93].

The assumption for the initial conditions has been that there is a certain amount of healthy host cells and small amounts of immune cells with equally distributed states of activity:

$$f_h(0, 1) = 100,$$

$$f_i(0, u) = 0.1, \quad \forall u \in [0, 1].$$

The following values of parameters have been used:

$$S_h(t) = 10, \forall t \geq 0, \quad H_{Max} = 100000,$$

$$p_h = 0.5, \quad d_{ht} = 0.2, \quad d_{hi} = 0.1, \quad d_{hd} = 0.8,$$

$$p_{iv} = 0.5, \quad d_i = 0.1, \quad c_i = 0.1,$$

$$p_v = 10.0, \quad d_{vi} = 4.0, \quad d_{vv} = 4.0,$$

and various values of parameter p_{id} describing the rate of production of immune cells due to damaged cells.

The numerical experiments have been aimed at studying the effects of damaging ability of the immune cells activated against healthy self-cells by viral infection.

First, the case without virus infection is considered setting $n_v(0) = 0$ and assuming that there is no damaged cells before the infection $f_h(0,0) = 0$ and $p_{id} = 0$. The dynamics of the concentrations of the healthy cells is shown in Figure 1. After initial growth, the healthy cells remains at high levels due to the absence of autoimmune reaction in this case.

In the further considerations it is assumed that initially small amounts of damaged host cells and viruses are present in the organism, by setting:

$$f_h(0,0) = 0.5, \quad n_v(0) = 0.1.$$

The results of numerical experiments show that for low values of the rate of production of immune cells due to damaged cells p_{id} the autoimmune reaction is weak and the concentration of healthy cells remains at high levels, see Figure 2 where $p_{id} = 0.00006$. In such cases autoimmune disorders are not observed or are very mild.

The results of numerical experiments show that for higher values of parameter p_{id} the cross-reaction of the immune system with self-antigens becomes very strong and damages a large amount of healthy cells. The result for $p_{id} = 0.00010$ is shown in Figure 3. One can see that in this case almost the half of the population of healthy cells is damaged. Such situation corresponds to a very severe autoimmune disease with possible lethal effect.

The performed numerical experiments show that p_{id} is a bifurcation parameter. It separates two different dynamics of the solutions. When the value of p_{id} is lower or equal to $p_{id} = 0.00016$, the solution is stable and tends to equilibrium while when it is greater than the value $p_{id} = 0.00016$, the solution is oscillating. In Figures 4 and 5 the concentrations of healthy cells for values $p_{id} = 0.00016$ and $p_{id} = 0.00040$ are shown for illustration.

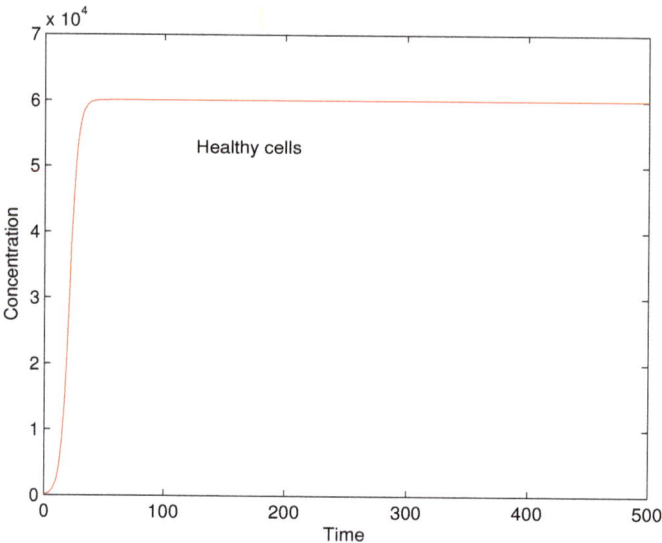

Figure 1. Concentrations of healthy cells for $n_v(0) = 0$, $f_h(0,0) = 0$ and $p_{id} = 0$.

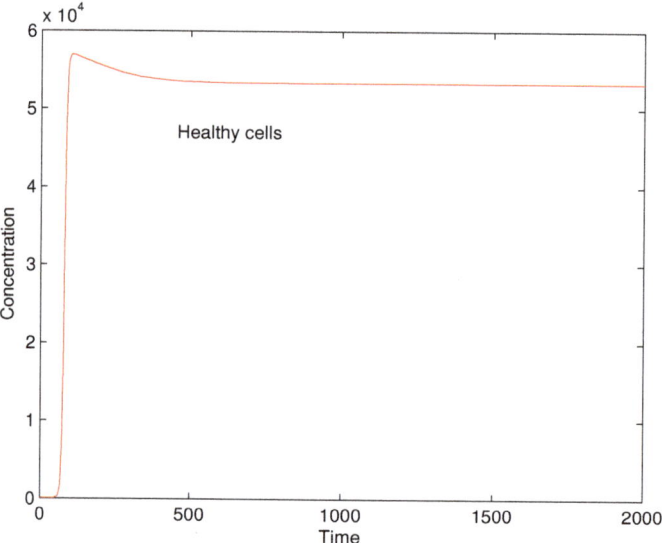

Figure 2. Concentrations of healthy cells for $p_{id} = 0.00006$.

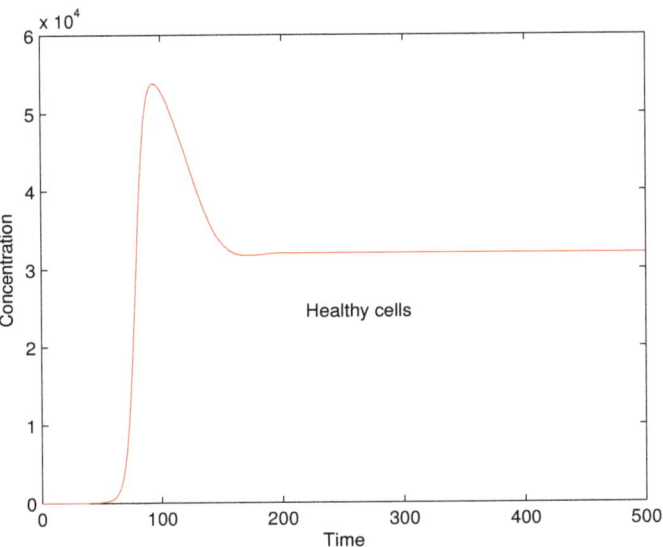

Figure 3. Concentrations of healthy cells for $p_{id} = 0.00010$.

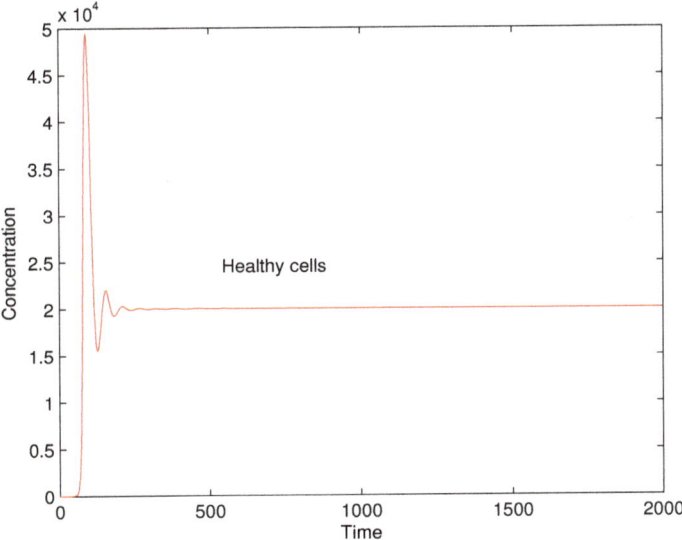

Figure 4. Concentrations of healthy cells for $p_{id} = 0.00016$.

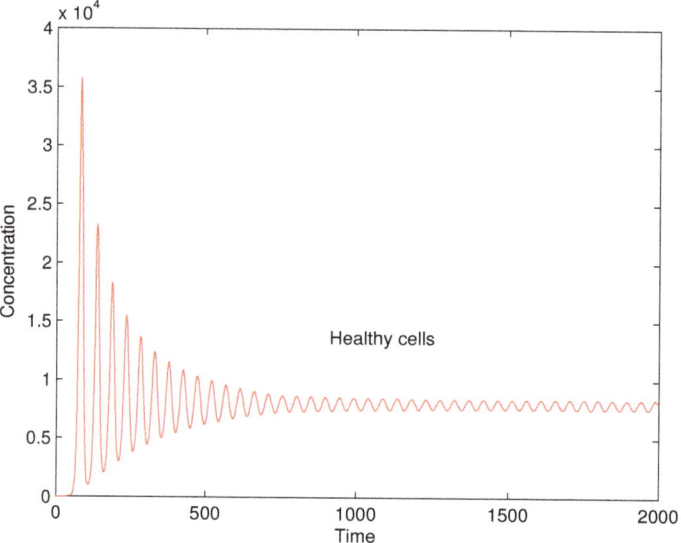

Figure 5. Concentrations of healthy cells for $p_{id} = 0.00040$.

The cases with periodic flare ups can be considered autoimmune diseases characterized with repeated periods of dormancy of the autoaggressive immune cells and their flare ups triggered by spreading pathogenic proteins, possible failure of regulatory immune cells to control the destructive reaction of effector immune cells against host constituents and so forth. An example of repeated flare ups is the MS [25,94]. The numerous repeated periods when large amounts of healthy cells are damaged can lead to serious tissue or organ destruction and even lethal result for the patients.

4. Concluding Remarks

In this paper, a kinetic type model of general autoimmune disease is presented. The modeled problem is solved numerically. The numerical solutions represent several typical dynamics of autoimmune diseases resulting from viral infection (absence of disease, mild symptoms, chronic disease, flare ups and severe autoimmune disease).

The presented results and further development of the model can be useful for the better understanding of the group of autoimmune diseases. The results confirm the evidence that the viral infections and damaged cells can be very potent pro-inflammatory triggers and sources for self-antigens for specific immune cells thus leading to chronic or severe autoimmune diseases [25,94].

Funding: This research received no external funding

Acknowledgments: The author wishes to express his gratitude to M. Lachowicz for the useful discussions on the mathematical model as well as to the anonymous referees for the helpful comments and remarks that led to the improvements in the presentation of the paper.

Conflicts of Interest: The author declares no conflict of interest.

References

1. Anderson, C.C.; Bretscher, P.; Corthay, A.; Dembic, Z.; Havele, C.; Nagy, Z.A.; Øynebraten, I. Immunological Tolerance: Part I of a Report of a Workshop on Foundational Concepts of Immune Regulation. *Scand. J. Immunol.* **2017**, *85*, 84–94. [CrossRef] [PubMed]
2. Janeway, C.; Travers, P.; Walport, M.; Shlomchik, M. *Immunobiology: The Immune System in Health and Disease*; Garland: New York, NY, USA, 2006.
3. Campbell, A.W. Autoimmunity and the Gut. *Autoimmune Dis.* **2014**, *2014*, 152428. [CrossRef] [PubMed]

4. Vojdani, A.A potential link between environmental triggers and autoimmunity. *Autoimmune Dis.* **2013**, *2013*, 437231. [CrossRef] [PubMed]
5. Ceccarelli, F.; Agmon-Levin, N.; Perricone, C. Genetic factors of autoimmune diseases 2017. *J. Immunol. Res.* **2017**, *2017*, 1–2. [CrossRef] [PubMed]
6. Pelajo, C.F.; Lopez-Benitez, J.M.; Miller, L.C. Vitamin D and autoimmune rheumatologic disorders. *Autoimmun. Rev.* **2010**, *9*, 507–510. [CrossRef] [PubMed]
7. Agmon-Levin, N.; Theodor, E.; Segal, R.; Shoenfeld, Y. Vitamin D in systemic and organ-specific autoimmune diseases. *Clin. Rev. Allergy Immunol.* **2013**, *45*, 256–266. [CrossRef] [PubMed]
8. Syage, J.; Kelly, C.; Dickason, M.; Ramirez, A.C.; Leon, F.; Dominguez, R.; Sealey-Voyksner, J. Determination of gluten consumption in celiac disease patients on agluten-free diet. *Am. J. Clin. Nutr.* **2018**, *107*, 201–207. [CrossRef]
9. Lebwohl, B.; Ludvigsson, J.F.; Green, P.H.R. Celiac disease and non-celiac gluten sensitivity. *Autoimmune Dis.* **2015**, *351*, h4347. [CrossRef]
10. Abbas, A.; Lichtman, A.; Pillai, S. *Cellular and Molecular Immunology*; Elsevier: Philadelphia, PA, USA, 2018.
11. Kivity, S.; Agmon-Levin, N.; Blank, M.; Shoenfeld, Y. Infections and autoimmunity: Friends or foes? *Trends Immunol.* **2009**, *30*, 409–414. [CrossRef]
12. Rose, N. Introduction. In *Infection and Autoimmunity*; Shoenfeld, Y., Agmon-Levin, N., Rose, N., Eds.; Academic Press: San Diego, CA, USA, 2015; pp. 1–12.
13. Richardson, S.J.; Willcox, A.; Bone, A.J.; Foulis, A.K.; Morgan, N.G. The prevalence of enteroviral capsid protein vp1 immunostaining in pancreatic islets in human type 1 diabetes. *Diabetologia* **2009**, *52*, 1143–1151. [CrossRef]
14. Martinelli, M.; Agmon-Levin, N.; Amital, H.; Shoenfeld, Y. Infections and autoimmune diseases: An interplay of pathogenic and protective links. In *Infection and Autoimmunity*; Shoenfeld, Y., Agmon-Levin, N., Rose, N., Eds.; Academic Press: San Diego, CA, USA, 2015; pp. 13–23.
15. Lidar, M.; Lipschitz, N.; Agmon-Levin, N.; Langevitz, P.; Barzilai, O.; Ram, M.; Porat-Katz, B.S.; Bizzaro, N.; Damoiseaux, J.; Tervaert, J.W.; et al. Infectious serologies and autoantibodies in hepatitis C and autoimmune disease-associated mixed cryoglobulinemia. *Clin. Rev. Allergy Immunol.* **2012**, *42*, 238–246. [CrossRef] [PubMed]
16. Fabrizi, F.; Dixit, V.; Messa, P. Antiviral therapy of symptomatic HCV-associated mixed cryoglobulinemia: Meta-analysis of clinical studies. *J. Med. Virol.* **2013**, *85*, 1019–1027. [CrossRef] [PubMed]
17. Harley, J.B.; Chen, X.; Pujato, M.; Miller, D.; Maddox, A.; Forney, C.; Magnusen, A.F.; Lynch, A.; Chetal, K.; Yukawa, M.; et al. Transcription factors operate across disease loci, with EBNA2 implicated in autoimmunity. *Nat. Genet.* **2018**, *50*, 699–706. [CrossRef] [PubMed]
18. Afrasiabi, A.; Parnell, G.P.; Fewings, N.; Schibeci, S.D.; Basuki, M.A.; Chandramohan, R.; Zhou, Y.; Taylor, B.; Brown, D.A.; Swaminathan, S.; et al. Evidence from genome wide association studies implicates reduced control of Epstein-Barr virus infection in multiplesclerosis susceptibility. *Genome Med.* **2019**, *11*, 26–38. [CrossRef]
19. Li, Z.X.; Zeng, S.; Wu, H.X.; Zhou, Y. The risk of systemic lupus erythematosus associated with Epstein–Barr virus infection: A systematic review and meta-analysis. *Clin. Exp. Med.* **2019**, *19*, 23–36. [CrossRef]
20. Strachan, D.P. Hay fever, hygiene, and household size. *BMJ* **1989**, *229*, 1259–1260. [CrossRef]
21. Davies, T.F. Infection and autoimmune thyroid disease. *J. Clin. Endocrinol. Metab.* **2008**, *93*, 6745–676. [CrossRef]
22. Navegantes, K.C.; de Souza Gomes, R.; Pereira, P.A.T.; Czaikoski, P.G.; Azevedo, C.H.M.; Monteiro, M.C. Immune modulation of some autoimmune diseases: The critical role of macrophages and neutrophils in the innate and adaptive immunity. *J. Transl. Med.* **2017**, *15*, 36. [CrossRef]
23. Ma, W.T.; Gao, F.; Gu, K.; Chen, D.K. The Role of Monocytes and Macrophages in Autoimmune Diseases: A Comprehensive Review. *Front. Immunol.* **2019**, *10*, 1140. [CrossRef]
24. McInnes, I.B.; Schett, G. The pathogenesis of rheumatoid arthritis. *New Engl. J. Med.* **2011**, *365*, 2205–2219. [CrossRef]
25. Iwami, S.; Takeuchi, Y.; Miura, Y.; Sasaki, T.; Kajiwara, T. Dynamical properties of autoimmune disease models: Tolerance, flare-up, dormancy. *J. Theor. Biol.* **2007**, *246*, 646–659. [CrossRef]
26. Iwami, S.; Takeuchi, Y.; Iwamoto, K.; Naruo, Y.; Yasukawa, T. A mathematical design of vector vaccine against autoimmune disease. *J. Theor. Biol.* **2009**, *256*, 382–392. [CrossRef] [PubMed]

27. Nicholson, D.; Kerr, E.; Jepps, O.; Nicholson, L. Modelling experimental uveitis: Barrier effects in autoimmune disease. *Inflamm. Res.* **2012**, *61*, 759–773. [CrossRef] [PubMed]
28. Delitala, M.; Dianzani, U.; Lorenzi, T.; Melensi, M. A mathematical model for immune and autoimmune response mediated by T-cells. *Comput. Math. Appl.* **2013**, *66*, 1010–1023. [CrossRef]
29. Landmann, S.; Preuss, N.; Behn, U. Self-tolerance and autoimmunity in a minimal model of the idiotypic network. *J. Theor. Biol.* **2017**, *426*, 17–39. [CrossRef] [PubMed]
30. Roy, A.K.; Roy, P.K.; Grigorieva, E. Mathematical insights on psoriasis regulation: Role of Th1 and Th2 cells. *Math. Biosci. Eng.* **2018**, *15*, 717–738.
31. Sakaguchi, S.; Wing, K.; Miyara, M. Regulatory T cells: A brief history and perspective. *Eur. J. Immunol.* **2007**, *37*, S116–S123. [CrossRef]
32. Romanczuk, P.; Bar, M.; Ebeling, W.; Linder, B.; Schimansky-Geier, L. Active Brownian particles, from individual to collective stochastic dynamics. *Eur. Phys. J.* **2012**, *202*, 1–162.
33. Bianca, C.; Bellomo, N. *Towards a Mathematical Theory of Multiscale Complex Biological Systems*; World Scientific: Singapore, 2011.
34. Bellomo, N.; Brezzi, F. Challenges in active particles methods: Theory and applications. *Math. Model. Methods Appl. Sci.* **2018**, *28*, 1627–1633. [CrossRef]
35. Bellomo, N.; Brezzi, F. Towards a multiscale vision of active particles. *Math. Model. Methods Appl. Sci.* **2019**, *29*, 581–588. [CrossRef]
36. Bianca, C. Thermostatted kinetic equations as models for complex systems in physics and life sciences. *Phys. Life Rev.* **2012**, *9*, 359–399. [CrossRef] [PubMed]
37. Bianca, C. Modeling complex systems by functional subsystems representation and thermostatted-KTAP methods. *Appl. Math. Inf. Sci.* **2012**, *6*, 495–499.
38. Bellomo, N.; Bianca, C.; Delitala, M. Complexity analysis and mathematical tools towards the modelling of living systems. *Phys. Life Rev.* **2009**, *6*, 144–175. [CrossRef] [PubMed]
39. Degond, P.; Pareschi, L.; Russo, G. (Eds.) *Modeling and Computational Methods for Kinetic Equations*; Springer Science+Bussiness Media: New York, NY, USA, 2004; pp. 219–258.
40. Bianca, C.; Dogbe, C. On the Boltzmann gas mixture equation: Linking the kinetic and fluid regimes. *Commun. Nonlinear Sci. Numer. Simulat.* **2015**, *29*, 240–256. [CrossRef]
41. Schiavo, M.L. The modelling of political dynamics by generalized kinetic (Boltzmann) models. *Math. Comput. Model.* **2003**, *37*, 261–281. [CrossRef]
42. Bertotti, M.L.; Delitala, M. Conservation laws and asymptotic behavior of a model of social dynamics. *Nonlinear Anal. Real World Appl.* **2008**, *9*, 183–196. [CrossRef]
43. Ajmone Marsan, G.; Bellomo, N.; Gibelli, L. Stochastic evolutionary differential games toward a systems theory of behavioral social dynamics. *Math. Model. Methods Appl. Sci.* **2016**, *26*, 1051–1093. [CrossRef]
44. Dolfin, M.; Leonida, L.; Outada, N. Modelling human behaviour in economics and social science. *Phys. Life Rev.* **2017**, *22–23*, 1–21. [CrossRef]
45. Bellomo, N.; Brezzi, F.; Pulvirenti, M. Modeling behavioral social systems. *Math. Model. Methods Appl. Sci.* **2017**, *27*, 1–11. [CrossRef]
46. Bianca, C.; Kombargi, A. On the inverse problem for thermostatted kinetic models with application to the financial market. *Appl. Math. Inf. Sci.* **2017**, *11*, 1463–1471. [CrossRef]
47. Bianca, C.; Kombargi, A. On the modeling of the stock market evolution by means of the information-thermostatted kinetic theory. *Nonlinear Stud.* **2017**, *24*, 935–944.
48. Bellomo, N.; Carbonaro, B. On the complexity of multiple interactions with additional reasoning about Kate, Jules and Jim. *Math. Comput. Model.* **2008**, *47*, 168–177. [CrossRef]
49. Bellomo, N.; Bianca, C.; Coscia, V. On the modeling of crowd dynamics: An overview and research perspectives. *Bol. Soc. Esp. Mat. Apl.* **2011**, *54*, 25–46. [CrossRef]
50. Bellomo, N.; Bellouquid, A. On multiscale models of pedestrian crowds - from mesoscopic to macroscopic. *Commun. Math. Sci.* **2015**, *13*, 1649–1664. [CrossRef]
51. Bellomo, N.; Gibelli, L. Toward a mathematical theory of behavioural-social dynamics for pedestrian crowds. *Math. Model. Methods Appl. Sci.* **2015**, *25*, 2417–2437. [CrossRef]
52. Elaiw, A.; Al-Turki, Y.; Alghamdi, M. A critical analysis of behavioural crowd dynamics—From a modelling strategy to kinetic theory methods. *Symmetry* **2019**, *11*, 851. [CrossRef]

53. Bellomo, N.; Gibelli, L. Behavioral human crowds. In *Crowd Dynamics, Volume 1—Theory, Models, and Safety Problems. Modeling and Simulation in Science, Engineering, and Technology*; Bellomo, N., Gibelli, L., Eds.; Birkhäuser: New York, NY, USA, 2018; pp. 1–14.
54. Bellomo, N.; Gibelli, L.; Outada, N. On the interplay between behavioral dynamics and social interactions in human crowds. *Kinet. Relat. Mod.* **2019**, *12*, 397–409. [CrossRef]
55. Cucker, F.; Smale, S. Emergent behavior in flocks. *IEEE Trans. Automat. Control* **2007**, *52*, 852–862. [CrossRef]
56. Ha, S.-Y.; Tadmor, E. From particle to kinetic and hydrodynamic description of flocking. *Kinet. Relat. Model.* **2008**, *1*, 415–435.
57. Bellomo, N.; Bellouquid, A.; Gibelli, L.; Outada, N. *A Quest Towards a Mathematical Theory of Living Systems*; Springer International Publishing: Cham, Switzerland, 2017.
58. Bellomo, N.; Ha, S.-Y. A quest toward a mathematical theory of the dynamics of swarms. *Math. Model. Methods Appl. Sci.* **2017**, *27*, 745–770. [CrossRef]
59. Degond, P.; Frouvelle, A.; Merino-Aceituno, S. A new flocking model through body attitude coordination. *Math. Model. Methods Appl. Sci.* **2017**, *27*, 1005–1049. [CrossRef]
60. Lachowicz, M.; Leszczynski, H.; Parisot, M. Blow-up and global existence for a kinetic equation of swarm formation. *Math. Model. Methods Appl. Sci.* **2017**, *27*, 1153–1175. [CrossRef]
61. Poyato, D.; Soler, J. Euler-type equations and commutators in singular and hyperbolic limits of kinetic Cucker–Smale models. *Math. Model. Methods Appl. Sci.* **2017**, *27*, 1089–1152. [CrossRef]
62. Delitala, M.; Tosin, A. Mathematical modelling of vehicular traffic: A discrete kinetic theory approach. *Math. Model. Methods Appl. Sci.* **2007**, *17*, 901–932. [CrossRef]
63. Klar, A.; Wegener, R. Traffic flow: Models and numerics. In *Modeling and Computational Methods for Kinetic Equations*; Degond, P., Pareschi, L., Russo, G., Eds.; Springer Science+Bussiness Media: New York, NY, USA, 2004; pp. 219–258.
64. Bianca, C.; Coscia, V. On the coupling of steady and adaptive velocity grids in vehicular traffic modelling. *Appl. Math. Lett.* **2011**, *24*, 149–155. [CrossRef]
65. Bianca, C. Mathematical modeling of crowds dynamics: Complexity and kinetic approach. *Nonlinear Stud.* **2012**, *19*, 345–354.
66. Bianca, C.; Dogbe, C. A mathematical model for crowd dynamics: Multiscale analysis, fluctuations and random noise. *Nonlinear Stud.* **2013**, *20*, 281–305.
67. Bianca, C.; Mogno, C. A thermostatted kinetic theory model for event-driven pedestrian dynamics. *Eur. Phys. J. Plus* **2018**, *133*, 213. [CrossRef]
68. Bianca, C.; Mogno, C. Modelling pedestrian dynamics into a metro station by thermostatted kinetic theory methods. *Math. Comput. Model. Dyn. Syst.* **2018**, *24*, 207–235. [CrossRef]
69. Bianca, C. Mathematical modelling for keloid formation triggered by virus: Malignant effects and immune system competition. *Math. Model. Methods Appl. Sci.* **2011**, *21*, 389–419. [CrossRef]
70. Bianca, C.; Fermo, L. Bifurcation diagrams for the moments of a kinetic type model of keloid-immune system competition. *Comput. Math. Appl.* **2011**, *61*, 277–288. [CrossRef]
71. Bianca, C.; Delitala, M. On the modelling of genetic mutations and immune system competition. *Comput. Math. Appl.* **2011**, *61*, 2362–2375. [CrossRef]
72. Bianca, C.; Riposo, J. Mimic therapeutic actions against keloid by thermostatted kinetic theory methods. *Eur. Phys. J. Plus* **2015**, *130*, 159. [CrossRef]
73. Ben Amar, M; Bianca, C. Towards a unified approach in the modelling of fibrosis: A review with research perspectives. *Phys. Life Rev.* **2016**, *16*, 61–85.
74. Ben Amar, M; Bianca, C. Multiscale modeling of fibrosis—What's next? *Phys. Life Rev.* **2016**, *16*, 118–123.
75. Bianca, C. How do mutative events modify moments evolution in thermostatted kinetic models? *Commun. Nonlinear Sci. Numer. Simul.* **2014**, *19*, 2155–2159. [CrossRef]
76. Bianca, C.; Dogbe, C.; Lemarchand, A. Mimic therapeutic actions against keloid by thermostatted kinetic theory methods. *Eur. Phys. J. Plus* **2016**, *131*, 41. [CrossRef]
77. Bianca, C.; Lemarchand, A. Miming the cancer-immune system competition by kinetic Monte Carlo simulations. *J. Chem. Phys.* **2016**, *145*, 154108. [CrossRef]
78. Masurel, L.; Bianca, C.; Lemarchand, A. On the learning control effects in the cancer-immune system competition. *Phys. A Stat. Mech. Its Appl.* **2018**, *506*, 462–475. [CrossRef]

79. Aylaj, B. Qualitative analysis and simulation of a nonlinear integro-differential system modeling tumor-immune cells competition. *Int. J. Biomath.* **2018**, *11*, 1850104. [CrossRef]
80. Bianca, C.; Lemarchand, A. Density evolution by the low-field limit of kinetic frameworks with thermostat and mutations. *Commun. Nonlinear Sci. Numer. Simul.* **2015**, *20*, 14–23. [CrossRef]
81. Bianca, C.; Brezin, L. Modeling the antigen recognition by B-cell and T-cell receptors through thermostatted kinetic theory methods. *Int. J. Biomath.* **2017**, *10*, 1750072. [CrossRef]
82. Bianca, C. Kinetic theory for active particles modelling coupled to Gaussian thermostats. *Appl. Math. Sci.* **2012**, *6*, 651–660.
83. Bianca, C. An existence and uniqueness theorem for the Cauchy problem for thermostatted-KTAP models. *Int. J. Math. Anal.* **2012**, *6*, 813–824.
84. Bianca, C. Thermostatted kinetic models for complex systems under microscopic external fields. *Math. Eng. Sci. Aerosp.* **2012**, *3*, 225–238.
85. Bianca, C. Onset of nonlinearity in thermostatted active particles models for complex systems. *Nonlinear Anal. Real World Appl.* **2012**, *13*, 2593–2608. [CrossRef]
86. Bianca, C. Thermostatted models - Multiscale analysis and tuning with real-world systems data. *J. Appl. Comput. Math.* **2012**, *9*, 418–425. [CrossRef]
87. Bianca, C.; Ferrara, M.; Guerrini, L. Thermostatted models - Multiscale analysis and tuning with real-world systems data. *J. Glob. Optim.* **2014**, *58*, 389–404. [CrossRef]
88. Bianca, C.; Dogbe, C. Kinetic models coupled with Gaussian thermostats: Macroscopic frameworks. *Nonlinearity* **2014**, *27*, 2771–2803. [CrossRef]
89. Bianca, C.; Menale, M. Existence and uniqueness of nonequilibrium stationary solutions in discrete thermostatted models. *Commun. Nonlinear Sci. Numer. Simulat.* **2019**, *73*, 25–34. [CrossRef]
90. Bianca, C.; Menale, M. On the interaction domain reconstruction in the weighted thermostatted kinetic framework. *Eur. Phys. J. Plus* **2019**, *134*, 143. [CrossRef]
91. Jäger, E.; Segel, L. On the distribution of dominance in a population of interacting anonymous organisms. *SIAM J. Appl. Math.* **1992**, *52*, 1442–1468. [CrossRef]
92. Kolev, M.; Nikolova, I. A mathematical model of some viral-induced autoimmune diseases. *Math. Appl.* **2018**, *46*, 97–108.
93. Shampine, L.F.; Reichelt, M.W. The Matlab ODE suite. *SIAM J. Sci. Comput.* **1997**, *18*, 1–22. [CrossRef]
94. Vadasz, Z.; Toubi, E. Acute and chronic infections: Their role in immune thrombocytopenia. In *Infection and Autoimmunity*; Shoenfeld, Y., Agmon-Levin, N., Rose, N., Eds.; Academic Press: San Diego, CA, USA, 2015; pp. 859–876.

© 2019 by the author. Licensee MDPI, Basel, Switzerland. This article is an open access article distributed under the terms and conditions of the Creative Commons Attribution (CC BY) license (http://creativecommons.org/licenses/by/4.0/).

Article
A Distributed Control Problem for a Fractional Tumor Growth Model

Pierluigi Colli [1], Gianni Gilardi [1] and Jürgen Sprekels [2,3]

1. Dipartimento di Matematica "F. Casorati", Università di Pavia and Research Associate at the IMATI—C.N.R. Pavia, via Ferrata 5, 27100 Pavia, Italy
2. Department of Mathematics, Humboldt-Universität zu Berlin, Unter den Linden 6, 10099 Berlin, Germany
3. Weierstrass Institute for Applied Analysis and Stochastics, Mohrenstrasse 39, 10117 Berlin, Germany
* Correspondence: pierluigi.colli@unipv.it

Received: 26 July 2019; Accepted: 23 August 2019; Published: 31 August 2019

Abstract: In this paper, we study the distributed optimal control of a system of three evolutionary equations involving fractional powers of three self-adjoint, monotone, unbounded linear operators having compact resolvents. The system is a generalization of a Cahn–Hilliard type phase field system modeling tumor growth that has been proposed by Hawkins–Daarud, van der Zee and Oden. The aim of the control process, which could be realized by either administering a drug or monitoring the nutrition, is to keep the tumor cell fraction under control while avoiding possible harm for the patient. In contrast to previous studies, in which the occurring unbounded operators governing the diffusional regimes were all given by the Laplacian with zero Neumann boundary conditions, the operators may in our case be different; more generally, we consider systems with fractional powers of the type that were studied in a recent work by the present authors. In our analysis, we show the Fréchet differentiability of the associated control-to-state operator, establish the existence of solutions to the associated adjoint system, and derive the first-order necessary conditions of optimality for a cost functional of tracking type.

Keywords: fractional operators; Cahn–Hilliard systems; well-posedness; regularity; optimal control; necessary optimality conditions

MSC: 35K55; 35Q92; 49J20; 92C50

---

1. Introduction

The recent paper [1] investigates the evolutionary system

$$\alpha \, \partial_t \mu + \partial_t \varphi + A^{2\rho} \mu = P(\varphi)(S - \mu), \tag{1}$$

$$\beta \, \partial_t \varphi + B^{2\sigma} \varphi + f(\varphi) = \mu, \tag{2}$$

$$\partial_t S + C^{2\tau} S = -P(\varphi)(S - \mu) + u, \tag{3}$$

where the equations are understood to hold in Ω, a bounded, connected and smooth domain in $\mathbb{R}^3$, and in the time interval $(0, T)$. In the above system, $A^{2\rho}$, $B^{2\sigma}$, and $C^{2\tau}$, with $\tau > 0$, $\sigma > 0$, $\rho > 0$, denote fractional powers of the self-adjoint, monotone, and unbounded, linear operators A, B and C, respectively, which are supposed to be densely defined in $H := L^2(\Omega)$ and to have compact resolvents. Moreover, α and β are positive real parameters.

System (1)–(3) is a generalization of a diffuse interface model for tumor growth. Such models, which are usually established in the framework of the Cahn–Hilliard model originating from the theory of phase transitions, have drawn increasing attention in the past years among mathematicians and applied scientists. We cite here just [2–8] as a sample of pioneering papers in this direction. In this connection, φ stands for an order parameter that should attain its values in the interval $[-1,1]$, where the values -1 and $+1$ indicate the healthy cell and tumor cell cases, respectively. The variable S represents the nutrient extra-cellular water concentration, u stands for a source term that acts as a control to monitor the evolution of the tumor cell fraction φ, and the nonlinearity P occurring in Equations (1) and (3) is a nonnegative and smooth function modeling a proliferation rate. Finally, μ represents the chemical potential, which acts as the driving thermodynamic force of the evolution and is obtained as the variational derivative with respect to the order parameter φ of a suitable free energy functional. In this connection, the nonlinearity f denotes the derivative of a double-well potential F which plays the role of a specific local free energy and yields the main contribution to the total free energy. Important examples for F are the so-called *classical regular potential* and the *logarithmic double-well potential*, given by the formulas

$$F_{reg}(r) := \frac{1}{4}(r^2-1)^2 \quad r \in \mathbb{R}, \quad \text{and} \tag{4}$$

$$F_{log}(r) := ((1+r)\ln(1+r) + (1-r)\ln(1-r)) - c_1 r^2 \quad r \in (-1,1), \tag{5}$$

respectively. In definition (5), the constant c_1 is larger than 1, so that F_{log} is nonconvex. Furthermore, the function P in (1) and (3) is nonnegative and smooth. Finally, the datum u appearing in (3) is given.

In the literature, the diffusional developments in the system have usually been modeled by the Laplacian, that is, the case $A^{2\rho} = B^{2\sigma} = C^{2\tau} = -\Delta$, accompanied by zero Neumann boundary conditions, was assumed, where two main classes of models were considered. The first class of models regards the tumor and healthy cells as inertialess fluids; in such models, special fluid effects can be incorporated by postulating a Darcy or Stokes–Brinkman law, see, e.g., the works [7,9–18], where we also refer to [19,20]. The other class of models, to which the model considered here belongs, neglects the velocity. Typical contributions in this direction were given in [21–27], to name just a few.

While the occurrence of more general diffusional regimes of fractional type has been studied for a long time in the mathematical literature, it was only recently (see, e.g., [28–36]) that fractional operators have been investigated in the framework of Cahn–Hilliard systems (for phase field systems of Caginalp type, see also [37]), and the only investigations of tumor growth models involving fractional diffusive regimes such as in the system (1)–(3) seem to be the recent papers [1,38] by the present authors. Concerning the motivation and biological meaningfulness of the systems (1)–(3), we point out that, in our approach, the three fractional operators, which may be considerably different one from the other, are used as major dynamics for the tumor growth and for the diffusion processes. These operators may be related and close to fractional Laplacians, but also to other elliptic operators, and can present different orders. Indeed, some components in tumor development, such as immune cells, do have anomalous diffusion dynamics observed by experiments [39], but other components, like chemical potential and nutrient concentration may be ruled by other fractional or non-fractional flows. However, taking all this into consideration, it is the case of emphasizing that fractional operators are more and more utilized in the world of biological applications. In this direction, let us mention some recent and related work: we quote [40] for growth of species and their movements; we quote [41] for the numerical study of a super-diffusive model for a benign brain tumor; we quote [39] for the application to cell movements, in particular for T cells migrating through chronically-infected brain tissue; we quote the monography [42] concerned with fractional diffusion processes in applied sciences; we quote [43] for a hyperbolic-parabolic model of chemotaxis related to tumor angiogenesis; we quote [44] for the study of the dynamics of growth bacteria; we quote [45] for the study of a space time fractional reaction–diffusion equation in Parkinson's disease; we quote [46] for the study of a model for collective cell movement due to chemical sensing; we quote [47] for applications to biological populations in competition; we quote [48] for the investigation of a fractional order tumor model; we quote [49] for the optimal control for a nonlinear mathematical model of tumor under immune suppression; we quote [50] for the study of a fractional reaction–diffusion equation with a nonlocal boundary condition modeling the invasion of a tumor and its growth.

Now, coming back to our system (1)–(3) and referring to the paper [1], it turns out that, under rather general assumptions on the operators and the potentials, well-posedness and regularity results for the initial value problem for (1)–(3) were established in the case $u = 0$, under the assumption that $\alpha > 0$ and $\beta > 0$. However, some remarks on more general cases including $u \in L^2(Q)$, where $Q := \Omega \times (0, T)$, have been given in [1]. In particular, under suitable assumptions on the initial data, for every $u \in L^2(Q)$, there exists at least a solution (μ, φ, S) in a proper functional space to a weak version of (1)–(3) (namely, Equation (2) is replaced by a variational inequality involving the convex part F_1 of F rather than f, since F_1 is not supposed to be differentiable). Moreover, the solution is unique if the domains of the fractional operators A^ρ and C^τ satisfy suitable embeddings of Sobolev type. Finally, if $B^{2\sigma}$ behaves like the Laplace operator with either Dirichlet or Neumann zero boundary conditions and f is single valued (like (4) and (5)), then the solution solves Equation (2) in a stronger sense, and it is even smoother under more restrictive assumptions on the initial data.

In this paper, we first establish similar results for system (1)–(3) by assuming $\alpha = 0$ and $\beta > 0$ (in fact, we take $\beta = 1$ without loss of generality). In particular, we extend some results shown for this case in the recent paper [38]. Then, we discuss a distributed control problem for the modified system. Namely, given nonnegative constants κ_i, $i = 1, \ldots, 5$, and functions $\varphi_Q, S_Q \in L^2(Q)$ and $\varphi_\Omega, S_\Omega \in L^2(\Omega)$, we consider the problem of minimizing the cost functional

$$\mathcal{J}(u, \varphi, S) := \frac{\kappa_1}{2} \int_Q |\varphi - \varphi_Q|^2 + \frac{\kappa_2}{2} \int_\Omega |\varphi(T) - \varphi_\Omega|^2$$
$$+ \frac{\kappa_3}{2} \int_Q |S - S_Q|^2 + \frac{\kappa_4}{2} \int_\Omega |S(T) - S_\Omega|^2 + \frac{\kappa_5}{2} \int_Q |u|^2,$$

where φ and S are the components of the solution (μ, φ, S) corresponding to the control u, which is supposed to vary under restrictions of the type $u_{min} \leq u \leq u_{max}$.

The choice of this tracking-type cost functional reflects the plan of a medical treatment via the application of drugs over some finite time interval $(0, T)$ with the aim of monitoring the evolution of the tumor fraction φ under the restriction that no harm be inflicted on the patient. We remark at this place that it would be desirable to minimize the duration, i.e., the time $T > 0$, of the medical treatment as well, in order to prevent that the tumor cells develop a resistance against the drug. However, such an approach, which was possible (see, e.g., [21]) in the special case when $A^{2\rho} = B^{2\sigma} = C^{2\tau} = -\Delta$, becomes very complicated in the situation considered here and was therefore not included.

The literature on optimal control problems for Cahn–Hilliard systems is still scarce. In this connection, we refer the reader to [31,51], where a number of references is given. Even less investigations have been made on optimal control problems for tumor growth models. About that, let us refer to the works [18,21,24,52–59], for various models involving the Laplacian. Concerning the optimal control of Cahn–Hilliard systems with fractional operators, we just can cite [31,32], and, to the authors' best knowledge, the present paper is the first contribution on the optimal control of the tumor growth model with fractional operators.

The remainder of the paper is organized as follows. In the next section, we list our assumptions and notations and present our results on the state system. Section 3 is devoted to the study of the control-to-state mapping and of its Fréchet differentiability. In the last section, we deal with the control problem. Namely, the existence of an optimal control is proved and the first order necessary conditions involving a proper adjoint system are derived.

2. The State System

In this section, we first introduce the notations and the assumptions needed for the analysis of the state system. Then, we present our results. We closely follow [1]. First, of all, the set $\Omega \subset \mathbb{R}^3$ is assumed to be bounded, connected and smooth, with volume $|\Omega|$ and outward unit normal vector field ν on $\Gamma := \partial\Omega$. Moreover, ∂_ν stands for the corresponding normal derivative. We set

$$H := L^2(\Omega) \tag{6}$$

and denote by $\|\cdot\|$ and $(\cdot,\cdot)$ the standard norm and inner product of H. As for the operators, we first postulate that

$$A : D(A) \subset H \to H, \quad B : D(B) \subset H \to H \quad \text{and} \quad C : D(C) \subset H \to H \quad \text{are}$$
unbounded monotone self-adjoint linear operators with compact resolvents. (7)

Therefore, there are sequences $\{\lambda_j\}$, $\{\lambda'_j\}$, $\{\lambda''_j\}$ and $\{e_j\}$, $\{e'_j\}$, $\{e''_j\}$ of eigenvalues and of corresponding eigenvectors satisfying

$$Ae_j = \lambda_j e_j, \quad Be'_j = \lambda'_j e'_j, \quad \text{and} \quad Ce''_j = \lambda''_j e''_j,$$
$$\text{with} \quad (e_i, e_j) = (e'_i, e'_j) = (e''_i, e''_j) = \delta_{ij} \quad \text{for } i, j = 1, 2, \ldots, \tag{8}$$

$$0 \leq \lambda_1 \leq \lambda_2 \leq \ldots, \quad 0 \leq \lambda'_1 \leq \lambda'_2 \leq \ldots \quad \text{and} \quad 0 \leq \lambda''_1 \leq \lambda''_2 \leq \ldots,$$
$$\text{with} \quad \lim_{j \to \infty} \lambda_j = \lim_{j \to \infty} \lambda'_j = \lim_{j \to \infty} \lambda''_j = +\infty, \tag{9}$$

$$\{e_j\}, \{e'_j\} \text{ and } \{e''_j\} \text{ are complete systems in } H. \tag{10}$$

As a consequence, we can define the powers of the above operators with arbitrary positive real exponents. As far as the first one is concerned, we have, for $\rho > 0$,

$$V_A^\rho := D(A^\rho) = \left\{ v \in H : \sum_{j=1}^\infty |\lambda_j^\rho(v, e_j)|^2 < +\infty \right\} \quad \text{and} \tag{11}$$

$$A^\rho v = \sum_{j=1}^\infty \lambda_j^\rho (v, e_j) e_j \quad \text{for } v \in V_A^\rho, \tag{12}$$

and we endow V_A^ρ with the graph norm

$$\|v\|_{A,\rho} := \left(\|v\|^2 + \|A^\rho v\|^2 \right)^{1/2} \quad \text{for every } v \in V_A^\rho. \tag{13}$$

Similarly, we set

$$V_B^\sigma := D(B^\sigma) \quad \text{and} \quad V_C^\tau := D(C^\tau), \tag{14}$$

with the graph norms

$$\|v\|_{B,\sigma} := \left(\|v\|^2 + \|B^\sigma v\|^2 \right)^{1/2} \quad \text{and} \quad \|v\|_{C,\tau} := \left(\|v\|^2 + \|C^\tau v\|^2 \right)^{1/2},$$
$$\text{for } v \in V_B^\sigma \text{ and } v \in V_C^\tau, \text{ respectively.} \tag{15}$$

From now on, we assume:

$$\rho, \sigma \text{ and } \tau \text{ are fixed positive real numbers.} \tag{16}$$

However, we need the further assumptions we list at once. It is understood that all of the embeddings below are assumed to be continuous.

The first eigenvalue λ_1 of A is strictly positive. (17)

$$V_A^{2\rho} \subset L^\infty(\Omega), \quad V_A^\rho \subset L^4(\Omega), \quad V_B^\sigma \subset L^4(\Omega), \quad \text{and} \quad V_C^\tau \subset L^4(\Omega). \tag{18}$$

$\psi(v) \in H$ and $(B^{2\sigma}v, \psi(v)) \geq 0$, for every $v \in V_B^{2\sigma}$ and every monotone
and Lipschitz continuous function $\psi : \mathbb{R} \to \mathbb{R}$ vanishing at the origin. (19)

Due to the continuus embeddings (18), there exists a constant $C_* > 0$ such that

$$\|v\|_\infty \leq C_*\|v\|_{A,2\rho} \quad \|v\|_4 \leq C_*\|v\|_{A,\rho} \quad \|v\|_4 \leq C_*\|v\|_{B,\sigma} \quad \text{and} \quad \|v\|_4 \leq C_*\|v\|_{C,\tau}$$

for every $v \in V_A^{2\rho}, v \in V_A^\rho, v \in V_B^\sigma$, and $v \in V_C^\tau$, respectively, (20)

where, for $p \in [1, +\infty]$, the symbol $\|\cdot\|_p$ denotes the norm in $L^p(\Omega)$. The same symbol will also be used for the norm in $L^p(Q)$ provided that no confusion can arise.

Remark 1. *We have to make some comments on (17)–(19). The first of these assumptions is satisfied if A is, e.g., the Laplace operator $-\Delta$ with zero Dirichlet (or Robin) boundary conditions, while the case of zero Neumann boundary conditions is excluded unless one adds to the Laplace operator, e.g., some zero-order term ensuring coerciveness. However, it is clear that A could be a much more general operator. By still considering the Laplace operator with (zero) Dirichlet boundary conditions as A, we can also discuss the first two embeddings in (18). By noting that $D(A) = H^2(\Omega) \cap H_0^1(\Omega)$ and Ω is smooth, it results that $D(A^{2\rho}) \subset H^{4\rho}(\Omega)$ and $D(A^\rho) \subset H^{2\rho}(\Omega)$. Hence, both embeddings hold true if $\rho \geq 3/8$, since Ω is three-dimensional. Finally, we make a comment on (19). Assume, for instance, that $B^{2\sigma} = -\Delta$ with zero Neumann boundary conditions. Then, $V_B^{2\sigma} = \{v \in H^2(\Omega) : \partial_\nu v = 0 \text{ on } \Gamma\}$ and, for every $v \in V_B^{2\sigma}$ and ψ as in (19), we have that $\psi(v) \in H^1(\Omega)$ (since $v \in H^1(\Omega)$) and*

$$(B^{2\sigma}v, \psi(v)) = \int_\Omega (-\Delta v) \psi(v) = \int_\Omega \nabla v \cdot \nabla \psi(v) = \int_\Omega \psi'(v)|\nabla v|^2 \geq 0.$$

The same argument works if we take the Dirichlet boundary conditions instead of the Neumann ones, since the functions ψ considered in (19) vanish at the origin. More generally, $B^{2\sigma}$ can be the principal part of an elliptic operator in divergence form with smooth coefficients. In particular, even though some restrictions on A, B, and C have to be imposed in order to fulfill the properties (17)–(19), no relationship between them is needed, and the three operators can be completely independent from each other.

Remark 2. *Assumption (17) allows us to consider an equivalent norm in V_A^ρ. Indeed, for every $v \in V_A^\rho$ we have that*

$$\|A^\rho v\|^2 = \sum_{j=1}^\infty |\lambda_j^\rho(v, e_j)|^2 \geq \lambda_1^{2\rho} \sum_{j=1}^\infty |(v, e_j)|^2 = \lambda_1^{2\rho} \|v\|^2. \tag{21}$$

Hence, since $\lambda_1 > 0$, we deduce that

$$\|v\| \leq \lambda_1^{-\rho} \|A^\rho v\| \quad \text{for every } v \in V_A^\rho, \tag{22}$$

so that the function $v \mapsto \|A^\rho v\|$ defines a norm in V_A^ρ that is equivalent to the graph norm (13).

For the nonlinear functions entering our system, we postulate the following properties:

$$D(F) \text{ is an open interval } (a,b) \text{ of the real line with } 0 \in (a,b). \tag{23}$$

$$F := D(F) \to \mathbb{R} \quad \text{is a } C^3 \text{ function}. \tag{24}$$

$$F(s) \geq C_1 s^2 - C_2 \quad \text{and} \quad F''(s) \geq -C_3$$
$$\text{for some constants } C_i > 0 \text{ and every } s \in D(F). \tag{25}$$

$$f := F' \text{ satisfies} \quad \lim_{s \searrow a} f(s) = -\infty \quad \text{and} \quad \lim_{s \nearrow b} f(s) = +\infty. \tag{26}$$

$$P : \mathbb{R} \to [0, +\infty) \quad \text{is bounded and Lipschitz continuous on } \mathbb{R}$$
$$\text{and of class } C^2 \text{ in } D(F). \tag{27}$$

Clearly, (23)–(26) are fulfilled by the significant potentials (4) and (5).

Remark 3. *The hypotheses (23)–(26) on F ensure that the conditions required in [1], i.e., $F = (\tilde{F}_1 + \tilde{F}_2)|_{(a,b)}$, where*

$$\tilde{F}_1 : \mathbb{R} \to [0, +\infty] \quad \text{is convex, proper, and l.s.c., with} \quad \tilde{F}_1(0) = 0, \tag{28}$$

$$\tilde{F}_2 : \mathbb{R} \to \mathbb{R} \quad \text{is of class } C^1 \text{ with a Lipschitz continuous first derivative}, \tag{29}$$

$$\tilde{F}_1^\lambda(s) + \tilde{F}_2(s) \geq -C_0 \quad \text{for some constant } C_0 \text{ and every } s \in \mathbb{R}, \tag{30}$$

are satisfied, as we show at once. We first split F by defining, for $s \in (a,b)$,

$$f_1(s) := \int_0^s (f'(s'))^+ \, ds', \quad F_1(s) := \int_0^s f_1(s') \, ds',$$

$$f_2(s) := F'(0) - \int_0^s (f'(s'))^- \, ds' \quad \text{and} \quad F_2(s) = F(0) + \int_0^s f_2(s') \, ds'.$$

Notice that F_1 is nonnegative and convex and that $F_1(0) = 0$. If $(a,b) \neq \mathbb{R}$, we properly extend these functions F_i to functions $\tilde{F}_i$ defined in the whole of $\mathbb{R}$. One can preserve the mentioned properties of F_1, including its lower semicontinuity, by setting

$$\tilde{F}_1(a) := \lim_{s \searrow a} F_1(s), \quad \tilde{F}_1(b) := \lim_{s \nearrow b} F_1(s) \quad \text{and} \quad \tilde{F}_1(s) := +\infty \quad \text{for } s \notin [a,b].$$

Moreover, one can ensure that the derivative of the extension $\tilde{F}_2$ is Lipschitz continuous, by noting that $F_2' = f_2$ already is Lipschitz continuous in (a,b) since its derivative $f_2' = -(f')^-$ is bounded by the assumption (25) on F''. The last condition that we have to check is (30), where $\tilde{F}_1^\lambda$ is the Moreau–Yosida approximation of $\tilde{F}_1$ at the level λ. We notice that this condition is not equivalent to an inequality of type $F(s) \geq -C_2$, which follows from (25) and looks rather natural in performing formal a priori estimates. On the other hand, one can prove that the inequality we need is implied by the full quadratic growth condition given in (25) (see (formula (3.1) [33] for some explanation). For this reason, we have postulated the latter.

Although some of the results to be presented will not require the whole set of hypotheses made so far, the statements will be greatly simplified if we do not each time recall the properties of the involved operators, spaces, and nonlinearities; we therefore make the following general assumption:

All of the assumptions made above on the structure are in force from now on. (31)

As mentioned in the Introduction, we only deal with the case $\alpha = 0$ and $\beta > 0$ of system (1)–(3). Clearly, we can take $\beta = 1$ without loss of generality. Hence, the Cauchy problem forming the state system under investigation reads as follows:

$$\partial_t \varphi + A^{2\rho} \mu = P(\varphi)(S - \mu), \tag{32}$$

$$\partial_t \varphi + B^{2\sigma} \varphi + f(\varphi) = \mu, \tag{33}$$

$$\partial_t S + C^{2\tau} S = -P(\varphi)(S - \mu) + u, \tag{34}$$

$$\varphi(0) = \varphi_0 \quad \text{and} \quad S(0) = S_0, \tag{35}$$

where φ_0 and S_0 are prescribed initial data that are supposed to satisfy

$$\varphi_0 \in V_B^{2\sigma}, \quad S_0 \in V_C^\tau, \quad \text{and} \quad a_0 \leq \varphi_0 \leq b_0 \quad \text{a.e. in } \Omega$$
$$\text{for some compact interval } [a_0\, b_0] \subset (a, b). \tag{36}$$

In fact, we could solve a weak form of the above problem under milder assumption on the initial data; however, in order to guarantee a sufficient regularity level of the solution, we need the whole of (36). Given a final time $T \in (0, +\infty)$, the regularity we can ensure (besides some boundedness to be discussed later on) is the following:

$$\mu \in L^\infty(0, T; V_A^{2\rho}), \tag{37}$$

$$\varphi \in W^{1,\infty}(0, T; H) \cap H^1(0, T; V_B^\sigma) \cap L^\infty(0, T; V_B^{2\sigma}), \tag{38}$$

$$f(\varphi) \in L^\infty(0, T; H), \tag{39}$$

$$S \in H^1(0, T; H) \cap L^\infty(0, T; V_C^\tau) \cap L^2(0, T; V_C^{2\tau}), \tag{40}$$

so that Equations (32)–(34) are satisfied a.e. in Q, where we recall that

$$Q := \Omega \times (0, T). \tag{41}$$

We notice at once that the first embedding in (18) yields that

$$\mu \in L^\infty(0, T; V_A^{2\rho}) \quad \text{implies that} \quad \mu \in L^\infty(Q). \tag{42}$$

At this point, we are ready to present our results. To this end, it is convenient to introduce the following variational formulation of (32)–(34):

$$(\partial_t \varphi(t), v) + (A^\rho \mu(t), A^\rho v) = (P(\varphi(t))(S(t) - \mu(t)), v)$$
$$\text{for every } v \in V_A^\rho \text{ and for a.a. } t \in (0, T), \tag{43}$$

$$(\partial_t \varphi(t), v) + (B^\sigma \varphi(t), B^\sigma v) + (f(\varphi(t)), v) = (\mu(t), v)$$
$$\text{for every } v \in V_B^\sigma \text{ and for a.a. } t \in (0, T), \tag{44}$$

$$(\partial_t S(t), v) + (C^\tau S(t), C^\tau v) = -(P(\varphi(t))(S(t) - \mu(t)), v) + (u(t), v)$$
$$\text{for every } v \in V_C^\tau \text{ and for a.a. } t \in (0, T). \tag{45}$$

This is based on obvious properties of the powers of the operators A, B and C, like the Green type formula $(A^{2\rho} v, w) = (A^\rho v, A^\rho w)$ for every $v \in V_A^{2\rho}$ and $w \in V_A^\rho$.

Before stating our well-posedness theorem, we prove some auxiliary results. The first one is a separation property enjoyed by any solution under our assumptions on the data.

Theorem 1. *Assume* (36) *and* $u \in L^2(0, T; H)$, *and let* (μ, φ, S) *be a solution to problem* (32)–(35) *satisfying* (37)–(40). *Then, it holds for every* $M > 0$ *that, if* $\|\mu\|_\infty < M$, *then there exists a compact interval* $[a_M, b_M] \subset (a, b)$ *such that*

$$a_M \leq \varphi \leq b_M \quad \text{a.e. in } Q. \tag{46}$$

This interval depends only on f, *the initial datum* φ_0, *and* M.

Proof. Notice that μ is bounded thanks to (42). Thus, we fix a constant M and assume that $\|\mu\|_\infty < M$. By the assumptions (26) on f and (36) on φ_0, we can choose $a_M \in (a, a_0]$ and $b_M \in [b_0, b)$ such that

$$f(z) < -M \quad \text{for all } z \in (a, a_M) \quad \text{and} \quad f(z) > M \quad \text{for all } z \in (b_M, b).$$

Now, we notice that, for a.a. $s \in (0, T)$, the value $\varphi(s)$ belongs to $V_B^{2\sigma}$ by (38). Moreover, the function $z \mapsto \psi(z) := (z - b_M)^+$ is monotone and Lipschitz continuous on $\mathbb{R}$ and vanishes at the origin. Hence, we have that $\psi(\varphi(s)) \in H$ by (19). Thus, we can multiply (33), written at the time s, by $\psi(\varphi(s))$ and integrate over $(0, t)$ with respect to s. By noting that $\psi(\varphi_0) = 0$ (since $\varphi_0 \leq b_0 \leq b_M$ a.e. in Ω), we obtain that

$$\frac{1}{2} \|\psi(\varphi(t))\|^2 + \int_0^t \left(B^{2\sigma} \varphi(s), \psi(\varphi(s))\right) ds = \int_0^t \left(\mu(s) - f(\varphi(s)), \psi(\varphi(s))\right) ds.$$

Thanks to the inequality (19), the second term on the left-hand side is nonnegative. Moreover, the right-hand side is nonpositive since $\psi(\varphi) = 0$, where $\varphi \leq b_M$, and $f(\varphi) \geq \mu$ whenever $\varphi > b_M$. Hence, we conclude that $\psi(\varphi) = 0$ a.e. in Q, i.e., that $\varphi \leq b_M$ a.e. in Q. By the same argument, with $\psi(z) := -(z + a_M)^-$, one obtains that $\varphi \geq a_M$ a.e. in Q. □

Theorem 2. *Under the assumptions* (36) *on the initial data, problem* (32)–(35) *has at most one solution satisfying* (37)–(40). *Moreover, if* $M > 0$, $u_i \in L^2(0, T; H)$, $i = 1, 2$, *and* (μ_i, φ_i, S_i) *are two corresponding solutions to* (32)–(35) *satisfying* (37)–(40) *and* $\|\mu_i\|_\infty < M$ *for* $i = 1, 2$, *then the estimate*

$$\|\mu_1 - \mu_2\|_{L^2(0,T;V_A^\rho)} + \|\varphi_1 - \varphi_2\|_{H^1(0,T;H) \cap L^\infty(0,T;V_B^\sigma)}$$
$$+ \|S_1 - S_2\|_{L^\infty(0,T;H) \cap L^2(0,T;V_C^\tau)} \leq K_M \|u_1 - u_2\|_{L^2(0,T;H)} \tag{47}$$

holds true with a constant K_M *that depends only on the structure of the system, the initial data, T and M*.

Proof. We show the uniqueness at the end and first prove the estimate (47), noting that the assumption $\|\mu_i\|_\infty < M$ is meaningful since the functions μ_i are bounded due to (42). We apply Theorem 1 and find a compact interval $[a_M, b_M]$ contained in (a, b) such that

$$a_M \leq \varphi_i \leq b_M \quad \text{a.e. in } Q, \quad \text{for } i = 1, 2.$$

Since f is (at least) a C^1 function on (a, b), it is Lipschitz continuous on $[a_M, b_M]$. Let L be the corresponding Lipschitz constant. After this preparation, we can start the proof. We set, for convenience, $u := u_1 - u_2$, $\mu := \mu_1 - \mu_2$, $\varphi := \varphi_1 - \varphi_2$, and $S := S_1 - S_2$, writing (32)–(34) for both solutions and multiplying the differences by μ, $\partial_t \varphi$, and S, respectively, in the inner product of H. Then, we sum up and integrate with respect to time over $(0, t)$. By noting that the terms involving $(\mu, \partial_t \varphi)$ cancel each other, and adding the same contributions $(1/2)\|\varphi(t)\|^2 = \int_0^t (\varphi(s), \partial_t \varphi(s)) ds$ and $\int_0^t \|S(s)\|^2 ds$ to both sides, we obtain the equation

$$\int_0^t \|A^\rho \mu(s)\|^2 ds + \int_0^t \|\partial_t \varphi(s)\|^2 + \frac{1}{2} \|\varphi(t)\|^2 + \frac{1}{2} \|B^\sigma \varphi(t)\|^2$$
$$+ \frac{1}{2} \|S(t)\|^2 + \int_0^t \|S(s)\|^2 ds + \int_0^t \|C^\tau S(s)\|^2 ds$$

$$= \int_0^t \Big(P(\varphi_1(s))(S_1(s) - \mu_1(s)) - P(\varphi_2(s))(S_2(s) - \mu_2(s)), \mu(s) - S(s) \Big) ds$$
$$- \int_0^t \big(f(\varphi_1(s)) - f(\varphi_2(s)), \partial_t \varphi(s) \big) ds$$
$$+ \int_0^t \big(\varphi(s), \partial_t \varphi(s) \big) ds + \int_0^t \big(u(s), S(s) \big) ds + \int_0^t \|S(s)\|^2 ds.$$

If we term I the first integral on the right-hand side, apply the Young inequality to the next three terms, use the Lipschitz continuity of f, and rearrange, we deduce that

$$\int_0^t \|A^\rho \mu(s)\|^2 ds + \frac{1}{2} \int_0^t \|\partial_t \varphi(s)\|^2 + \frac{1}{2} \|\varphi(t)\|^2_{B,\sigma} + \frac{1}{2} \|S(t)\|^2 + \int_0^t \|S(s)\|^2_{C,\tau} ds$$
$$\leq I + (L^2 + 1) \int_0^t \|\varphi(s)\|^2 ds + \frac{1}{2} \int_0^t \|u(s)\|^2 + \frac{3}{2} \int_0^t \|S(s)\|^2 ds.$$

Now, we rewrite I as the sum of two terms. The first of these is nonpositive, since P is nonnegative, and we estimate the other one recalling that P' is bounded because P is Lipschitz continuous. By applying the Hölder inequality, and recalling (20), we have for every $\delta > 0$ that

$$I = \int_0^t \big(P(\varphi_1(s))(S(s) - \mu(s)), \mu(s) - S(s) \big) ds$$
$$+ \int_0^t \big((P(\varphi_1(s))) - P(\varphi_2(s)))(S_2(s) - \mu_2(s)), \mu(s) - S(s) \big) ds$$
$$\leq \sup |P'| \int_0^t \|\varphi(s)\|_4 \|S_2(s) - \mu_2(s)\|_4 \|S(s) - \mu(s)\| ds$$
$$\leq \delta \int_0^t \big(\|S(s)\| + \|\mu(s)\| \big)^2 ds$$
$$+ \frac{\sup |P'|^2 C_*^4}{4\delta} \int_0^t \big(\|S_2(s)\|_{C,\tau} + \|\mu_2(s)\|_{A,\rho} \big)^2 \|\varphi(s)\|^2_{B,\sigma} ds.$$

We notice that the function $s \mapsto \big(\|S_2(s)\|_{C,\tau} + \|\mu_2(s)\|_{A,\rho} \big)^2$ belongs to $L^\infty(0,T)$, thanks to the regularity (37) and (40) of μ_2 and S_2, respectively. Therefore, by choosing $\delta > 0$ small enough, and applying the Gronwall lemma, we obtain the estimate (47) with a constant K_M whose dependence on the data agrees with that specified in the statement.

We now come back to uniqueness. As both $u_i \in L^2(0,T;H)$ and the corresponding solutions (μ_i, φ_i, S_i) are arbitrary in the above argument (since no restriction on M is made), we conclude that $(\mu_1, \varphi_1, S_1) = (\mu_2, \varphi_2, S_2)$ if $u_1 = u_2$, which shows the uniqueness for the solution to problem (32)–(35). With this, the proof is complete. □

Finally, we can state our well-posedness and stability result. Here, and later on in this paper, we use the notation
$$\mathcal{B}_R := \{ u \in L^2(0,T;H) : \|u\|_{L^2(0,T;H)} < R \}, \tag{48}$$
where R is a positive real parameter.

Theorem 3. *Under the assumptions* (36) *on the initial data φ_0 and S_0, problem* (32)–(35) *has, for every $u \in L^2(0,T;H)$, a unique solution (μ, φ, S) that satisfies* (37)–(40). *In particular, μ is bounded.*

Moreover, for every $R > 0$, there exist a constant $K_1(R)$ and a compact interval $[a_R, b_R] \subset (a, b)$, which depend only on the structure of the system, the initial data, T and R, such that both the estimate

$$\|\mu\|_{L^\infty(0,T;V_A^{2\rho})} + \|\mu\|_\infty + \|P(\varphi)(S - \mu)\|_{L^2(0,T;H)}$$
$$+ \|\varphi\|_{W^{1,\infty}(0,T;H) \cap H^1(0,T;V_B^\sigma) \cap L^2(0,T;V_B^{2\sigma})} + \|f(\varphi)\|_{L^\infty(0,T;H)}$$
$$+ \|S\|_{H^1(0,T;H) \cap L^\infty(0,T;V_C^\tau) \cap L^2(0,T;V_C^{2\tau})}$$
$$\leq K_1(R) \tag{49}$$

and the separation property

$$a_R \leq \varphi \leq b_R \quad \text{a.e. in } Q \tag{50}$$

hold true for every $u \in \mathcal{B}_R$ and the corresponding solution (μ, φ, S). Finally, if $R > 0$, $u_i \in \mathcal{B}_R$, $i = 1, 2$, and (μ_i, φ_i, S_i) are the corresponding solutions, then the estimate

$$\|\mu_1 - \mu_2\|_{L^2(0,T;V_A^\rho)} + \|\varphi_1 - \varphi_2\|_{H^1(0,T;H) \cap L^\infty(0,T;V_B^\sigma)}$$
$$+ \|S_1 - S_2\|_{L^\infty(0,T;H) \cap L^2(0,T;V_C^\tau)} \leq K_2(R) \|u_1 - u_2\|_{L^2(0,T;H)} \tag{51}$$

holds true with a constant $K_2(R)$ that depends only on the structure of the system, the initial data, T, and R.

Proof. Uniqueness follows from Theorem 2. Let us come to the existence of a solution and to the estimates (49) and (51). However, we do not give a complete proof. Indeed, one can adapt the arguments of [1] on account of Remark 3, and we briefly explain the reason for this. The procedure used there is based on the Yosida regularization of the nonlinearity f, a time discretization of the regularized system, and the derivation of suitable a priori estimates, and the same line of argumentation can be followed in our situation. Here, is the main remark: in [1], some estimates for μ have been derived from estimates of $\partial_t \mu$, and this term is missing in (32), in contrast to (1). In the present case, an estimate of the norm of μ, e.g., in $L^2(0, T; H)$, can be deduced from an estimate of $A^\rho \mu$ in the same space as shown in Remark 2, by using the assumption (17) on the first eigenvalue λ_1 of A. Hence, we do not repeat the arguments of [1] with the corresponding modifications. However, for the reader's convenience, we sketch the formal proofs of the estimates that would be obtained step by step in the rigorous procedure in order to prove the existence of a solution. We assume $u \in \mathcal{B}_R$ from the very beginning, so that these estimates eventually lead to (49) as well.

In order to simplify notation, we use the same symbol c without any subscript for possibly different constants that depend only on the structure of our system, the initial data and T, but neither on u nor on R. Moreover, the symbol c_R stands for (possibly different) constants that depend on the constant R, in addition, but still not on u. Thus, it is understood that the actual values of such constants may vary from line to line and even in the same chain of inequalities. Notice that the notations used for constants we want to refer to (like, e.g., those used in (25)) are different.

First a priori estimate. We test (43), (44) and (45), written at the time s, by $\mu(s)$, $\partial_t \varphi(s)$, and $S(s)$, respectively, in the scalar product of H. Then, we sum up and integrate over $(0, t)$, where $t \in (0, T)$ is arbitrary, noting that the terms involving the product $\mu \partial_t \varphi$ cancel each other. By also adding $|\Omega| C_2$ to both sides (see (25)), we obtain the identity

$$\int_0^t \|A^\rho \mu(s)\|^2 \, ds + \int_0^t \|\partial_t \varphi(s)\|^2 \, ds + \frac{1}{2} \|B^\sigma \varphi(t)\|^2 + \int_\Omega (F(\varphi(t)) + C_2)$$
$$+ \frac{1}{2} \|S(t)\|^2 + \int_0^t \|C^\tau S(s)\|^2 \, ds + \int_{Q_t} P(\varphi)(S - \mu)^2$$
$$= \frac{1}{2} \|B^\sigma \varphi_0\|^2 + \int_\Omega (F(\varphi_0) + C_2) + \frac{1}{2} \|S_0\|^2 + \int_0^t (u(s), S(s)) \, ds \, .$$

Recalling the consequence (22) of (17), taking (25) into account and applying the Gronwall lemma, we conclude that

$$\|\mu\|_{L^2(0,T;V_A^\rho)} + \|\varphi\|_{H^1(0,T;H)\cap L^\infty(0,T;V_B^\sigma)}$$
$$+ \|S\|_{L^\infty(0,T;H)\cap L^2(0,T;V_C^\tau)} + \|P^{1/2}(\varphi)(S-\mu)\|_{L^2(0,T;H)} \leq c_R. \tag{52}$$

Consequence. We can also infer that

$$\|P(\varphi)(S-\mu)\|^2_{L^2(0,T;H)} + \|F(\varphi)\|_{L^\infty(0,T;L^1(\Omega))} \leq c_R.$$

Indeed, the estimate is right for the first term, since P is bounded by (27); for the bound of the second term, we can argue as in (Section 4.4 [1]).

Second a priori estimate. We would like to test (43) by $\partial_t \mu$ even though $\partial_t \mu$ does not appear in the equation (in contrast to (1)). In fact, the estimate we derive here by a formal procedure should be performed rigorously at the level of the discrete scheme, which can contain that time derivative multiplied by a viscosity coefficient that tends to zero at some point of the procedure. Thus, we test (32) and (34) formally by $\partial_t \mu$ and $\partial_t S$, respectively. At the same time, we formally differentiate (33) with respect to time and test the resulting equality by $\partial_t \varphi$. Then, we sum up and integrate with respect to time, as usual. Since the terms involving the product $\partial_t \mu \, \partial_t \varphi$ cancel each other, we obtain the identity

$$\frac{1}{2}\|A^\rho \mu(t)\|^2 + \frac{1}{2}\|\partial_t \varphi(t)\|^2 + \int_0^t \|B^\sigma \partial_t \varphi(s)\|^2 \, ds$$
$$+ \int_0^t \|\partial_t S(s)\|^2 \, ds + \frac{1}{2}\|C^\tau S(t)\|^2$$
$$= \frac{1}{2}\|A^\rho \mu(0)\|^2 + \frac{1}{2}\|\partial_t \varphi(0)\|^2 + \frac{1}{2}\|C^\tau S_0\|^2$$
$$+ \int_0^t (u(s), \partial_t S(s)) \, ds - \int_0^t (f'(s)\partial_t \varphi(s), \partial_t \varphi(s)) \, ds$$
$$+ \int_0^t (P(\varphi(s))(S(s) - \mu(s)), \partial_t \mu(s) - \partial_t S(s)) \, ds. \tag{53}$$

The integral containing u can be handled using Young's inequality, and the one involving f' is easily treated using the second inequality in (25) and (52). We postpone the estimate of the last integral and first deal with the initial values appearing on the right-hand side of (53). By using the initial conditions for φ and S, we write (32) and (33) at the time $t = 0$ in the following way:

$$(A^{2\rho} + P(\varphi_0))\mu(0) = P(\varphi_0)S_0 \quad \text{and} \quad \partial_t \varphi(0) = \mu(0) - B^{2\sigma}\varphi_0 - f(\varphi_0). \tag{54}$$

Since P is nonnegative, by multiplying the first identity in (54) by $\mu(0)$, we have that (see (22))

$$(P(\varphi_0)S_0, \mu(0)) = ((A^{2\rho} + P(\varphi_0))\mu(0), \mu(0)) \geq \|A^\rho \mu(0)\|^2 \geq \lambda_1^{2\rho}\|\mu(0)\|^2$$

whence

$$\|\mu(0)\| \leq \lambda_1^{-2\rho}\|P(\varphi_0)S_0\| \leq c, \quad \|A^\rho \mu(0)\|^2 \leq \|\mu(0)\|\,\|P(\varphi_0)S_0\| \leq c \tag{55}$$

and, on account of (36), we also deduce from the second identity in (54) that

$$\|\partial_t \varphi(0)\| \leq c + \|B^{2\sigma}\varphi_0\| + \|f(\varphi_0)\| \leq c.$$

Finally, we deal with the last integral on the right-hand side of (53), which we term I for brevity. We perform an integration by parts in time, recall that P' is bounded by (27), and invoke (55). By recalling that $\|\cdot\|_p$ denotes the norm in $L^p(\Omega)$, we then have

$$\begin{aligned}I &= -\int_0^t\!\!\int_\Omega P(\varphi)(S-\mu)\partial_t(S-\mu)\\ &= -\frac{1}{2}\int_\Omega P(\varphi(t))\bigl(S(t)-\mu(t)\bigr)^2 + \frac{1}{2}\int_\Omega P(\varphi_0)\bigl(S_0-\mu(0)\bigr)^2\\ &\quad + \frac{1}{2}\int_0^t\!\!\int_\Omega P'(\varphi)\partial_t\varphi(S-\mu)^2\\ &\le c + c\int_0^t \|\partial_t\varphi(s)\|_2\bigl(\|S(s)\|_4^2 + \|\mu(s)\|_4^2\bigr)\,ds\,.\end{aligned}$$

Finally, we notice that

$$\int_0^T \bigl(\|S(s)\|_4^2 + \|\mu(s)\|_4^2\bigr)\,ds \le c_R$$

by (52) and some of the embeddings in (18). This allows us to apply Gronwall's lemma, whence we conclude that

$$\|\mu\|_{L^\infty(0,T;V_A^\rho)} + \|\partial_t\varphi\|_{L^\infty(0,T;H)\cap L^2(0,T;V_B^\sigma)} + \|S\|_{H^1(0,T;H)\cap L^\infty(0,T;V_C^\tau)} \le c_R\,. \tag{56}$$

Third a priori estimate. By taking $v = \mu(t)$ in (43) and recalling that P is nonnegative and bounded, we obtain the following inequality for a.a. $t \in (0,T)$

$$\|A^\rho \mu(t)\|^2 \le \bigl(P(\varphi(t))S(t) - \partial_t\varphi(t), \mu(t)\bigr) \le c\bigl(\|S\|_{L^\infty(0,T;H)} + \|\partial_t\varphi\|_{L^\infty(0,T;H)}\bigr)\|\mu(t)\|\,.$$

By accounting for Remark 2, we deduce that

$$\|\mu\|_{L^\infty(0,T;V_A^\rho)} \le c_R\,. \tag{57}$$

In particular, the norm of μ in $L^\infty(0,T;H)$ is bounded by some constant c_R. Since the same holds for S due to (52), we infer that

$$\|P(\varphi)(S-\mu)\|_{L^\infty(0,T;H)} \le c_R\,. \tag{58}$$

Consequence. By comparison in (32), we deduce that

$$\|A^{2\rho}\mu\|_{L^\infty(0,T;H)} \le \|P(\varphi)(S-\mu)\|_{L^\infty(0,T;H)} + \|\partial_t\varphi\|_{L^\infty(0,T;H)}.$$

Combining this with (58) and (56), we conclude that

$$\|\mu\|_{L^\infty(0,T;V_A^{2\rho})} \le c_R\,. \tag{59}$$

Then, the first embedding in (18) yields that μ is bounded (as claimed in the statement) and that

$$\|\mu\|_\infty \le c_R\,. \tag{60}$$

By comparison in (34), we also deduce that

$$\|S\|_{L^2(0,T;V_C^{2\tau})} \le c_R\,. \tag{61}$$

No better estimate for S is available, since $u \in L^2(0,T;H)$ only.

Fourth a priori estimate. We recall that Remark 3 provides a splitting of f as $f_1 + f_2$, with f_1 monotone and vanishing at the origin and f_2 Lipschitz continuous. Thus, we can write (33) for a.a. $t \in (0, T)$ in the form

$$B^{2\sigma} \varphi(t) + f_1(\varphi(t)) = \mu(t) - \partial_t \varphi(t) - f_2(\varphi(t)),$$

and test this identity by $f_1(\varphi(t))$. More precisely, in the correct argument, f_1 is replaced by its Yosida regularization, which is monotone and Lipschitz continuous and vanishes at the origin, and the equation itself is replaced by a scheme, which is obtained by discretizing time differentiation and for which the analogue of $\varphi(t)$ belongs to $V_B^{2\sigma}$. Hence, assumption (19) can actually be applied. Here, we formally apply it to the above identity with $v = \varphi(t)$ and $\psi = f_1$. We obtain that

$$\|f_1(\varphi(t))\|^2 \leq (B^{2\sigma} \varphi(t) + f_1(\varphi(t)), f_1(\varphi(t))) = (\mu(t) - \partial_t \varphi(t) - f_2(\varphi(t)), f_1(\varphi(t)))$$
$$\leq \|\mu - \partial_t \varphi - f_2(\varphi)\|_{L^\infty(0,T;H)} \|f_1(\varphi(t))\|.$$

On account of the previous estimates, and by a comparison in (33), we conclude that

$$\|f_1(\varphi)\|_{L^\infty(0,T;H)} + \|\varphi\|_{L^\infty(0,T;V_B^{2\sigma})} \leq c_R. \tag{62}$$

Conclusion. This concludes the formal proof of the existence part of Theorem 3 and of estimate (49). As already said, in the rigorous argument, the above bounds are established for the solution to an approximating problem, and one has to perform some limiting procedure. The estimates provide convergence of weak and weak-star type. However, even strong convergence in $L^2(0, T; H)$ for the approximations of φ and S is obtained. Indeed, the embeddings $V_B^\sigma \subset H$ and $V_C^\tau \subset H$ are compact due to (7), so that one can apply the Aubin–Lions lemma (see, e.g., (Thm. 5.1, p. 58 [60])). Therefore, the nonlinear terms can be correctly managed.

Separation. Let us come to estimate (50). This is a trivial consequence of the above estimate and Theorem 1. Indeed, this theorem can be applied with $M := c_R + 1$, where c_R is the constant that appears in (60). The corresponding compact interval $[a_R, b_R]$ of the statement is nothing but the interval $[a_M, b_M]$ considered in (46), which depends only on the structure of the system, the initial data, T, and R.

Continuous dependence. In addition, (51) is a trivial consequence of a fact already proved, namely, of Theorem 2. Indeed, if $u_i \in \mathcal{B}_R$, $i = 1, 2$, then the L^∞ bound for the corresponding μ_i is ensured by (49), and Theorem 2 can be applied with $M = K_1(R) + 1$. Hence, we can take as $K_2(R)$ the constant K_M that appears in (47). In addition, this constant depends only on the structure of the system, the initial data, T and R. □

Remark 4. *The existence part of Theorem 2.6 is closely connected to the existence result proved in (Theorem 3.4 [38]), where, however, no statement concerning separation or uniqueness was proved. For purposes of control theory, however, it is indispensable to have uniqueness, since otherwise no control-to-state operator can be defined, and this seems to be available only under the assumptions made here.*

3. The Control-to-State Mapping

The results of the previous section ensure that we can correctly define a control-to-state mapping to be used in the control problem under investigation. Taking into account that the cost functional to be minimized depends only on the components φ and S of the solution corresponding to a given u, we set

$$\mathcal{Y}_1 := L^2(0, T; V_A^\rho), \quad \mathcal{Y}_2 := H^1(0, T; H) \cap L^\infty(0, T; V_B^\sigma),$$
$$\mathcal{Y}_3 := C^0([0, T]; H) \cap L^2(0, T; V_C^\tau), \quad \text{and} \quad \mathcal{Y} := \mathcal{Y}_2 \times \mathcal{Y}_3, \tag{63}$$

and define
$$\mathcal{S}_i : L^2(0,T;H) \to \mathcal{Y}_i \quad i=1,2,3, \quad \text{and} \quad \mathcal{S} : L^2(0,T;H) \to \mathcal{Y},$$
by setting, for $u \in L^2(0,T;H)$,
$$\mathcal{S}_1(u) := \mu, \quad \mathcal{S}_2(u) := \varphi, \quad \mathcal{S}_3(u) := S, \quad \text{and} \quad \mathcal{S}(u) := (\varphi, S),$$
where (μ, φ, S) is the solution to (32)–(35) corresponding to u. (64)

More precisely, we need to consider the restriction of these maps to $\mathcal{B}_R$ for any given radius $R > 0$. The choice of the space $\mathcal{Y}$ mainly is due to the following fact: the inequality (51) implies that
$$\|\mathcal{S}(u_1) - \mathcal{S}(u_2)\|_\mathcal{Y} \leq K_2(R) \|u_1 - u_2\|_{L^2(0,T;H)} \quad \text{for every } u_1, u_2 \in \mathcal{B}_R. \tag{65}$$

A very important consequence of the separation property (50) and of the regularity of f ensured by (24) is the following global boundedness condition:
$$\|f^{(k)}(\mathcal{S}_2(u))\|_\infty \leq K_3(R) \quad \text{for } k = 0,1,2, \text{ and every } u \in \mathcal{B}_R, \tag{66}$$
where $K_3(R)$ depends only on the structure of the system, the initial data, T, and R.

The Fréchet differentiability of the maps $\mathcal{S}$ is strictly related to the properties of the linearized problem we introduce now. To this end, we fix $\bar{u} \in L^2(0,T;H)$. The linearized system associated with $\bar{u}$ and the variation $h \in L^2(0,T;H)$ is the following:

$$\partial_t \xi + A^{2\rho}\eta = P(\bar{\varphi})(\zeta - \eta) + P'(\bar{\varphi})\xi(\bar{S} - \bar{\mu}), \tag{67}$$
$$\partial_t \xi + B^{2\sigma}\xi + f'(\bar{\varphi})\xi = \eta, \tag{68}$$
$$\partial_t \zeta + C^{2\tau}\zeta = -P(\bar{\varphi})(\zeta - \eta) - P'(\bar{\varphi})\xi(\bar{S} - \bar{\mu}) + h, \tag{69}$$
$$\xi(0) = 0 \quad \text{and} \quad \zeta(0) = 0, \tag{70}$$

where $\bar{\mu} := \mathcal{S}_1(\bar{u})$, $\bar{\varphi} := \mathcal{S}_2(\bar{u})$, and $\bar{S} := \mathcal{S}_3(\bar{u})$. We also write the weak formulation of (67)–(69): the identities

$$(\partial_t \xi(t), v) + (A^\rho \eta(t), A^\rho v)$$
$$= (P(\bar{\varphi}(t))(\zeta(t) - \eta(t)), v) + (P'(\bar{\varphi}(t))\xi(t)(\bar{S}(t) - \bar{\mu}(t)), v) \tag{71}$$

$$(\partial_t \xi(t), v) + (B^\sigma \xi(t), B^\sigma v) + (f'(\bar{\varphi}(t))\xi(t), v)$$
$$= (\eta(t), v) \tag{72}$$

$$(\partial_t \zeta(t), v) + (C^\tau \zeta(t), C^\tau v)$$
$$= -(P(\bar{\varphi}(t))(\zeta(t) - \eta(t)), v)$$
$$\quad - (P'(\bar{\varphi}(t))\xi(t)(\bar{S}(t) - \bar{\mu}(t)), v) + (h(t), v) \tag{73}$$

have to hold true for every $v \in V_A^\rho$, $v \in V_B^\sigma$, and $v \in V_C^\tau$, respectively, and for a.a. $t \in (0,T)$. We have the following results.

Theorem 4. *Suppose that the assumptions* (36) *on the initial data of problem* (32)–(35) *are fulfilled, and let* $\overline{u} \in L^2(0,T;H)$ *and* $h \in L^2(0,T;H)$. *Then, the linearized problem* (67)–(70) *has a unique solution* (η, ξ, ζ) *satisfying the regularity requirements*

$$\eta \in L^2(0, T; V_A^{2\rho}), \tag{74}$$

$$\xi \in H^1(0, T; H) \cap L^\infty(0, T; V_B^\sigma) \cap L^2(0, T; V_B^{2\sigma}), \tag{75}$$

$$\zeta \in H^1(0, T; H) \cap L^\infty(0, T; V_C^\tau) \cap L^2(0, T; V_C^{2\tau}). \tag{76}$$

Moreover, if $R > 0$ and $\overline{u} \in \mathcal{B}_R$, then this solution satisfies the estimate

$$\|(\xi, \zeta)\|_{\mathcal{Y}} \leq K_4(R) \, \|h\|_{L^2(0,T;H)}, \tag{77}$$

where the constant $K_4(R)$ depends only on the structure of the system (32)–(34), *the initial data φ_0 and S_0, T, and R.*

Proof. We notice that the coefficients $P(\overline{\varphi})$, $P'(\overline{\varphi})$, $f'(\overline{\varphi})$, as well as $\overline{\mu}$, are bounded functions. Moreover, if $\overline{u}$ belongs to some $\mathcal{B}_R$, then the L^∞ bounds are uniform, i.e., they just depend on R and not on $\overline{u}$. On the contrary, $\overline{S}$ might be unbounded. However, as stated in Theorem 3, it is smooth. Thus, the linear system is not worse than the nonlinear one and can be solved by the same argument (which we do not repeat here) based on time discretization that has been used in [1] (see also [31] for the linearized system associated with the Cahn–Hilliard equations). This first leads to a solution to the variational problem (71)–(73) and then to the strong formulation (67)–(69). However, we perform at least some formal estimates that can justify both the regularity asserted in the statement and the validity of estimate (77). To this end, we fix $R > 0$ and assume that $\overline{u} \in \mathcal{B}_R$ at once.

In addition, in this section, i.e., in this proof and later on, we adopt a convention on the constants similar to the one used in the previous section: c stands for possibly different constants depending only on the structure, the data, and T, while the notation c_R indicates an additional dependence on R.

First a priori estimate. We formally test (71)–(73) by η, $\partial_t \xi$, and ζ, respectively, sum up and integrate over $(0,t)$. Moreover, we add the same quantities $(1/2)\|\xi(t)\|^2 = \int_0^t (\xi(s), \partial_t \xi(s))\, ds$ and $\int_0^t \|\zeta(s)\|^2$ to both sides of the resulting identity, in order to recover the full norms in V_B^σ and V_C^τ on the left-hand side. In addition, in this case, a cancellation occurs, and we have that

$$\int_0^t \|A^\rho \eta(s)\|^2\, ds + \int_0^t \|\partial_t \xi(s)\|^2\, ds + \frac{1}{2}\|\xi(t)\|_{B,\sigma}^2 + \frac{1}{2}\|\zeta(t)\|^2 + \int_0^t \|\zeta(s)\|_{C,\tau}^2\, ds$$
$$= \int_0^t (P(\overline{\varphi}(s))(\zeta(s) - \eta(s)), \eta(s) - \zeta(s))\, ds$$
$$+ \int_0^t (P'(\overline{\varphi}(s))\, \xi(s)\, (\overline{S}(s) - \overline{\mu}(s)), \eta(s) - \zeta(s))\, ds$$
$$+ \int_0^t (\xi(s) - f'(\overline{\varphi}(s))\, \xi(s), \partial_t \xi(s))\, ds + \int_0^t (h(s) + \zeta(s), \zeta(s))\, ds. \tag{78}$$

The first integral on the right-hand side is nonpositive, while the next one, which we term I, needs some treatment. By using the Hölder inequality, two of the inequalities (20), Remark 2, and the Young inequality, we obtain that

$$I \leq c \int_0^t \|\xi(s)\|_2 \, \|\overline{S}(s) - \overline{\mu}(s)\|_4 \, \|\eta(s) - \zeta(s)\|_4\, ds$$
$$\leq c \int_0^t \|\xi(s)\| \, (\|\overline{S}(s)\|_{C,\tau} + \|\overline{\mu}(s)\|_{A,\rho})\, (\|\eta(s)\|_{A,\rho} + \|\zeta(s)\|_{C,\tau})\, ds$$
$$\leq \frac{1}{2} \int_0^t \|A^\rho \eta(s)\|^2\, ds + \frac{1}{2} \int_0^t \|\zeta(s)\|_{C,\tau}^2\, ds + c \int_0^t (\|\overline{S}(s)\|_{C,\tau}^2 + \|\overline{\mu}(s)\|_{A,\rho}^2)\, \|\xi(s)\|^2\, ds,$$

where we notice that the function $s \mapsto \|\overline{S}(s)\|_{C,\tau}^2 + \|\overline{\mu}(s)\|_{A,\rho}^2$ belongs to $L^\infty(0,T)$ and that its norm is bounded by a constant like in (49), due to Theorem 3 applied to $\overline{u}$. By treating the last terms of (78) using the Schwarz and Young inequalities, and applying Gronwall's lemma, we conclude that

$$\|\eta\|_{L^2(0,T;V_A^\rho)} + \|\xi\|_{H^1(0,T;H) \cap L^\infty(0,T;V_B^\sigma)} + \|\zeta\|_{L^\infty(0,T;H) \cap L^2(0,T;V_C^\tau)}$$
$$\leq c_R \|h\|_{L^2(0,T;H)}. \tag{79}$$

Second a priori estimate. We estimate the right-hand side of (67). On account of the embeddings (18), we have a.e. in $(0,T)$ that

$$\|P(\overline{\varphi})(\zeta - \eta) + P'(\overline{\varphi})\xi(\overline{S} - \overline{\mu})\| \leq c(\|\zeta\| + \|\eta\|) + c\|\xi\|_4 \|\overline{S} - \overline{\mu}\|_4$$
$$\leq c(\|\zeta\| + \|\eta\|) + c\|\xi\|_{B,\sigma}(\|\overline{S}\|_{C,\tau} + \|\overline{\mu}\|_{A,\rho}).$$

On account of (79), we conclude that

$$\|P(\overline{\varphi})(\zeta - \eta) + P'(\overline{\varphi})\xi(\overline{S} - \overline{\mu})\|_{L^2(0,T;H)} \leq c_R \|h\|_{L^2(0,T;H)}. \tag{80}$$

Since (79) also yields an estimate for $\partial_t \xi$, a (formal) comparison in (67) allows us to conclude that

$$\|A^{2\rho}\eta\|_{L^2(0,T;H)} \leq c_R \|h\|_{L^2(0,T;H)} \quad \text{i.e.,} \quad \|\eta\|_{L^2(0,T;V_A^{2\rho})} \leq c_R \|h\|_{L^2(0,T;H)}. \tag{81}$$

Third a priori estimate. We test (69) by $\partial_t \zeta$ and integrate in time, as usual. On account of (80) and the Young inequality, we obtain that

$$\int_0^t \|\partial_t \zeta(s)\|^2 \, ds + \frac{1}{2} \|C^\tau \zeta(t)\|^2 \leq \frac{1}{2} \int_0^t \|\partial_t \zeta(s)\|^2 \, ds + c_R \|h\|_{L^2(0,T;H)}^2.$$

We thus deduce that

$$\|\partial_t \zeta\|_{L^2(0,T;H)} + \|\zeta\|_{L^\infty(0,T;V_C^\tau)} \leq c_R \|h\|_{L^2(0,T;H)}. \tag{82}$$

Now that $\partial_t \zeta$ is estimated, a comparison in (69) provides a bound for $C^{2\tau}\zeta$. Hence, we conclude that

$$\|\zeta\|_{H^1(0,T;H) \cap L^\infty(0,T;V_C^\tau) \cap L^2(0,T;V_C^{2\tau})} \leq c_R \|h\|_{L^2(0,T;H)}. \tag{83}$$

This ends the list of the formal estimates and formally leads to a strong solution satisfying (77). Even though uniqueness formally follows by taking $h = 0$, we remark that it can be proved rigorously. Indeed, by assuming the regularity (74)–(76), the procedure used to obtain the above estimates is justified. □

Theorem 5. *Assume* (36) *for the initial data of problem* (32)–(35). *Then, the control-to-state mapping $\mathcal{S}$ defined in* (64) *is Fréchet differentiable at every point in $L^2(0,T;H)$. More precisely, if $\overline{u} \in L^2(0,T;H)$ and $h \in L^2(0,T;H)$, then the value $(D\mathcal{S})(\overline{u})[h]$ of the Fréchet derivative $(D\mathcal{S})(\overline{u})$ in the direction h is given by the pair (ξ, ζ), where (η, ξ, ζ) is the solution to the linearized problem* (67)–(70) *associated with $\overline{u}$ and h.*

Proof. Fix any $\overline{u} \in L^2(0,T;H)$, and let $(\overline{\mu}, \overline{\varphi}, \overline{S})$ be the corresponding state. For every $h \in L^2(0,T;H)$, let (μ^h, φ^h, S^h) be the state corresponding to $\overline{u} + h$. Finally, let (η, ξ, ζ) be the solution to the linearized problem (67)–(70) associated with $\overline{u}$ and h. We set, for convenience,

$$\eta^h := \mu^h - \overline{\mu} - \eta, \quad \xi^h := \varphi^h - \overline{\varphi} - \xi, \quad \zeta^h := S^h - \overline{S} - \zeta. \tag{84}$$

According to the definitions of differentiability and derivative in the sense of Fréchet, we have to prove that the (linear) map $h \mapsto (\xi, \zeta)$ is continuous from $L^2(0, T; H)$ into $\mathcal{Y}$ and that there exist a real number $\bar{h} > 0$ and a function $\Lambda : (0, \bar{h}) \to \mathbb{R}$ satisfying

$$\|(\xi^h, \zeta^h)\|_{\mathcal{Y}} \leq \Lambda(\|h\|_{L^2(0,T;H)}) \quad \text{and} \quad \lim_{s \searrow 0} \frac{\Lambda(s)}{s} = 0. \tag{85}$$

The first fact is ensured by (77) once R is chosen larger than $\|\bar{u}\|_{L^\infty(0,T;H)}$. Hence, we fix $R > \|\bar{u}\|_{L^2(0,T;H)}$ once and for all. As for the construction of Λ, we set $\bar{h} := R - \|\bar{u}\|_{L^2(0,T;H)}$, and we assume that $\|h\|_{L^2(0,T;H)} < \bar{h}$. This implies that $\bar{u}$ and $\bar{u} + h$ belong to $\mathcal{B}_R$, so that Theorem 3 can be applied to both of them. We thus derive uniform estimates for the corresponding states, hence for the coefficients of the corresponding linearized systems. This entails uniform estimates for the corresponding solutions. In order to establish (85), we observe that (η^h, ξ^h, ζ^h) satisfies the regularity properties

$$\eta^h \in L^\infty(0, T; V_A^{2\rho}),$$
$$\xi^h \in H^1(0, T; H) \cap L^\infty(0, T; V_B^\sigma) \cap L^2(0, T; V_B^{2\sigma}),$$
$$\zeta^h \in H^1(0, T; H) \cap L^\infty(0, T; V_C^\tau) \cap L^2(0, T; V_C^{2\tau})$$

and solves the problem (in a strong form, i.e., the equations are satisfied a.e. in Q, since all the contributions are L^2 functions)

$$\partial_t \xi^h + A^{2\rho} \eta^h = Q_1^h, \tag{86}$$
$$\partial_t \xi^h + B^{2\sigma} \xi^h + Q_2^h = \eta^h, \tag{87}$$
$$\partial_t \zeta^h + C^{2\tau} \zeta^h = -Q_1^h, \tag{88}$$
$$\eta^h(0) = 0, \quad \xi^h(0) = 0 \quad \text{and} \quad \zeta^h(0) = 0, \tag{89}$$

where Q_1^h and Q_2^h are defined by

$$Q_1^h := P(\varphi^h)(S^h - \mu^h) - P(\bar{\varphi})(\bar{S} - \bar{\mu}) - P(\bar{\varphi})(\zeta - \eta) - P'(\bar{\varphi}) \xi (\bar{S} - \bar{\mu}),$$
$$Q_2^h := f(\varphi^h) - f(\bar{\varphi}) - f'(\bar{\varphi}) \xi.$$

It is convenient to rewrite the functions Q_i^h by accounting for the Taylor expansions of P and f. Using the formula with integral remainder, it is immediately checked that

$$Q_2^h = f'(\bar{\varphi})\xi^h + R_2^h (\varphi^h - \bar{\varphi})^2, \quad \text{where} \quad R_2^h := \int_0^1 (1 - \theta) f''(\bar{\varphi} + \theta(\varphi^h - \bar{\varphi})) \, d\theta \tag{90}$$

while it is more complicated to find a convenient representation of Q_1^h. However, simple algebraic manipulations show that

$$Q_1^h = P(\bar{\varphi})(\zeta^h - \eta^h) + (P(\varphi^h) - P(\bar{\varphi}))[(S^h - \bar{S}) - (\mu^h - \bar{\mu})]$$
$$+ P'(\bar{\varphi})(\bar{S} - \bar{\mu}) \xi^h + (\bar{S} - \bar{\mu}) R_1^h (\varphi^h - \bar{\varphi})^2, \tag{91}$$

where

$$R_1^h := \int_0^1 (1 - \theta) P''(\bar{\varphi} + \theta(\varphi^h - \bar{\varphi})) \, d\theta.$$

Notice that, by (66), both R_1^h and R_2^h are bounded uniformly with respect to h:

$$\|R_1^h\|_\infty + \|R_2^h\|_\infty \leq c_R. \tag{92}$$

After this preparation, we start estimating. To this end, we test (86), (87), and (88), by η^h, $\partial_t \xi^h$, and ζ^h, respectively. Then, we sum up and integrate in time. There is a usual cancellation. By adding the same contributions to both sides similarly as in the previous proof, we obtain

$$\int_0^t \|A^\rho \eta^h(s)\|^2 \, ds + \int_0^t \|\partial_t \xi^h(s)\|^2 \, ds + \frac{1}{2} \|\xi^h(t)\|_{B,\sigma}^2$$
$$+ \frac{1}{2} \|\zeta^h(t)\|^2 + \int_0^t \|\zeta^h(s)\|_{C,\tau}^2 \, ds$$
$$= \int_0^t \left(Q_1^h(s), \eta^h(s) - \zeta^h(s)\right) ds - \int_0^t \left(Q_2^h(s), \partial_t \xi^h(s)\right) ds$$
$$+ \int_0^t \left(\xi^h(s), \partial_t \xi^h(s)\right) ds + \int_0^t \|\zeta^h(s)\|^2 \, ds. \tag{93}$$

We have to estimate only the integrals involving Q_1^h and Q_2^h. The first term produces four integrals, termed I_j for $j = 1, \ldots, 4$ for brevity, which correspond to the four summands, in that order, of (91). Clearly, I_1 is nonpositive. As for I_2, we use the Hölder inequality and the embeddings (18) as well as the estimate (51) applied with $u_1 = \overline{u} + h$ and $u_2 = \overline{u}$. Hence, by omitting the integration variable s to shorten the lines, we have, for every $\delta > 0$,

$$I_2 \leq c \int_0^t \|\varphi^h - \overline{\varphi}\|_4 \left(\|S^h - \overline{S}\| + \|\mu^h - \overline{\mu}\|\right) \left(\|\eta^h\|_4 + \|\zeta^h\|_4\right) ds$$
$$\leq c \int_0^t \|\varphi^h - \overline{\varphi}\|_{B,\sigma} \left(\|S^h - \overline{S}\| + \|\mu^h - \overline{\mu}\|\right) \left(\|\eta^h\|_{A,\rho} + \|\zeta^h\|_{C,\tau}\right) ds$$
$$\leq \delta \int_0^t \left(\|\eta^h\|_{A,\rho}^2 + \|\zeta^h\|_{C,\tau}^2\right) ds$$
$$+ \frac{c}{\delta} \|\varphi^h - \overline{\varphi}\|_{L^\infty(0,T;V_B^\sigma)}^2 \left(\|S^h - \overline{S}\|_{L^2(0,T;H)}^2 + \|\mu^h - \overline{\mu}\|_{L^2(0,T;H)}^2\right)$$
$$\leq \delta \int_0^t \left(\|\eta^h\|_{A,\rho}^2 + \|\zeta^h\|\right)_{C,\tau}^2 ds + \frac{c_R}{\delta} \|h\|_{L^2(0,T;H)}^4.$$

Next, by also accounting for (49) applied to $\overline{\mu}$ and $\overline{S}$, we similarly have that

$$I_3 \leq c \int_0^t \|\overline{S} - \overline{\mu}\|_4 \|\xi^h\|_4 \left(\|\eta^h\| + \|\zeta^h\|\right) ds$$
$$\leq c \int_0^t \left(\|\overline{S}\|_{C,\tau} + \|\overline{\mu}\|_{A,\rho}\right) \|\xi^h\|_{B,\sigma} \left(\|\eta^h\| + \|\zeta^h\|\right) ds$$
$$\leq \delta \int_0^t \|\eta^h\|^2 \, ds + c_R (1 + \delta^{-1}) \int_0^t \left(\|\xi^h\|_{B,\sigma}^2 + \|\zeta^h\|^2\right) ds$$

and for the fourth contribution, thanks to (92), we obtain that

$$I_4 \leq c_R \int_0^t \|\overline{S} - \overline{\mu}\|_4 \|\varphi^h - \overline{\varphi}\|_4^2 \|\eta^h - \zeta^h\|_4 \, ds$$
$$\leq c_R \int_0^t \left(\|\overline{S}\|_{C,\tau} + \|\overline{\mu}\|_{A,\rho}\right) \|\varphi^h - \overline{\varphi}\|_{B,\sigma}^2 \left(\|\eta^h\|_{A,\rho} + \|\zeta^h\|_{C,\tau}\right) ds$$
$$\leq \delta \int_0^t \left(\|\eta^h\|_{A,\rho}^2 + \|\zeta^h\|_{C,\tau}^2\right) ds + \frac{c_R}{\delta} \|h\|_{L^2(0,T;H)}^4.$$

Finally, we estimate the term involving Q_2^h by accounting for (90) and (92) in this way:

$$-\int_0^t (Q_2^h, \partial_t \zeta^h) \, ds = -\int_0^t (f'(\overline{\varphi})\zeta^h + R_2^h (\varphi^h - \overline{\varphi})^2, \partial_t \zeta^h) \, ds$$

$$\leq c_R \int_0^t \|\zeta^h\| \|\partial_t \zeta^h\| \, ds + c_R \int_0^t \|\varphi^h - \overline{\varphi}\|_4^2 \|\partial_t \zeta^h\| \, ds$$

$$\leq \delta \int_0^t \|\partial_t \zeta^h\|^2 \, ds + \frac{c_R}{\delta} \int_0^t \|\zeta^h\|^2 \, ds + \frac{c_R}{\delta} \|h\|_{L^2(0,T;H)}^4 \,.$$

By treating the last two terms of (93) in a trivial way, recalling all the inequalities derived above, choosing $\delta > 0$ small enough, and applying the Gronwall lemma, we conclude that

$$\|\eta^h\|_{L^2(0,T;V_A^\rho)} + \|\xi^h\|_{H^1(0,T;H) \cap L^\infty(0,T;V_B^\sigma)} + \|\zeta^h\|_{L^\infty(0,T;H) \cap L^2(0,T;V_C^\tau)} \leq c_R \|h\|_{L^2(0,T;H)}^2 \,.$$

If we term c_R the value of the constant c_R of the last inequality, then we obtain (85) with Λ defined on $(0, \overline{h})$ by $\Lambda(s) := c_R s^2$. This completes the proof. □

4. The Control Problem

As announced in the Introduction, the main aim of this paper is the discussion of a control problem for the state system studied in the previous sections. For this problem, we assume that

$$\kappa_i \geq 0, \quad \text{for } i = 1, \dots, 5, \quad \varphi_Q, S_Q \in L^2(Q), \quad \text{and} \quad \varphi_\Omega, S_\Omega \in L^2(\Omega), \tag{94}$$

$$u_{min}, u_{max} \in L^\infty(Q), \quad \text{and} \quad u_{min} \leq u_{max} \quad \text{a.e. in } Q. \tag{95}$$

Then, the cost functional $\mathcal{J}$ and the set $\mathcal{U}_{ad}$ of the admissible controls are defined by

$$\mathcal{J}(u, \varphi, S) := \frac{\kappa_1}{2} \int_Q |\varphi - \varphi_Q|^2 + \frac{\kappa_2}{2} \int_\Omega |\varphi(T) - \varphi_\Omega|^2$$

$$+ \frac{\kappa_3}{2} \int_Q |S - S_Q|^2 + \frac{\kappa_4}{2} \int_\Omega |S(T) - S_\Omega|^2 + \frac{\kappa_5}{2} \int_Q |u|^2 \tag{96}$$

$$\mathcal{U}_{ad} := \{u \in L^2(0,T;H) : u_{min} \leq u \leq u_{max} \text{ a.e. in } Q\} \tag{97}$$

and the control problem is the following:

$$\text{Minimize } \mathcal{J}(u, \varphi, S) \text{ under the constraints that } u \in \mathcal{U}_{ad} \text{ and}$$
$$(\mu, \varphi, S) \text{ is the solution to (32)–(35) corresponding to } u. \tag{98}$$

For the above problem, we prove the existence of an optimal control, and we derive the first-order necessary conditions for optimality. This involves an adjoint problem for which we prove a well-posedness result. We recall that the control-to-state mapping $\mathcal{S}$ is defined in (64) and state our first result.

Theorem 6. *Under the assumptions (94)–(95), the control problem has at least one solution, that is, there is some $\overline{u} \in \mathcal{U}_{ad}$ satisfying the following condition: for every $v \in \mathcal{U}_{ad}$ we have that $\mathcal{J}(\overline{u}, \overline{\varphi}, \overline{S}) \leq \mathcal{J}(v, \varphi, S)$, where $(\overline{\varphi}, \overline{S}) = \mathcal{S}(\overline{u})$ and $(\varphi, S) = \mathcal{S}(v)$.*

Proof. Since $\mathcal{U}_{ad}$ is nonempty, the infimum of $\mathcal{J}$ under the constraints given in (98) is a well-defined real number $d \geq 0$, and we can pick a minimizing sequence $\{u_n\} \subset \mathcal{U}_{ad}$. Hence, denoting by (μ_n, φ_n, S_n) the state corresponding to u_n for $n \in \mathbb{N}$, we have that $\mathcal{J}(u_n, \varphi_n, S_n) \to d$ as $n \to \infty$.

Since $\mathcal{U}_{ad}$ is bounded and closed in $L^2(0,T;H)$ (in fact, it is even bounded and closed in $L^\infty(Q)$), we can assume that

$$u_n \to \bar{u} \quad \text{weakly in } L^2(0,T;H) \tag{99}$$

for some $\bar{u} \in \mathcal{U}_{ad}$. Moreover, we can choose some $R > 0$ such that $\mathcal{U}_{ad} \subset \mathcal{B}_R$. Therefore, we can apply Theorem 3 to u_n and deduce that (μ_n, φ_n, S_n) satisfies the estimate (49), as well as the separation and global boundedness properties (50) and (66), for all $n \in \mathbb{N}$. Hence, for a subsequence indexed again by n, we have that

$$\mu_n \to \bar{\mu} \quad \text{weakly star in } L^\infty(0,T;V_A^{2\rho}), \tag{100}$$
$$\varphi_n \to \bar{\varphi} \quad \text{weakly star in } W^{1,\infty}(0,T;H) \cap H^1(0,T;V_B^\sigma) \cap L^2(0,T;V_B^{2\sigma}), \tag{101}$$
$$S_n \to \bar{S} \quad \text{weakly star in } H^1(0,T;H) \cap L^\infty(0,T;V_C^\tau) \cap L^2(0,T;V_C^{2\tau}). \tag{102}$$

It follows that the initial conditions (35) are satisfied by the limiting pair $(\bar{\varphi}, \bar{S})$. Moreover, thanks to the compact embedding $V_B^\sigma \subset H$ ensured by (7), and consequently of $H^1(0,T;V_B^\sigma)$ into $L^2(0,T;H)$, we deduce that

$$\varphi_n \to \bar{\varphi} \quad \text{strongly in } L^2(0,T;H).$$

Since f and P are Lipschitz continuous in $[a_R, b_R]$, we also infer that

$$f(\varphi_n) \to f(\bar{\varphi}) \quad \text{and} \quad P(\varphi_n) \to P(\bar{\varphi}), \quad \text{strongly in } L^2(0,T;H).$$

It follows that $(\bar{\mu}, \bar{\varphi})$ solves (33). From the above strong convergence and the weak convergence of $\{\mu_n\}$ and $\{S_n\}$ at least in $L^2(0,T;H)$, we deduce that $\{P(\varphi_n)(S_n - \mu_n)\}$ converges to $P(\bar{\varphi})(\bar{S} - \bar{\mu})$ weakly in $L^1(Q)$. Hence, the limiting triplet $(\bar{\mu}, \bar{\varphi}, \bar{S})$ satisfies Equations (32) and (34) as well, i.e., $(\bar{\mu}, \bar{\varphi}, \bar{S})$ is the state corresponding to the control $\bar{u}$. On the other hand, we have that

$$\mathcal{J}(\bar{u}, \bar{\varphi}, \bar{S}) \leq \liminf_{n \nearrow \infty} \mathcal{J}(u_n, \varphi_n, S_n) = d$$

by semicontinuity. We conclude that $\bar{u}$ is an optimal control. $\square$

The rest of the section is devoted to the derivation of the first-order necessary conditions for optimality. Hence, we fix an optimal control $\bar{u} \in \mathcal{U}_{ad}$ and the corresponding $(\bar{\mu}, \bar{\varphi}, \bar{S})$ once and for all. If we introduce the so-called reduced cost functional $\tilde{\mathcal{J}}$ by setting

$$\tilde{\mathcal{J}}(u) := \mathcal{J}(u, \mathcal{S}_2(u), \mathcal{S}_3(u)) \quad \text{for } u \in L^2(0,T;H),$$

we immediately find from the convexity of $\mathcal{U}_{ad}$ that the Fréchet derivative $(D\tilde{\mathcal{J}})(\bar{u}) \in \mathcal{L}(L^2(0,T;H);\mathbb{R})$ must satisfy

$$(D\tilde{\mathcal{J}})(\bar{u})[v - \bar{u}] \geq 0 \quad \text{for every } v \in \mathcal{U}_{ad},$$

provided that it exists. However, this is the case due to the obvious differentiability of the quadratic functional $\mathcal{J}$ and the differentiability of the operator $\mathcal{S}$, which takes its values in $\mathcal{Y} \subset (C^0([0,T];H))^2$. Hence, by accounting for the full statement of Theorem 5, we can even apply the chain rule and rewrite the above inequality as

$$\kappa_1 \int_Q (\bar{\varphi} - \varphi_Q)\xi + \kappa_2 \int_\Omega (\bar{\varphi}(T) - \varphi_\Omega)\xi(T) + \kappa_3 \int_Q (\bar{S} - S_Q)\zeta$$
$$+ \kappa_4 \int_\Omega (\bar{S}(T) - S_\Omega)\zeta(T) + \kappa_5 \int_Q \bar{u}(v - \bar{u}) \geq 0 \quad \text{for every } v \in \mathcal{U}_{ad}, \tag{103}$$

where ξ and ζ are the components of the solution (η, ξ, ζ) to the linearized system (67)–(70) associated with $\overline{u}$ and $h = v - \overline{u}$.

As usual in control problems, a condition of this sort is not satisfactory, since it requires solving the linearized problem for infinitely many choices of $h \in L^2(0, T; H)$ because v is arbitrary in $\mathcal{U}_{ad}$. Therefore, we have to eliminate ξ and ζ from (103), which can be done by introducing and solving a proper adjoint problem. This is a backward-in-time problem for the adjoint state variables (q, p, r) that formally reads as follows:

$$A^{2\rho} q - p + P(\overline{\varphi})(q - r) = 0, \tag{104}$$

$$-\partial_t (q + p) + B^{2\sigma} p + f'(\overline{\varphi}) p - P'(\overline{\varphi})(\overline{S} - \overline{\mu})(q - r) = \kappa_1 (\overline{\varphi} - \varphi_Q), \tag{105}$$

$$-\partial_t r + C^{2\tau} r - P(\overline{\varphi})(q - r) = \kappa_3 (\overline{S} - S_Q), \tag{106}$$

$$(q + p)(T) = \kappa_2 (\overline{\varphi}(T) - \varphi_\Omega) \quad \text{and} \quad r(T) = \kappa_4 (\overline{S}(T) - S_\Omega). \tag{107}$$

However, in order to give this system a proper meaning according to the regularity that we will prove, we need some preliminaries. First, due to the density of V_B^σ in H, we can identify H with a subspace of the dual space $V_B^{-\sigma} := (V_B^\sigma)^*$ of V_B^σ in such a way that $\langle v, w \rangle = (v, w)$ for every $v \in H$ and $w \in V_B^\sigma$, where $\langle \cdot, \cdot \rangle$ denotes the duality pairing between $V_B^{-\sigma}$ and V_B^σ. Now, thanks to the obvious formula $(B^{2\sigma} v, w) = (B^\sigma v, B^\sigma w)$, which holds for every $v \in V_B^{2\sigma}$ and $w \in V_B^\sigma$ and, owing to the above identification, can also be read in the form $\langle B^{2\sigma} v, w \rangle = (B^\sigma v, B^\sigma w)$, one can extend the operator $B^{2\sigma} : V_B^{2\sigma} \to H$ to a continuous linear operator, still termed $B^{2\sigma}$, from V_B^σ to $V_B^{-\sigma}$ by means of the above formula, namely,

$$\langle B^{2\sigma} v, w \rangle = (B^\sigma v, B^\sigma w) \quad \text{for every } v, w \in V_B^\sigma. \tag{108}$$

At this point, it is meaningful to postulate the following regularity for the adjoint variables:

$$q \in L^\infty(0, T; V_A^{2\rho}), \tag{109}$$

$$p \in L^2(0, T; V_B^\sigma) \quad \text{and} \quad \partial_t (q + p) \in L^2(0, T; V_B^{-\sigma}), \tag{110}$$

$$r \in H^1(0, T; H) \cap L^\infty(0, T; V_C^\tau) \cap L^2(0, T; V_C^{2\tau}). \tag{111}$$

Indeed, then all of the equations, as well as the final conditions, have a precise meaning, by also accounting for the properties of the other ingredients which we recall for the reader's convenience: $P(\overline{\varphi})$, $P'(\overline{\varphi})$, $f'(\overline{\varphi})$, and $\overline{\mu}$ are bounded, and $\overline{S} \in H^1(0, T; H) \cap L^\infty(0, T; V_B^\sigma)$, whence, in particular, $\overline{S} \in L^\infty(0, T; L^4(\Omega))$. However, we also consider a variational formulation of the adjoint system, which makes sense in a much weaker regularity setting for (q, p, r), namely,

$$q \in L^\infty(0, T; V_A^\rho), \quad p \in L^2(0, T; V_B^\sigma), \quad \text{and} \quad r \in H^1(0, T; H) \cap L^2(0, T; V_C^\tau). \tag{112}$$

We require that

$$\int_0^T \{(A^\rho q, A^\rho v) - (p, v) + (P(\overline{\varphi})(q-r), v)\} \, ds = 0$$

for every $v \in L^2(0, T; V_A^\rho)$, (113)

$$\int_0^T \{(q+p, \partial_t v) + (B^\sigma p, B^\sigma v) + (f'(\overline{\varphi}) p - P'(\overline{\varphi})(\overline{S} - \overline{\mu})(q-r), v)\} \, ds$$
$$= \int_0^T (g_1, v) \, ds + (g_2, v(T))$$

for every $v \in H^1(0, T; H) \cap L^2(0, T; V_B^\sigma)$ vanishing at $t = 0$, (114)

$$\int_0^T \{(-\partial_t r, v) + (C^\tau r, C^\tau v) - (P(\overline{\varphi})(q-r), v)\} \, ds$$
$$= \int_0^T (g_3, v) \, ds \quad \text{for every } v \in L^2(0, T; V_C^\tau),$$ (115)

$$r(T) = g_4,$$ (116)

where we have introduced the abbreviating notation

$$g_1 := \kappa_1(\overline{\varphi} - \varphi_Q), \ g_2 := \kappa_2(\overline{\varphi}(T) - \varphi_\Omega), \ g_3 := \kappa_3(\overline{S} - S_Q), \ g_4 := \kappa_4(\overline{S}(T) - S_\Omega).$$ (117)

In addition, for brevity, and in order to shorten the exposition, we have omitted the integration time variable termed s. We will do the same in the following.

Clearly, (109)–(111) and (104)–(107) imply (112) and (113)–(116). In fact, these problems are equivalent. The proof given below makes use of the Leibniz rule proved in (Lem. 4.5 [30]) (and well known under slightly different assumptions), which we here state as a lemma.

Lemma 1. *Let $(\mathcal{V}, \mathcal{H}, \mathcal{V}^*)$ be a Hilbert triplet, and assume that*

$$y \in H^1(0, T; \mathcal{H}) \cap L^2(0, T; \mathcal{V}) \quad \text{and} \quad z \in H^1(0, T; \mathcal{V}^*) \cap L^2(0, T; \mathcal{H}).$$ (118)

Then, the function $t \mapsto (y(t), z(t))_\mathcal{H}$ is absolutely continuous on $[0, T]$, and its derivative is given by

$$\frac{d}{dt}(y, z)_\mathcal{H} = (y', z)_\mathcal{H} + {}_{\mathcal{V}^*}\langle z', y \rangle_\mathcal{V} \quad \text{a.e. in } (0, T),$$ (119)

where $(\cdot, \cdot)_\mathcal{H}$ and ${}_{\mathcal{V}^*}\langle \cdot, \cdot \rangle_\mathcal{V}$ denote the inner product in $\mathcal{H}$ and the dual pairing between $\mathcal{V}^*$ and $\mathcal{V}$, respectively.

Lemma 2. *Assume that (112) and (113)–(116) are valid. Then, (109)–(111) and (104)–(107) hold true as well.*

Proof. We first notice that (113) implies the pointwise variational inequality

$$(A^\rho q, A^\rho v) = (p - P(\overline{\varphi})(q-r), v) \quad \text{for every } v \in V_A^\rho \text{ and a.e. in } (0, T).$$

On the other hand, the conditions $w \in V_A^\rho$, $g \in H$, and $(A^\rho w, A^\rho v) = (g, v)$ for every $v \in V_A^\rho$, imply that $w \in V_A^{2\rho}$ and $A^{2\rho}w = g$, as one immediately sees by using the spectral representation. Hence, we obtain (109) and (104). The same argument can be used to deduce that r belongs to $L^2(0, T; V_C^{2\tau})$ and solves (106), since even $\partial_t r$ belongs to $L^2(0, T; H)$ by assumption. The last condition $r \in L^\infty(0, T; V_C^\tau)$ in (111) then follows from interpolation.

Much more work has to be done for the second equations. First, for the same test functions v as in (114), we deduce that

$$\int_0^T \{(q+p, \partial_t v) + \langle B^{2\sigma} p, v \rangle\} \, ds = \int_0^T (g, v) \, ds + (g_2, v(T)),$$ (120)

where, for brevity, we have set

$$g := g_1 - f'(\overline{\varphi})\, p + P'(\overline{\varphi})(\overline{S} - \overline{\mu})(q - r).$$

We immediately infer that

$$\left| \int_0^T (q + p, \partial_t v)\, ds \right| \leq \|p\|_{L^2(0,T;V_B^\sigma)} \|v\|_{L^2(0,T;V_B^\sigma)}$$
$$+ \|g\|_{L^2(0,T;H)} \|v\|_{L^2(0,T;H)} + \|g_2\| \|v(T)\|.$$

In particular, we have for some constant $c > 0$ that

$$\left| \int_0^T (q + p, \partial_t v)\, ds \right| \leq c\, \|v\|_{L^2(0,T;V_B^\sigma)} \quad \text{for every } v \in C_c^\infty(0, T; V_B^\sigma).$$

This exactly means that $\partial_t(q+p) \in (L^2(0,T;V_B^\sigma))^* = L^2(0,T;V_B^{-\sigma})$. Thus, we can replace the expression $(q+p, \partial_t v)$ by $-\langle \partial_t(q+p), v\rangle$ in (120), provided that $v \in C_c^\infty(0,T;V_B^\sigma)$. The variational equation we obtain is just (105) understood in the sense of $V_B^{-\sigma}$.

It remains to derive the first of the final conditions (107). To this end, we also assume that $v(0) = 0$ and exploit (120) once more. Moreover, we can apply Lemma 1 with $\mathcal{V} = V_B^\sigma$, $\mathcal{H} = H$, $y = v$, and $z = q + p$, since $v \in H^1(0,T;H) \cap L^2(0,T;V_B^\sigma)$ and $q + p \in H^1(0,T;V_B^{-\sigma}) \cap L^2(0,T;H)$. Finally, we account for the already proved Equation (105). We then obtain that

$$\int_0^T (g,v)\, ds + (g_2, v(T)) = \int_0^T (q + p, \partial_t v)\, ds + \int_0^T (B^\sigma p, B^\sigma v)\, ds$$
$$= \int_0^T \{-\langle \partial_t(q+p), v\rangle + \langle B^{2\sigma} p, v\rangle\}\, ds + \langle (q+p)(T), v(T)\rangle$$
$$= \int_0^T (g,v)\, ds + \langle (q+p)(T), v(T)\rangle.$$

Therefore, we have that $((q+p)(T), v(T)) = (g_2, v(T))$ for every v with the required properties, and the desired final condition obviously follows. □

Thus, we can choose between the strong form (104)–(107) and the weak formulation (113)–(116), according to our convenience, in proving a well-posedness result, which is our next goal. We prepare the existence part by introducing a Faedo–Galerkin scheme with viscosity that looks like an approximation of (104)–(107). We recall (8) on the eigenvalues and the eigenvectors of the operators and set, for every integer $n \geq 1$,

$$V_A^{(n)} := \mathrm{span}\{e_1, \ldots, e_n\}, \quad V_B^{(n)} := \mathrm{span}\{e_1', \ldots, e_n'\}, \quad \text{and} \quad V_C^{(n)} := \mathrm{span}\{e_1'', \ldots, e_n''\}.$$

Then, we look for a triplet (q^n, p^n, r^n) satisfying

$$q^n \in H^1(0,T;V_A^{(n)}), \quad p^n \in H^1(0,T;V_B^{(n)}), \quad \text{and} \quad r^n \in H^1(0,T;V_C^{(n)}), \tag{121}$$

and solving the system

$$\left(-\tfrac{1}{n}\partial_t q^n + A^{2\rho} q^n - p^n + P(\overline{\varphi})(q^n - r^n), v\right)$$
$$= 0 \quad \text{for every } v \in V_A^{(n)} \text{ and a.e. in } (0,T), \tag{122}$$
$$\left(-\partial_t(q^n + p^n) + (B^{2\sigma} p^n + f'(\overline{\varphi})p^n - P'(\overline{\varphi})(\overline{S} - \overline{\mu})(q^n - r^n), v\right)$$
$$= (g_1, v) \quad \text{for every } v \in V_B^{(n)} \text{ and a.e. in } (0,T), \tag{123}$$

$$(-\partial_t r^n + C^{2\tau} r^n - P(\overline{\varphi})(q^n - r^n), v)$$
$$= (g_3, v) \quad \text{for every } v \in V_C^{(n)} \text{ and a.e. in } (0, T), \tag{124}$$

as well as the final conditions

$$(q^n(T), v) = 0, \quad ((q^n + p^n)(T), v) = (g_2, v), \quad \text{and} \quad (r^n(T), v) = (g_4, v),$$
$$\text{for every } v \in V_A^{(n)}, v \in V_B^{(n)}, \text{ and } v \in V_C^{(n)}, \text{ respectively.} \tag{125}$$

The following result holds true.

Proposition 1. *System (122)–(125) has a unique solution (q^n, p^n, r^n) satisfying the conditions (121).*

Proof. The requirements (121) mean that

$$q^n(t) = \sum_{j=1}^n q_j^n(t) e_j \quad p^n(t) = \sum_{j=1}^n p_j^n(t) e_j' \quad \text{and} \quad r^n(t) = \sum_{j=1}^n r_j^n(t) e_j''$$

for a.a. $t \in (0, T)$ and some functions $q_j^n, p_j^n, r_j^n \in H^1(0, T)$. Moreover, an equivalent system is obtained by taking for $i = 1, \ldots, n$ just $v = e_i$, $v = e_i'$, and $v = e_i''$, in the three variational equations, respectively. Hence, (122)–(124) become an ODE system having the column vectors $q_n := (q_j^n)$, $p_n := (p_j^n)$, and $r_n := (r_j^n)$, as unknowns. This system reads as follows:

$$\sum_{j=1}^n \left\{ -\frac{1}{n}(e_j, e_i) \frac{d}{dt} q_j^n + \lambda_j^{2\rho}(e_j, e_i) q_j^n - (e_j', e_i) p_j^n \right.$$
$$\left. + (P(\overline{\varphi}) e_j, e_i) q_j^n - (P(\overline{\varphi}) e_j'', e_i) r_j^n \right\} = 0$$

$$\sum_{j=1}^n \left\{ -(e_j, e_i') \frac{d}{dt} q_j^n - (e_j', e_i') \frac{d}{dt} p_j^n + (\lambda_j')^{2\sigma}(e_j', e_i') p_j^n + (f'(\overline{\varphi}) e_j', e_i') p_j^n \right.$$
$$\left. - (P'(\overline{\varphi})(\overline{S} - \overline{\mu}) e_j, e_i') q_j^n + (P'(\overline{\varphi})(\overline{S} - \overline{\mu}) e_j'', e_i') r_j^n \right\} = (g_1, e_i')$$

$$\sum_{j=1}^n \left\{ -(e_j'', e_i'') \frac{d}{dt} r_j^n + (\lambda_j'')^{2\tau}(e_j'', e_i'') r_j^n \right.$$
$$\left. - (P(\overline{\varphi}) e_j, e_i'') q_j^n + (P(\overline{\varphi}) e_j'', e_i'') r_j^n \right\} = (g_3, e_i''),$$

where the index i runs over $\{1, \ldots, n\}$ in all of the equations, which are understood to hold a.e. in $(0, T)$. Thus, thanks to the orthogonality conditions in (8), it takes the form

$$-\frac{1}{n} q_n' + M_1 q_n + M_2 p_n + M_3 r_n = 0$$
$$M_4 q_n' - p_n' + M_5 q_n + M_6 p_n + M_7 r_n = b_n'$$
$$-r_n' + M_8 q_n + M_9 r_n = b_n''$$

for some (possibly time dependent, but bounded) $(n \times n)$ matrices M_k, $k = 1, \ldots, 9$, and column vectors $b_n', b_n'' \in L^2(0, T; \mathbb{R}^n)$. Therefore, one can solve the first equation for q_n' and replace q_n' in the second one by the resulting expression. At the same time, one multiplies the first equation by n and keeps the third one as it is. This procedure leads to an equivalent system of the form $-y' + My = b$ for some matrix $M \in L^\infty(0, T; \mathbb{R}^{3n \times 3n})$ and some vector $b \in L^2(0, T; \mathbb{R}^{3n})$ in the unknown $y \in H^1(0, T; \mathbb{R}^{3n})$ obtained by rearranging the triplet $(q_n \, p_n \, r_n)$ as a $3n$-column vector. On the other hand, the final conditions (125) provide a final condition for y. Hence, standard results for ODEs show the unique solvability. □

At this point, we are ready to solve the adjoint problem. We need, however, the following additional compatibility condition:

$$\text{It holds} \quad \kappa_4 S_\Omega \in V_C^\tau. \tag{126}$$

Remark 5. *The compatibility condition (126) is satisfied if either $\kappa_4 = 0$ or $S_\Omega \in V_C^\tau$. Obviously, $\kappa_4 = 0$ means that we do not have a tracking of the solution variable S at the final time T; while this is not desirable, it is not too much of a restriction, since one is rather interested in monitoring the final tumor fraction $\varphi(T)$ than $S(T)$. On the other hand, the assumption $S_\Omega \in V_C^\tau$ is not overly restrictive in view of the fact that $S \in H^1(0,T;H) \cap L^2(0,T;V_C^{2\tau})$, whence it follows that $S \in C^0([0,T];V_C^\tau)$ by continuous embedding, and thus $S(T) \in V_C^\tau$. To assume the same regularity for S_Ω is certainly not unreasonable.*

We have the following result.

Theorem 7. *Suppose that also (126) is fulfilled. Then, the adjoint system (104)–(107) has a unique solution satisfying the regularity conditions (109)–(111).*

Proof. In order to prove the existence of a solution, we start from the finite-dimensional problem (122)–(125), perform an a priori estimate, and let n tend to infinity. In addition, in this section, we simplify the notation as far as constants are concerned and use the same symbol c for different constants that can depend only on the structure, the data, T, the optimal control $\bar{u}$, and the corresponding state $(\bar{\mu}, \bar{\varphi}, \bar{S})$.

A priori estimate. We write the Equations (122)–(124) at the time s and test them by $-\partial_t q^n(s)$, $p^n(s)$, and $-\partial_t r^n(s)$, respectively. Then, we sum up, integrate over (t,T) with respect to s, and notice that the terms involving the product $p^n \partial_t q^n$ cancel each other. Moreover, we add the same quantities $\int_t^T \|p^n\|^2 \, ds$ and $(1/2)\|r^n(t)\|^2 = \int_t^T (r^n, \partial_t r^n) \, ds$ to both sides in order to recover the full norms in the spaces V_B^σ and V_C^τ. We then obtain the identity

$$\frac{1}{n} \int_t^T \|\partial_t q^n\|^2 \, ds + \frac{1}{2} \|A^\rho q^n(t)\|^2 + \frac{1}{2} \|p^n(t)\|^2 + \int_t^T \|p^n\|_{B,\sigma}^2 \, ds$$
$$+ \int_t^T \|\partial_t r^n\|^2 \, ds + \frac{1}{2} \|r^n(t)\|_{C,\tau}^2$$
$$= \int_t^T (P(\bar{\varphi})(q^n - r^n) \partial_t (q^n - r^n)) \, ds + \int_t^T (P'(\bar{\varphi})(\bar{S} - \bar{\mu})(q^n - r^n), p^n) \, ds$$
$$- \int_t^T (f'(\bar{\varphi})p^n, p^n) \, ds + \int_t^T (g_1, p^n) \, ds - \int_t^T (g_3, \partial_t r^n) \, ds$$
$$+ \frac{1}{2} \|A^\rho q^n(T)\|^2 + \frac{1}{2} \|p^n(T)\|^2 + \frac{1}{2} \|r^n(T)\|_{C,\tau}^2$$
$$+ \int_t^T \|p^n\|^2 \, ds + \int_t^T (r^n, \partial_t r^n) \, ds. \tag{127}$$

At first, we exploit the endpoint conditions (125). Obviously, $q^n(T) = 0$, which entails that $A^\rho q^n(T) = 0$, as well as $(p^n(T), v) = (g_2, v)$ for all $v \in V_B^{(n)}$. The latter identity just means that $p^n(T)$ is the H-orthogonal projection of g_2 onto $V_B^{(n)}$, which implies that $\|p^n(T)\| \leq \|g_2\|$ for all $n \in \mathbb{N}$. By the same token, we can infer that $\|r^n(T)\| \leq \|g_4\|$ for all $n \in \mathbb{N}$. Finally, we insert $v = C^{2\tau} r^n(T) \in V_C^{(n)}$ in the last identity in (125). Recalling that $\bar{S} \in C^0([0,T]; V_C^\tau)$, and by virtue of (126), we infer that $g_4 \in V_C^\tau$. We thus find that

$$\|C^\tau r^n(T)\|^2 = (r^n(T), C^{2\tau} r^n(T)) = (g_4, C^{2\tau} r^n(T)) = (C^\tau g_4, C^\tau r^n(T)),$$

whence we infer that $\|C^\tau r^n(T)\| \leq \|C^\tau g_4\|$. In conclusion, we have shown the estimate $\|r^n(T)\|_{C,\tau} \leq \|g_4\|_{C,\tau}$ for all $n \in \mathbb{N}$.

Next, we consider the first two terms on the right-hand side, which we denote by Y_1 and Y_2. We only need to estimate these terms, since the remaining other ones can easily be handled using Young's inequality and, eventually, Gronwall's lemma. As for Y_1, we first integrate by parts, and one of the important terms we obtain is nonpositive. Then, we account for the Hölder and Young inequalities, the equivalence of norms in V_A^ρ related to (22), and the embeddings (18) as follows:

$$Y_1 = \frac{1}{2}\int_t^T\int_\Omega P(\overline{\varphi})\,\partial_t|q^n - r^n|^2\,dx\,ds = \frac{1}{2}\int_\Omega P(\overline{\varphi}(T))\,|q^n(T) - r^n(T)|^2$$
$$- \frac{1}{2}\int_\Omega P(\overline{\varphi}(t))\,|q^n(t) - r^n(t)|^2 - \int_t^T\int_\Omega P'(\overline{\varphi})\,\partial_t\overline{\varphi}\,|q^n - r^n|^2$$
$$\leq c + c\int_t^T \|\partial_t\overline{\varphi}\|_4\,(\|q^n\|_4 + \|r^n\|_4)^2\,ds \leq c + c\int_t^T (\|A^\rho q^n\|^2 + \|r^n\|_{C,\tau}^2)\,ds.$$

Concerning Y_2, we have that

$$Y_2 \leq c\int_t^T \|\overline{S} - \overline{\mu}\|_4\,\|q^n - r^n\|_4\,\|p^n\|_4\,ds$$
$$\leq \frac{1}{2}\int_t^T \|p^n\|_{B,\sigma}^2\,ds + c\int_t^T (\|A^\rho q^n\|^2 + \|r^n\|_{C,\tau}^2)\,ds.$$

By treating the remaining terms on the right-hand side of (127) as announced before, and applying the Gronwall lemma, we conclude that

$$\frac{1}{\sqrt{n}}\|\partial_t q^n\|_{L^2(0,T;H)}^2 + \|q^n\|_{L^\infty(0,T;V_A^\rho)}$$
$$+ \|p^n\|_{L^\infty(0,T;H)\cap L^2(0,T;V_B^\sigma)} + \|r^n\|_{H^1(0,T;H)\cap L^\infty(0,T;V_C^\tau)} \leq c. \tag{128}$$

Existence. The above estimate ensures that, for a subsequence again indexed by n,

$$\frac{1}{n}\partial_t q^n \to 0 \quad \text{strongly in } L^2(0,T;H), \tag{129}$$
$$q^n \to q \quad \text{weakly star in } L^\infty(0,T;V_A^\rho), \tag{130}$$
$$p^n \to p \quad \text{weakly star in } L^\infty(0,T;H) \cap L^2(0,T;V_B^\sigma), \tag{131}$$
$$r^n \to r \quad \text{weakly star in } H^1(0,T;H) \cap L^\infty(0,T;V_C^\tau). \tag{132}$$

We aim at proving that (q,p,r) is the desired solution to the weak form (113)–(116) of the adjoint problem. Clearly, (116) is satisfied, and we have to prove that the variational equations are satisfied as well. We confine ourselves to the second equation, which is the most complicated one. To this end, we write an integrated version of (123). We fix any integer $m > 1$, take any $v \in H^1(0,T;H) \cap L^2(0,T;V_B^{(m)})$ vanishing at $t = 0$, and assume that $n \geq m$. Then $V_A^{(m)} \subset V_A^{(n)}$, so that $v(s)$ is admissible in (123) written at the time s, and we can test the equation in the inner product of H. Moreover, we replace $(B^{2\sigma} p^n(s), v(s))$ by $(B^\sigma p^n(s), B^\sigma v(s))$. Then, we integrate over $(0,T)$ with respect to s. Now, we observe that $v(T)$ is admissible in the second identity of (125). Thus, by an integration by parts, we obtain that

$$\int_0^T \{(q^n + p^n, \partial_t v) + (B^\sigma p^n, B^\sigma v) + (f'(\overline{\varphi})\,p^n - P'(\overline{\varphi})(\overline{S} - \overline{\mu})(q^n - r^n), v)\}\,ds$$
$$= \int_0^T (g_1, v)\,ds + (g_2, v(T)).$$

Since $n \geq m$ is arbitrary, we can let n tend to infinity by using (130)–(132). Concerning, e.g., the worst term, we recall that $\overline{S}$ and $\overline{\mu}$ belong to $L^\infty(0,T;L^4(\Omega))$ and observe that q^n and r^n converge to q and r, respectively, also weakly in $L^2(0,T;L^4(\Omega))$. Hence, we have that

$$P'(\overline{\varphi})(\overline{S}-\overline{\mu})(q^n - r^n) \to P'(\overline{\varphi})(\overline{S}-\overline{\mu})(q - r) \quad \text{weakly in } L^2(0,T;H).$$

As the other terms are easier, we conclude that (114) holds for such a function v. At this point, we fix any $v \in H^1(0,T;H) \cap L^2(0,T;V_B^\sigma)$ vanishing at $t=0$ and define v_m by setting

$$v_m(t) := \sum_{j=1}^m (v(t), e_j') e_j' \quad \text{for } t \in [0,T].$$

Then, v_m belongs to $H^1(0,T;H) \cap L^2(0,T;V_B^{(m)})$ and vanishes at $t=0$. Hence, we can use it in the equality just obtained. As m is arbitrary, we can take the limit as $m \to \infty$. By noting that v_m converges even strongly to v in $H^1(0,T;H) \cap L^2(0,T;V_B^\sigma)$, we conclude that (114) is satisfied for such a v. By similarly reasoning for the other equations, we can conclude. Hence, the existence part of the statement is proved.

Uniqueness. By linearity, we can assume that all the right-hand sides of the strong formulation (104)–(107) vanish, so that the problem becomes

$$A^{2\rho}q - p + P(\overline{\varphi})(q-r) = 0, \tag{133}$$

$$-\partial_t(q+p) + B^{2\sigma}p + f'(\overline{\varphi})p - P'(\overline{\varphi})(\overline{S}-\overline{\mu})(q-r) = 0, \tag{134}$$

$$-\partial_t r + C^{2\tau} r - P(\overline{\varphi})(q-r) = 0, \tag{135}$$

$$(q+p)(T) = 0 \quad \text{and} \quad r(T) = 0. \tag{136}$$

We cannot adapt the argument used to arrive at (128), since no information for $\partial_t q$ is available now. Thus, we proceed in a different way. With the notation

$$(1 * v)(t) := \int_t^T v(s)\,ds \quad \text{for a.a. } t \in (0,T) \text{ and every } v \in L^1(0,T;H),$$

we integrate (for a.a. $t \in (0,T)$) (134) over (t,T) and obtain a.e. in $(0,T)$

$$q + p + B^{2\sigma}(1 * p) = 1 * \left(P'(\overline{\varphi})(\overline{S}-\overline{\mu})(q-r)\right) - 1 * \left(f'(\overline{\varphi})p\right). \tag{137}$$

At this point, we test (133) by q, (137) by p, and (135) by r, sum up, and integrate over (t,T). The terms involving the product pq cancel each other. We also add the same quantities $(1/2)\|(1*p)(t)\|^2 = \int_t^T p(1*p)\,ds$ and $\int_t^T \|r\|^2\,ds$ to both sides and obtain

$$\int_t^T \|A^\rho q\|^2\,ds + \int_t^T (P(\overline{\varphi})(q-r), q-r)\,ds + \int_t^T \|p\|^2\,ds + \frac{1}{2}\|(1*p)(t)\|_{B,\sigma}^2$$

$$+ \frac{1}{2}\|r(t)\|^2 + \int_t^T \|r\|_{C,\tau}^2\,ds$$

$$= \int_t^T (1 * (P'(\overline{\varphi})(\overline{S}-\overline{\mu})(q-r) - f'(\overline{\varphi})p), p)\,ds$$

$$+ \int_t^T p(1*p)\,ds + \int_t^T \|r\|^2\,ds. \tag{138}$$

All of the terms on the left-hand side are nonnegative. Now, we treat the first integral on the right-hand side, which we term Y. We first integrate by parts. Then, we owe to Young's inequality and to the obvious inequality $\|(1*v)(t)\|^2 \leq T \int_t^T \|v(s)\|^2 ds$, which holds true for every $t \in [0,T]$ and every $v \in L^2(0,T;H)$. We set, for brevity, $w := P'(\overline{\varphi})(\overline{S} - \overline{\mu})(q - r) - f'(\overline{\varphi})p$ and observe that

$$Y \leq \frac{1}{4} \int_t^T \|p\|^2 ds + \int_t^T \|1*w\|^2 ds$$
$$\leq \frac{1}{4} \int_t^T \|p\|^2 ds + \int_t^T T \int_s^T \|w\|^2 ds' \, ds.$$

On the other hand, we recall (18), (22) and the regularity $L^\infty(0,T;L^4(\Omega))$ of $\overline{S}$ and $\overline{\mu}$. We thus deduce that

$$\int_s^T \|w\|^2 ds' \leq \|P'(\overline{\varphi})\|_\infty^2 \int_s^T \|\overline{S} - \overline{\mu}\|_4^2 \|q - r\|_4^2 ds' + \|f'(\overline{\varphi})\|_\infty^2 \int_s^T \|p\|^2 ds'$$
$$\leq c \int_s^T \left(\|A^\rho q\|^2 + \|r\|_{C,\tau}^2 + \|p\|^2 \right) ds',$$

whence

$$Y \leq \frac{1}{4} \int_t^T \|p\|^2 ds + \int_t^T c \left(\int_s^T \left(\|A^\rho q\|^2 + \|p\|^2 + \|r\|_{C,\tau}^2 \right) ds' \right) ds. \qquad (139)$$

Therefore, coming back to (138) and estimating the second integral on the right-hand side as

$$\int_t^T p(1*p) \, ds \leq \frac{1}{4} \int_t^T \|p\|^2 ds + \int_t^T \|1*p\|^2 ds$$

and then applying the Gronwall lemma, we easily conclude that $(q,p,r) = (0,0,0)$. □

Now that the adjoint problem is solved, we can rewrite the variational inequality (103) in a much better form. Indeed, we have the following result.

Theorem 8. *Under the assumptions (94)–(95) and (126), let $\overline{u} \in \mathcal{U}_{ad}$ be an optimal control, and let (q,p,r) be the solution to the associated adjoint problem (104)–(107). Then, it holds*

$$\int_Q (r + \kappa_5 \overline{u})(v - \overline{u}) \geq 0 \quad \text{for every } v \in \mathcal{U}_{ad}. \qquad (140)$$

In particular, if $\kappa_5 > 0$, then $\overline{u}$ is the projection of $-r/\kappa_5$ on $\mathcal{U}_{ad}$ in the sense of the space $L^2(Q)$ with its standard inner product. That is, it is given by

$$\overline{u} = \min\{u_{max}, \max\{u_{min}, -r/\kappa_5\}\} \quad \text{a.e. in } Q.$$

Proof. We fix $v \in \mathcal{U}_{ad}$ and consider the linearized system with $h = v - \overline{u}$. Now, we observe that the regularity (109)–(111) is suitable for integrating over $(0,T)$ the equations (67), (68), and (69), tested by $q(t)$, $p(t)$, and $r(t)$, respectively. By doing this, rearranging and summing up, we obtain (as before in this section, we omit the integration variable, which we term s for uniformity)

$$\int_0^T \left\{ (\partial_t \xi, q) + (A^\rho \eta, A^\rho q) - (P(\overline{\varphi})(\zeta - \eta), q) - (P'(\overline{\varphi}) \xi (\overline{S} - \overline{\mu}), q) \right\} ds$$
$$+ \int_0^T \left\{ (\partial_t \xi, p) + (B^\sigma \xi, B^\sigma p) + (f'(\overline{\varphi}) \xi, p) - (\eta, p) \right\} ds$$
$$+ \int_0^T \left\{ (\partial_t \zeta, r) + (C^\tau \zeta, C^\tau r) + (P(\overline{\varphi})(\zeta - \eta), r) + (P'(\overline{\varphi}) \xi (\overline{S} - \overline{\mu}), r) \right\} ds$$
$$= \int_0^T (v - \overline{u}, r) \, ds.$$

At the same time, we take $v = -\eta$ in (113), $v = -\xi$ in (114), $v = -\zeta$ in (115), respectively, and note that all the three test functions are admissible in their equations. Then, we sum up, rearrange, and get

$$\int_0^T \left\{ -(A^\rho q, A^\rho \eta) + (p, \eta) - (P(\overline{\varphi})(q-r), \eta) \right\} ds$$
$$+ \int_0^T \left\{ -(q+p, \partial_t \xi) - (B^\sigma p, B^\sigma \xi) - (f'(\overline{\varphi})p - P'(\overline{\varphi})(\overline{S} - \overline{\mu})(q-r), \xi) \right\} ds$$
$$+ \int_0^T \left\{ (\partial_t r, \zeta) - (C^\tau r, C^\tau \zeta) + (P(\overline{\varphi})(q-r), \zeta) \right\} ds$$
$$= -\int_0^T (g_1, \xi) \, ds - (g_2, \xi(T)) - \int_0^T (g_3, \zeta) \, ds.$$

Next, we add the identities just obtained to each other. Several cancellations occur, and what remains is just the following identity:

$$\int_0^T \{(\partial_t \zeta, r) + (\partial_t r, \zeta)\} \, ds = \int_0^T (v - \overline{u}, r) \, ds - \int_0^T (g_1, \xi) \, ds - (g_2, \xi(T)) - \int_0^T (g_3, \zeta) \, ds.$$

At this point, we observe that the left-hand side equals $(g_4, \zeta(T))$ by (116), so that the above identity becomes

$$\int_0^T (g_1, \xi) \, ds + (g_2, \xi(T)) + \int_0^T (g_3, \zeta) \, ds + (g_4, \zeta(T)) = \int_0^T (v - \overline{u}, r) \, ds.$$

Hence, by recalling the notation (117), and comparing with (103), we obtain (140). □

5. Conclusions

In this paper, we analyzed a system of three evolution equations involving fractional operators and nonlinearities. This system turns out to be a generalization of a Cahn–Hilliard type phase field system modeling tumor growth. Actually, we considered the distributed control problem in terms of the control in the third equation, by dealing with a tracking-type cost functional. We discussed separation properties, continuous dependence with respect to the control, and well-posedness results for the state system (cf. Theorems 1–3). Then, we introduced the linearized problem and showed its solvability by Theorem 4. The Fréchet differentiability of the control-to-state operator, with the Fréchet derivative mapping the increment into the solution of the linearized problem, has been established by Theorem 5. The existence of solutions to the optimal control problem has been the subject of Theorem 6. Finally, the associated adjoint system has been introduced and its well-posedness has been proved (see Theorem-7), then the first-order necessary conditions of optimality have been derived in a simple and clear form, as stated in Theorem 8.

Author Contributions: Conceptualization, P.C., G.G. and J.S.; Methodology, P.C., G.G. and J.S.; Software, P.C., G.G. and J.S.; Validation, P.C., G.G. and J.S.; Formal Analysis, P.C., G.G. and J.S.; Investigation, P.C., G.G. and J.S.; Resources, P.C., G.G. and J.S.; Data Curation, P.C., G.G. and J.S.; Writing—Original Draft Preparation, P.C., G.G. and J.S.; Writing—Review & Editing, P.C., G.G. and J.S.; Visualization, P.C., G.G. and J.S.; Supervision, P.C., G.G. and J.S.; Project Administration, P.C., G.G. and J.S.; Funding Acquisition, P.C., G.G. and J.S.

Funding: This research received no external funding.

Acknowledgments: This research was partly supported by the Italian Ministry of Education, University and Research (MIUR): Dipartimenti di Eccellenza Program (2018–2022)—Dept. of Mathematics "F. Casorati", University of Pavia. In addition, P.C. and C.G. gratefully acknowledge some other support from the GNAMPA (Gruppo Nazionale per l'Analisi Matematica, la Probabilità e le loro Applicazioni) of INdAM (Istituto Nazionale di Alta Matematica).

Conflicts of Interest: The authors declare no conflict of interest.

References

1. Colli, P.; Gilardi, G.; Sprekels, J. Well-posedness and regularity for a fractional tumor growth model. *Adv. Math. Sci. Appl.* **2019**, *28*, 343–375.
2. Cristini, V.; Li, X.; Lowengrub, J.S.; Wise, S.M. Nonlinear simulations of solid tumor growth using a mixture model: Invasion and branching. *J. Math. Biol.* **2009**, *58*, 723–763. [CrossRef] [PubMed]
3. Cristini, V.; Lowengrub, J.S. *Multiscale Modeling of Cancer: An Integrated Experimental and Mathematical Modeling Approach*; Cambridge University Press: Leiden, The Netherlands, 2010.
4. Hawkins-Daarud, A.; Prudhomme, S.; van der Zee, K.G.; Oden, J.T. Bayesian calibration, validation, and uncertainty quantification of diffuse interface models of tumor growth. *J. Math. Biol.* **2013**, *67*, 1457–1485. [CrossRef] [PubMed]
5. Hawkins-Daruud, A.; van der Zee, K.G.; Oden, J.T. Numerical simulation of a thermodynamically consistent four-species tumor growth model. *Int. J. Numer. Math. Biomed. Eng.* **2012**, *28*, 3–24. [CrossRef]
6. Oden, J.T.; Hawkins, A.; Prudhomme, S. General diffuse-interface theories and an approach to predictive tumor growth modeling. *Math. Models Methods Appl. Sci.* **2010**, *20*, 477–517. [CrossRef]
7. Wise, S.M.; Lowengrub, J.S.; Frieboes, H.B.; Cristini, V. Three-dimensional multispecies nonlinear tumor growth-I: Model and numerical method. *J. Theor. Biol.* **2008**, *253*, 524–543. [CrossRef]
8. Wu, X.; van Zwieten, G.J.; van der Zee, K.G. Stabilized second-order splitting schemes for Cahn–Hilliard models with applications to diffuse-interface tumor-growth models, *Int. J. Numer. Meth. Biomed. Eng.* **2014**, *30*, 180–203. [CrossRef]
9. Dai, M.; Feireisl, E.; Rocca, E.; Schimperna, G.; Schonbek, M. Analysis of a diffuse interface model of multi-species tumor growth. *Nonlinearity* **2017**, *30*, 1639–1658. [CrossRef]
10. Ebenbeck, M.; Garcke, H. Analysis of a Cahn–Hilliard–Brinkman model for tumour growth with chemotaxis. *J. Differ. Equ.* **2019**, *266*, 5998–6036. [CrossRef]
11. Frigeri, S.; Lam, K.F.; Rocca, E.; Schimperna, G. On a multi-species Cahn–Hilliard–Darcy tumor growth model with singular potentials. *Commun. Math. Sci.* **2018**, *16*, 821–856. [CrossRef]
12. Garcke, H.; Lam, K.F. Global weak solutions and asymptotic limits of a Cahn–Hilliard–Darcy system modelling tumour growth. *AIMS Math.* **2016**, *1*, 318–360. [CrossRef]
13. Garcke, H.; Lam, K.F. Well-posedness of a Cahn–Hilliard system modelling tumour growth with chemotaxis and active transport. *Eur. J. Appl. Math.* **2017**, *28*, 284–316. [CrossRef]
14. Garcke, H.; Lam, K.F. Analysis of a Cahn–Hilliard system with non–zero Dirichlet conditions modeling tumor growth with chemotaxis. *Discret. Contin. Dyn. Syst.* **2017**, *37*, 4277–4308. [CrossRef]
15. Garcke, H.; Lam, K.F. On a Cahn–Hilliard–Darcy system for tumour growth with solution dependent source terms. In *Trends on Applications of Mathematics to Mechanics*; Rocca, E., Stefanelli, U., Truskinovski, L., Visintin, A., Eds.; Springer INdAM Series; Springer: Cham, Switzerland, 2018; Volume 27, pp. 243–264.
16. Garcke, H.; Lam, K.F.; Nürnberg, R.; Sitka, E. A multiphase Cahn–Hilliard–Darcy model for tumour growth with necrosis. *Math. Models Methods Appl. Sci.* **2018**, *28*, 525–577. [CrossRef]
17. Garcke, H.; Lam, K.F.; Sitka, E.; Styles, V. A Cahn–Hilliard–Darcy model for tumour growth with chemotaxis and active transport. *Math. Models Methods Appl. Sci.* **2016**, *26*, 1095–1148. [CrossRef]
18. Sprekels, J.; Wu, H. Optimal distributed control of a Cahn–Hilliard–Darcy system with mass sources. *Appl. Math. Optim.* **2019**, 1–42. [CrossRef]
19. Conti, M.; Giorgini, A. The three-dimensional Cahn–Hilliard–Brinkman system with unmatched densities. *HAL* **2018**, 1–34.
20. della Porta, F.; Giorgini, A.; Grasselli, M. The nonlocal Cahn–Hilliard–Hele–Shaw system with logarithmic potential. *Nonlinearity* **2018**, *31*, 4851–4881. [CrossRef]
21. Cavaterra, C.; Rocca, E.; Wu, H. Long-time dynamics and optimal control of a diffuse interface model for tumor growth. *Appl. Math. Optim.* **2019**, 1–49. [CrossRef]
22. Colli, P.; Gilardi, G.; Hilhorst, D. On a Cahn–Hilliard type phase field system related to tumor growth. *Discret. Contin. Dyn. Syst.* **2015**, *35*, 2423–2442. [CrossRef]
23. Colli, P.; Gilardi, G.; Rocca, E.; Sprekels, J. Vanishing viscosities and error estimate for a Cahn–Hilliard type phase field system related to tumor growth. *Nonlinear Anal. Real World Appl.* **2015**, *26*, 93–108. [CrossRef]
24. Colli, P.; Gilardi, G.; Rocca, E.; Sprekels, J. Optimal distributed control of a diffuse interface model of tumor growth. *Nonlinearity* **2017**, *30*, 2518–2546. [CrossRef]

25. Colli, P.; Gilardi, G.; Rocca, E.; Sprekels, J. Asymptotic analyses and error estimates for a Cahn–Hilliard type phase field system modelling tumor growth. *Discret. Contin. Dyn. Syst. Ser. S* **2017**, *10*, 37–54. [CrossRef]
26. Frigeri, S.; Grasselli, M.; Rocca, E. On a diffuse interface model of tumor growth. *Eur. J. Appl. Math.* **2015**, *26*, 215–243. [CrossRef]
27. Miranville, A.; Rocca, E.; Schimperna, G. On the long time behavior of a tumor growth model. *J. Differ. Equ.* **2019**, *267*, 2616–2642. [CrossRef]
28. Ainsworth, M.; Mao, Z. Analysis and approximation of a fractional Cahn–Hilliard equation. *SIAM J. Numer. Anal.* **2017**, *55*, 1689–1718. [CrossRef]
29. Akagi, G.; Schimperna, G.; Segatti, A. Fractional Cahn–Hilliard, Allen–Cahn, and porous medium equations. *J. Differ. Equ.* **2016**, *261*, 2935–2985. [CrossRef]
30. Colli, P.; Gilardi, G.; Sprekels, J. Well-posedness and regularity for a generalized fractional Cahn–Hilliard system. *Atti Accad. Naz. Lincei Rend. Lincei Mat. Appl.* **2019**, *30*, 437–478.
31. Colli, P.; Gilardi, G.; Sprekels, J. Optimal distributed control of a generalized fractional Cahn–Hilliard system, *Appl. Math. Optim.* **2019**, 1–39. [CrossRef]
32. Colli, P.; Gilardi, G.; Sprekels, J. Deep quench approximation and optimal control of general Cahn–Hilliard systems with fractional operators and double-obstacle potentials, *arXiv* 2018, arXiv:1812.01675.
33. Colli, P.; Gilardi, G.; Sprekels, J. Longtime behavior for a generalized Cahn–Hilliard system with fractional operators. *arXiv* 2019, arXiv:1904.00931.
34. Gal, C.G. On the strong-to-strong interaction case for doubly nonlinear Cahn–Hilliard equations. *Discret. Contin. Dyn. Syst.* **2017**, *37*, 131–167. [CrossRef]
35. Gal, C.G. Non-local Cahn–Hilliard equations with fractional dynamic boundary conditions. *Eur. J. Appl. Math.* **2017**, *28*, 736–788. [CrossRef]
36. Gal, C.G. Doubly nonlinear Cahn–Hilliard equations. *Ann. Inst. H. Poincaré Anal. Non Linéaire* **2018**, *35*, 357–392. [CrossRef]
37. Colli, P.; Gilardi, G. Well-posedness, regularity and asymptotic analyses for a fractional phase field system. *Asymptot. Anal.* **2019**, *114*, 93–128. [CrossRef]
38. Colli, P.; Gilardi, G.; Sprekels, J. Asymptotic analysis of a tumor growth model with fractional operators. *arXiv* 2019, arXiv:1908.10651.
39. Estrada-Rodriguez, G.; Gimperlein, H.; Painter, K.J.; Stocek, J. Space-time fractional diffusion in cell movement models with delay. *Math. Models Methods Appl. Sci.* **2019**, *29*, 65–88. [CrossRef]
40. Baeumer, B.; Kovács, M.; Meerschaert, M.M. Numerical solutions for fractional reaction-diffusion equations. *Comput. Math. Appl.* **2008**, *55*, 2212–2226. [CrossRef]
41. Chandra, S.K.; Bajpai, M.K. Mesh free alternate directional implicit method based three-dimensional super-diffusive model for benign brain tumor segmentation. *Comput. Math. Appl.* **2019**, *77*, 3212–3223. [CrossRef]
42. Evangelista, L.R.; Lenzi, E.K. *Fractional Diffusion Equations and Anomalous Diffusion*; Cambridge University Press: Cambridge, UK, 2018.
43. Granero-Belinchón, R. Global solutions for a hyperbolic-parabolic system of chemotaxis. *J. Math. Anal. Appl.* **2017**, *449*, 872–883. [CrossRef]
44. Ibrahim, R.W.; Nashine, H.K.; Kamaruddin, N. Hybrid time-space dynamical systems of growth bacteria with applications in segmentation. *Math. Biosci.* **2017**, *292*, 10–17. [CrossRef] [PubMed]
45. Joshi, H.; Jha, B.K. Fractionally delineate the neuroprotective function of calbindin-D28k in Parkinson's disease. *Int. J. Biomath.* **2018**, *11*, 1850103. [CrossRef]
46. Karlsen, K.H.; Ulusoy, S. On a hyperbolic Keller–Segel system with degenerate nonlinear fractional diffusion. *Netw. Heterog. Media* **2016**, *11*, 181–201. [CrossRef]
47. Massaccesi, A.; Valdinoci, E. Is a nonlocal diffusion strategy convenient for biological populations in competition? *J. Math. Biol.* **2017**, *74*, 113–147. [CrossRef] [PubMed]
48. Sohail, A.; Arshad, S.; Javed, S.; Maqbool, K. Numerical analysis of fractional-order tumor model. *Int. J. Biomath.* **2015**, *8*, 1550069. [CrossRef]
49. Sweilam, N.H.; Al-Mekhlafi, S.M. Optimal control for a nonlinear mathematical model of tumor under immune suppression: A numerical approach. *Optim. Control Appl. Methods* **2018**, *39*, 1581–1596. [CrossRef]
50. Zhou, Y.; Shangerganesh, L.; Manimaran, J.; Debbouche, A. A class of time-fractional reaction-diffusion equation with nonlocal boundary condition. *Math. Methods Appl. Sci.* **2018**, *41*, 2987–2999. [CrossRef]

51. Colli, P.; Gilardi, G.; Sprekels, J. Optimal velocity control of a viscous Cahn–Hilliard system with convection and dynamic boundary conditions, *SIAM J. Control Optim.* **2018**, *56*, 1665–1691. [CrossRef]
52. Colli, P.; Gilardi, G.; Marinoschi, G.; Rocca, E. Sliding mode control for a phase field system related to tumor growth. *Appl. Math. Optim.* **2019**, *79*, 647–670. [CrossRef]
53. Ebenbeck, M.; Knopf, P. Optimal medication for tumors modeled by a Cahn–Hilliard–Brinkman equation. *arXiv* **2018**, arXiv:1811.07783.
54. Ebenbeck, M.; Knopf, P. Optimal control theory and advanced optimality conditions for a diffuse interface model of tumor growth. *arXiv* **2019**, arXiv:1903.00333.
55. Garcke, H.; Lam, K.F.; Rocca, E. Optimal control of treatment time in a diffuse interface model of tumor growth. *Appl. Math. Optim.* **2018**, *78*, 495–544. [CrossRef]
56. Signori, A. Optimal distributed control of an extended model of tumor growth with logarithmic potential. *Appl. Math. Optim.* **2019**, 1–33. [CrossRef]
57. Signori, A. Optimality conditions for an extended tumor growth model with double obstacle potential via deep quench approach. *arXiv* **2018**, arXiv:1811.08626.
58. Signori, A. Optimal treatment for a phase field system of Cahn–Hilliard type modeling tumor growth by asymptotic scheme. *arXiv* **2019**, arXiv:1902.01079.
59. Signori, A. Vanishing parameter for an optimal control problem modeling tumor growth. *arXiv* **2019**, arXiv:1903.04930.
60. Lions, J.-L. *Quelques Méthodes de Résolution des Problèmes aux Limites non Linéaires*; Gauthier-Villars: Paris, France, 1969.

© 2019 by the authors. Licensee MDPI, Basel, Switzerland. This article is an open access article distributed under the terms and conditions of the Creative Commons Attribution (CC BY) license (http://creativecommons.org/licenses/by/4.0/).

Article

Analysis of a Model for Coronavirus Spread

Youcef Belgaid [1], Mohamed Helal [1] and Ezio Venturino [2,*,†]

[1] Laboratory of Biomathematics, Univ. Sidi Bel Abbes, P.B. 89, Sidi Bel Abbes 22000, Algeria; youcef.belgaid@univ-sba.dz (Y.B.); mohamed.helal@univ-sba.dz (M.H.)
[2] Dipartimento di Matematica "Giuseppe Peano", Università di Torino, via Carlo Alberto 10, 10123 Torino, Italy
* Correspondence: ezio.venturino@unito.it; Tel.: +39-011-670-2833
† Member of the INdAM research group GNCS.

Received: 3 April 2020; Accepted: 15 May 2020; Published: 19 May 2020

Abstract: The spread of epidemics has always threatened humanity. In the present circumstance of the Coronavirus pandemic, a mathematical model is considered. It is formulated via a compartmental dynamical system. Its equilibria are investigated for local stability. Global stability is established for the disease-free point. The allowed steady states are an unlikely symptomatic-infected-free point, which must still be considered endemic due to the presence of asymptomatic individuals; and the disease-free and the full endemic equilibria. A transcritical bifurcation is shown to exist among them, preventing bistability. The disease basic reproduction number is calculated. Simulations show that contact restrictive measures are able to delay the epidemic's outbreak, if taken at a very early stage. However, if lifted too early, they could become ineffective. In particular, an intermittent lock-down policy could be implemented, with the advantage of spreading the epidemics over a longer timespan, thereby reducing the sudden burden on hospitals.

Keywords: dynamical systems; compartment model; epidemics; basic reproduction number; stability

MSC: 92D30; 92D25

---

1. Introduction

The coronavirus infection has been spreading for a few months. Authorities in several countries have relied on scientific tools for fighting the epidemics. With the lack of a vaccine, distancing methods have been forced on populations to avoid the transmission by direct contact. In laboratories, possible vaccines are being developed, but at the moment they are still at the experimental stage [1]. Meanwhile several models, mathematical, statistical and computer-science-based, are being developed to study the disease and contribute to fighting it.

Models for the spread of epidemics are classic, and an excellent presentation is [2]. In general, the total population is partitioned into at least two classes, susceptibles and infectives, with migrations from the former to the latter by disease propagation through direct or indirect contact, if the disease is transmissible. Additionally, if it can be overcome but causes relapses, the infected can become susceptible again, after maybe going through an intermediate class of being recovered. More sophisticated versions include quarantined and exposed individuals. Some of these classes will be considered also in the present study and illustrated in detail before the model formulation process.

In [3] the disease evolution forecast in several of the most affected countries is attempted, using for that purpose parameter estimation techniques to calibrate the model. The involved compartments are susceptibles, asymptomatic individuals and symptomatic ones, which in turn are partitioned into reported and unreported cases. In [4] a simple SIRI model is considered, in which the recovered could still contribute to the disease spreading. The model is then extended to account for a possible vaccine, which, unfortunately, at present is not yet available, although several laboratories worldwide are trying to develop and test it, as mentioned above. In [5] a dynamic model for the diffusion of Covid-19 has been proposed. The transmission network is made by the bats–hosts–reservoir–people compartments; compare also [1]. As it amounts to about 14 differential equations and 25 parameters, it is rather complex. From it, the authors have obtained a simplified version, consisting of six compartments and 13 parameters. Then, the disease basic reproduction number has been calculated.

Our aim here is the mathematical analysis of a slightly modified version of the simpler model in [5]. The most important change accounts for the fact that asymptomatic people may indeed turn into fully symptomatic and infectious individuals. This feature also distinguishes the system introduced here from the one studied in [6], which, however, contains more compartments. The main aim of that study is the forecast of the epidemic spread in various cities in China, considering, additionally, weather data, which finally indicate that higher humidity favors the containment of a coronavirus epidemic. Our focus in the first part of this investigation is the theoretical analysis of the proposed system, and then we perform some preliminary simulations with realistic parameter values. More extended simulations will be devoted to a subsequent study.

The analysis of dynamical systems usually considers the possible equilibria that can be attained, assessing their feasibility and stability, and possible connections between them. For more details on these issues we refer the reader to classical texts, such as [7–9].

The paper is organized as follows. The main findings are outlined in the next section, which also discusses the results of numerical simulations. Section 3 contains an evaluation of their implications under various distancing policies. We formulate the model in Section 4, where we also analyze it mathematically, showing boundedness of the trajectories, establishing an expression for the disease basic reproduction number, finding its equilibria and assessing their local stability, and global stability is established just for the disease-free equilibrium. The section ends with the details on the numerical simulations.

2. Results

2.1. Theoretical Findings

The main analytical findings of this investigation are summarized in Tables 1 and 2. The expressions of B_T, C_T, H_T, D_T and R_0 are given by Equations (3) and (6).

The model (1) allows only three possible equilibria; the disease-free state C_0, where only susceptibles thrive; an equilibrium without symptomatic infected, which occurs only for a very particular case, when the exposed individuals become all asymptomatic infected; and finally, the endemic equilibrium C^*. All these equilibria are locally asymptotically stable, if suitable, rather complicated conditions, hold. Among the endemic and the disease-free equilibrium bistability is impossible, since they are related via a transcritical bifurcation.

Table 1. Equilibria of the model (1) and their feasibility conditions.

Equilibrium	Populations	Feasibility
$C_0 = (S_0, 0, 0, 0, 0)$	$S_0 = \dfrac{\Lambda}{d_p}$	–
$C_I = (S_I, E_I, 0, A_I, R_I)$	$S_I = \dfrac{\Lambda - B_T E_I}{d_p}$	
	$E_I = \dfrac{1}{B_T}\left(\Lambda - \dfrac{d_p B_T C_T H_T}{\beta_I \omega'_p D_T}\right)$	$\Lambda > \dfrac{d_p B_T C_T H_T}{\beta_I \omega'_p D_T}$
	$A_I = \left(\dfrac{\omega'_p}{H_T}\right) E_I$	(resp. $R_0 > 1$)
	$R_I = \dfrac{\gamma'_p}{d_p}\left(\dfrac{\omega'_p}{H_T}\right) E_I$	$\alpha = 1$ and $\xi = 0$

Table 1. *Cont.*

Equilibrium	Populations	Feasibility
$C^* = (S^*, E^*, I^*, A^*, R^*)$	$S^* = \dfrac{\Lambda - B_T E^*}{d_p}$	$\Lambda > \dfrac{d_p B_T C_T H_T}{\beta_I \left[(1-\alpha)\omega_p H_T + \alpha\omega'_p D_T\right]}$
	$E^* = \dfrac{1}{B_T}\left(\Lambda - \dfrac{d_p B_T C_T H_T}{\beta_I\left[(1-\alpha)\omega_p H_T + \alpha\omega'_p D_T\right]}\right)$	(resp. $R_0 > 1$)
	$I^* = \left(\dfrac{(1-\alpha)\omega_p H_T + \alpha\omega'_p \xi}{C_T H_T}\right) E^*$	$\alpha \neq 1$ or $\xi \neq 0$
	$A^* = \left(\dfrac{\alpha\omega'_p}{H_T}\right) E^*$	
	$R^* = \left[\dfrac{\gamma_p}{d_p}\left(\dfrac{(1-\alpha)\omega_p H_T + \alpha\omega'_p \xi}{C_T H_T}\right) + \dfrac{\gamma'_p}{d_p}\left(\dfrac{\alpha\omega'_p}{H_T}\right)\right] E^*$	

Table 2. Stability conditions of the equilibria of the model (1).

Point	Coefficients	Stability
C_0	$a_2 = B_T + C_T + H_T$	
	$a_1 = H_T[B_T + C_T] + B_T C_T - \beta_I S_0((1-\alpha)\omega_p + k\alpha\omega'_p)$	$\Lambda < \dfrac{d_p B_T C_T H_T}{\beta_I\left[(1-\alpha)\omega_p H_T + \alpha\omega'_p D_T\right]}$
	$a_0 = B_T C_T H_T - \beta_I S_0\left((1-\alpha)\omega_p H_T + \alpha\omega'_p D_T\right)$	(resp. $R_0 < 1$)
C_I	$b_3 = \beta_I k A_I + d_p + H_T + B_T + C_T$,	
	$b_2 = [\beta_I k A_I + d_p](H_T + B_T + C_T)$	$\Lambda > \dfrac{d_p B_T C_T H_T}{\beta_I \omega'_p D_T}$
	$+ B_T C_T + H_T(B_T + C_T) - \omega'_p k \beta_I S_I$	
	$b_1 = [\beta_I k A_I + d_p][B_T C_T + H_T(B_T + C_T)]$	(resp. $R_0 > 1$)
	$+ H_T B_T C_T - \omega'_p(kd_p + D_T)\beta_I S_I$	
	$b_0 = [\beta_I k A_I + d_p]H_T B_T C_T - d_p \omega'_p D_T \beta_I S^*$	$\alpha = 1$ and $\xi = 0$
C^*	$c_3 = \beta_I(I^* + kA^*) + d_p + H_T + B_T + C_T$,	
	$c_2 = [\beta_I(I^* + kA^*) + d_p](H_T + B_T + C_T)$	$\Lambda > \dfrac{d_p B_T C_T H_T}{\beta_I\left[(1-\alpha)\omega_p H_T + \alpha\omega'_p D_T\right]}$
	$+ B_T C_T + H_T(B_T + C_T) - [\alpha\omega'_p k + (1-\alpha)\omega_p]\beta_I S^*$	(resp. $R_0 > 1$)
	$c_1 = [\beta_I(I^* + kA^*) + d_p][B_T C_T + H_T(B_T + C_T)] + H_T B_T C_T$	$\alpha \neq 1$ or $\xi \neq 0$
	$- [\alpha\omega'_p(kd_p + D_T) + (1-\alpha)\omega_p(d_p + H_T)]\beta_I S^*$	
	$c_0 = [\beta_I(I^* + kA^*) + d_p]H_T B_T C_T$	
	$- d_p[\alpha\omega'_p D_T + (1-\alpha)\omega_p H_T]\beta_I S^*$	

2.2. Simulations Results

We have performed some simulations with the parameter values listed in Table 3, to simulate various implementations of the distancing policy, which actually is in current use in several countries. The simulations may not be fully realistic, but our point is to investigate their qualitative behavior, not to give quantitative forecasts.

Table 3. Parameters values.

Parameter	Value	Parameter	Value
Λ	500	d_p	8.2×10^{-6}
γ_p	1.764	γ'_p	0.6024
ξ	0.1	k	$0.1 \in [0:005; 0:2]$
μ	0.001	α	$0.15 \in [0.01, 0.3]$
ω_p	0.1	ω'_p	0.1
ν	0		

We look at the influence that the time of starting the restrictive measures has on the disease spread, while keeping fixed the time of their lifting. We next investigate the effect of the time at which the restricting measures are lifted.

Now comparing the results for the start of implementation at $t_1 = 1$ and $t_1 = 10$, and ending them at the same time, it is seen that the earlier the measures are taken, the better it is, because the epidemic's outbreak is kept in check. In Figure 1 the epidemic outbreak starts around time 30, immediately after lifting the restrictions, while in Figure 2 the initial surge before the measures are implemented is damped by their implementation, and after their lifting the outbreak occurs. Both figures use $t_2 = 30$. The same result is seen using $t_2 = 90$ as the time for removing the restrictions; compare Figures 3 and 4. In Figure 3 nothing apparently happens until time 100 because of the restrictions. When they are lifted, the epidemic spreads. In Figure 4 there is a small peak at the onset of the contagion, immediately curbed by the containment measures, lasting as long as they are in use. In spite of their longer implementation, the outbreak occurs nevertheless with the peak at the same time as in Figure 3.

The investigation of different timings for introducing and relaxing the distancing measures shows that a late implementation has no effect, as the peak of the epidemic occurs and then these measures are ineffective, independently of how long they are implemented. An earlier implementation followed by their subsequent lifting leads to a secondary peak at some time later, the occurrence of which seems to be related to the time for which the measures are implemented; the longer the latter, the longer the delay in the secondary outbreak. However, the number of affected people remains the same.

Unfortunately, the result of the simulations indicates that essentially the whole population gets affected by the disease. Only the timings differ, if distancing measures are taken.

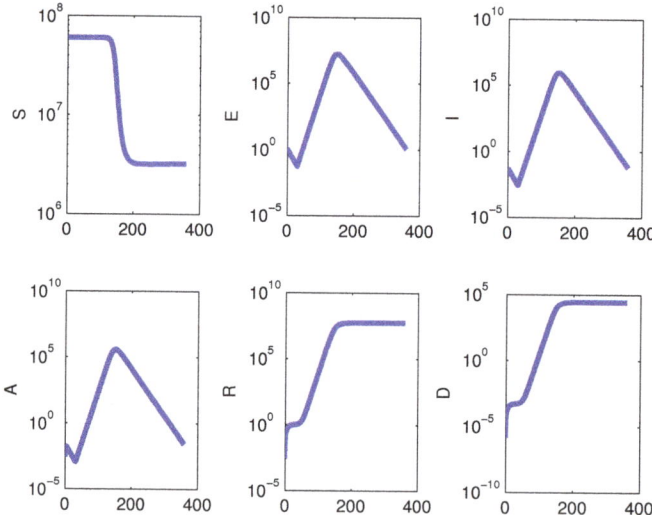

Figure 1. Using a semilogarithmic scale for the vertical axis, we show the results of starting the restrictions at time $t_1 = 1$, using $\beta_I = 10^{-10}$ and lifting them at time $t_2 = 30$, returning to $\beta_I = 10^{-7}$ one month later, over a one year timespan for the model with demographics (1). Left to right and top to bottom, the subpopulations are: S, E, I, A, R and D, the disease-related deceased class.

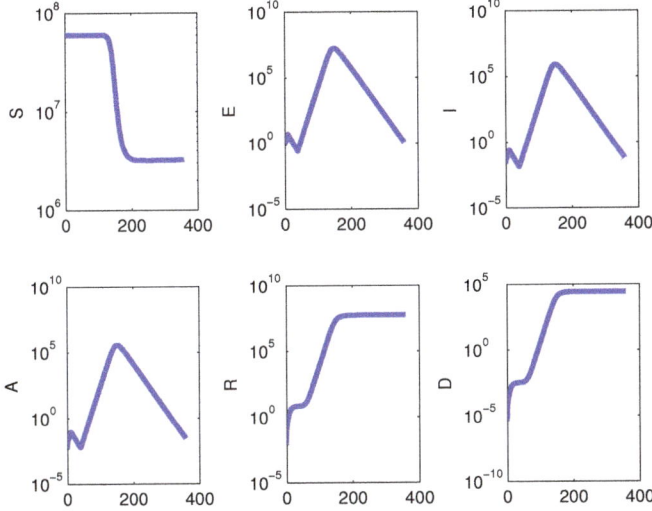

Figure 2. Using a semilogarithmic scale for the vertical axis, we show the results of starting the restrictions at time $t_1 = 10$, using $\beta_I = 10^{-10}$ and lifting them at time $t_2 = 30$, returning to $\beta_I = 10^{-7}$ one month later, over a one year timespan for the model with demographics (1). Left to right and top to bottom, the subpopulations are: S, E, I, A, R and D, the disease-related deceased class.

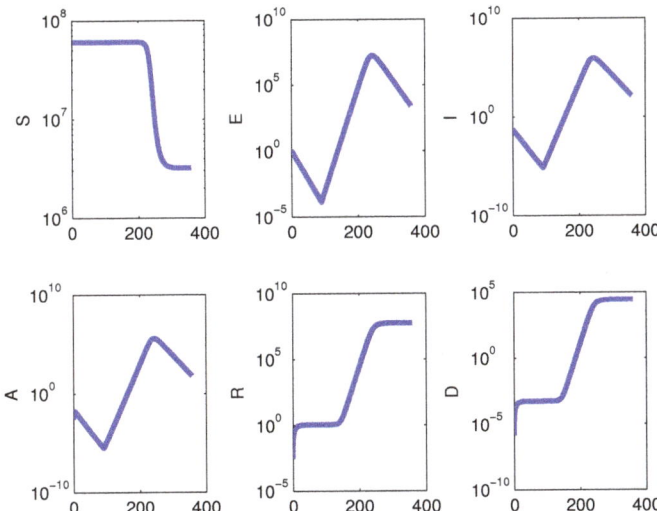

Figure 3. Using a semilogarithmic scale for the vertical axis, we show the results of starting the restrictions at time $t_1 = 1$, using $\beta_I = 10^{-10}$ and lifting them at time $t_2 = 90$, returning to $\beta_I = 10^{-7}$ three months later, over a one year timespan for the model with demographics (1). Left to right and top to bottom, the subpopulations are: S, E, I, A, R and D, the disease-related deceased class.

Thus, if the measures are implemented too late, independently of the time at which they are removed, the outbreak occurs and their subsequent application becomes, therefore, irrelevant, as it cannot be kept in check any longer; compare Figures 2 and 4. On the other hand, by implementing them at the early stages of the contagion process, the outbreak can be delayed, as long as these measures are implemented, as can be seen from Figures 1 and 3. If they are lifted, the final results of the epidemic's outbreak are essentially the same as if they were not at all implemented, in terms of the number of people being affected by the disease and with possible ultimate fatal consequences; compare the peaks of all the infected classes in Figures 1–5 also with the results in Figure 6 where no measures are taken to prevent the epidemic from spreading.

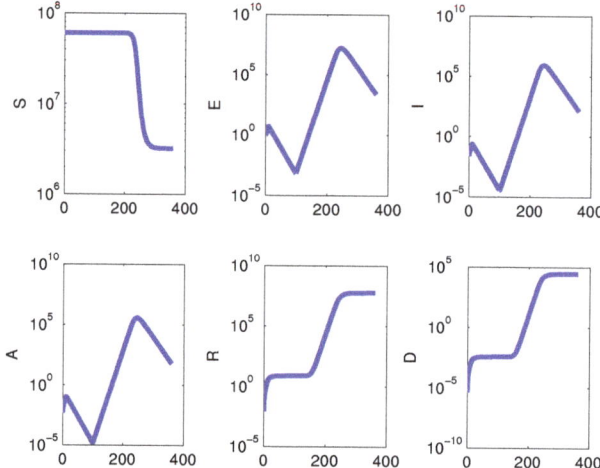

Figure 4. Using a semilogarithmic scale for the vertical axis, we show the results of starting the restrictions at time $t_1 = 10$, using $\beta_I = 10^{-10}$ and lifting them at time $t_2 = 90$, returning to $\beta_I = 10^{-7}$ three months later, over a one year timespan for the model with demographics (1). Left to right and top to bottom, the subpopulations are: S, E, I, A, R and D, the disease-related deceased class.

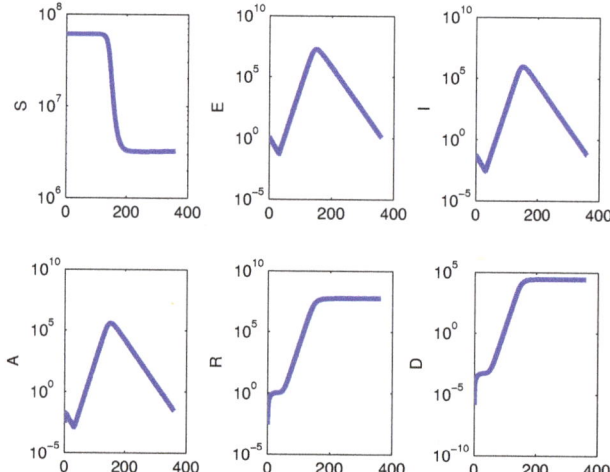

Figure 5. Using a semilogarithmic scale for the vertical axis, we show the results of absolute isolation, starting at time $t_1 = 1$ setting $\beta_I = 0$ and lifting it at time $t_2 = 30$, returning to $\beta_I = 10^{-7}$ one month later, over a one year timespan for the model with demographics (1). Left to right and top to bottom, the subpopulations are: S, E, I, A, R and D, the disease-related deceased class.

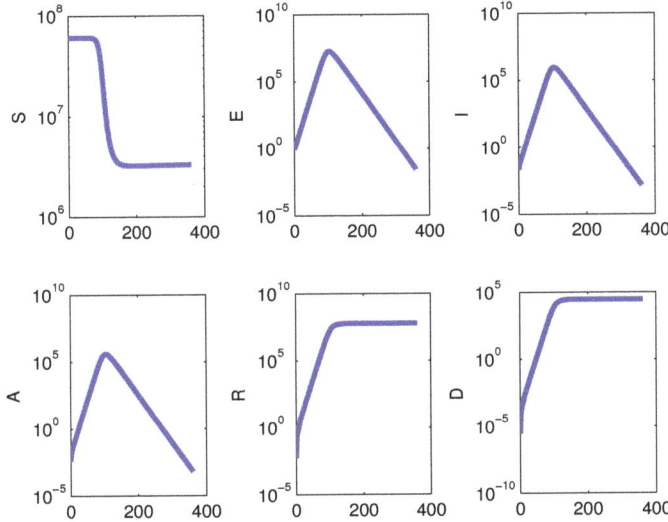

Figure 6. The epidemic's effect on the population in the absence of measures for $\beta_I = 10^{-7}$, on a semilogarithmic scale, over a period of one year. Left to right and top to bottom: S, E, I, A, R and D, the disease-related deceased class.

An intermittent lock-down policy, simulated as an alternative way of coping with the outbreak, might be important to render the burden on hospitalizations smaller, as it tends to spread the epidemic over a longer timespan.

For the particular situation in Italy, note that patient number 1 was diagnosed on 21 February, and the distancing measures in the area were in place starting the following days up to about two weeks later, and then extended to the whole country. Incidentally, patient number 0, the initial carrier of the disease, has never been identified, although there are some speculations. However, in the current news, it is reported that the virus was already circulating yet not known of in Northern Italy in January, which means that additional time had elapsed before the restrictions were applied.

Thus, apparently, these results are negative as for the possibility of containing the spread in the long run, in line with what is hinted in [10], with the exception of the intermittent distancing measures policy, which may spread the epidemic's effects over longer timespans. However, there are some assumptions in the model that make it too crude, so that we plan a deeper subsequent study. In particular, here the results depend on homogeneous mixing, which for a large country is hardly the case. Secondly, this is a continuous model, for which the compartments are depleted only asymptotically. Thus it is not possible to prevent the class of infectives from vanishing in finite time, so that even a small residual in it would start the epidemic's outbreak again. Therefore the somewhat negative results obtained might hopefully be better off in practice. Suitable modifications of the model along these lines will be the subject of a further investigation.

3. Discussion

We have investigated a simple model for the coronavirus pandemic. The steady states, apart from a symptomatic-infected-free point, which is unlikely to exist, are the disease-free equilibrium and the endemic state. The model differs from other current models that are being studied for a few features. From the simplified model that appears in [5], because our formulation contains less equations, it does not consider the viruses compartment, and above all, we allow disease-related mortality, which apparently is missing in the cited paper. Furthermore, we allow the progression of asymptomatic individuals to the class of fully symptomatic. This feature certainly distinguishes it also from [6], where asymptomatics recover or become diagnosed with the disease, but do not spread it any longer. In the present situation in Italy our assumption is very realistic.

There is no possibility of bistability in our situation, as the two fully meaningful equilibria are related to each other via a transcritical bifurcation. The disease-free equilibrium is also globally asymptotically stable, if it is locally asymptotically stable. An expression for the basic reproduction number is established, with a possibly realistic numerical value [11,12].

The simulations show that containment measures could be effective in delaying the epidemic's outbreaks if taken at a very early stage, but when lifted the outbreaks would occur anyway and affect almost the whole population. However, this last statement should be mitigated by the drawbacks inherent in the model's assumptions, as mentioned in the previous section, thereby leaving hope that in practice it will not occur, if the measures are properly implemented.

We next discuss in detail the various different restriction policies that we have simulated.

3.1. Epidemic with a Lock-Down

In this case, in particular, assuming for the disease transmission coefficient the reference value $\beta_I = 10^{-7}$, we reduce it to $\beta_I = 10^{-10}$ during the interval $[t_1, t_1 + t_2]$ and reinstate the standard value afterwards; we monitor the epidemic's evolution over six months. Figures 1–5 show the results of different choices for t_1 and t_2. Containment measures are effective as long as they are implemented, if they are taken early enough, before the epidemic attains its peak.

Since reducing the transmission by one order of magnitude means that to infect a susceptible with rate β_I, it is necessary for only one infected; with $\beta_I/10$, 10 infected would be necessary. Thus since the lock-down is not perfect, as for instance, some essential activities like food production are still going on, a hypothetical reasonable estimate for the contact reduction is three orders of magnitude. A comparison with a different, milder reduction, $\beta_I = 10^{-8}$ is made, showing essentially no difference in the results, see Figure 7.

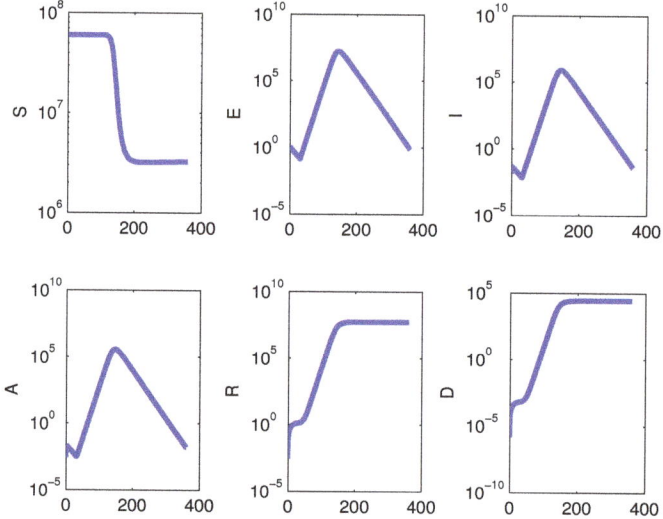

Figure 7. On a semilogarithmic scale, the total populations with the lock-down policy, implemented from time 1 up to time 30, with the milder reduced contact rate $\beta_I = 10^{-8}$, after which $\beta_I = 10^{-7}$ resumes. The simulation runs over a one year timespan for the simplified model (1). Left to right and top to bottom, the subpopulations are: S, E, I, A, R and D, the disease-related deceased class.

3.2. Epidemic with Total Isolation

We changed also the policy to an improbable absolute confinement of every individual in the population, reducing the transmission to exactly zero. The results show no change with respect to those of the lock-down policy. We report only Figure 5, which is identical to Figure 1. The same occurs in the cases contemplated by Figures 2–4.

3.3. The Simplified No-Demographics Model

We repeated the simulations for the model (1) in which we set $\Lambda = d_p = 0$. In the simulations we observed some small changes in the susceptibles behavior, with respect to the full model with vital dynamics. Figures 8 and 9 are the counterparts of the Figures 1 and 2. The ultimate impact of the epidemic is essentially the same; compare in particular, the curves of recovered and deceased. For the total isolation case, Figure 10 shows the same features; compare it with Figure 5. Similar considerations hold for the various remaining cases, and therefore, the pictures are not reported.

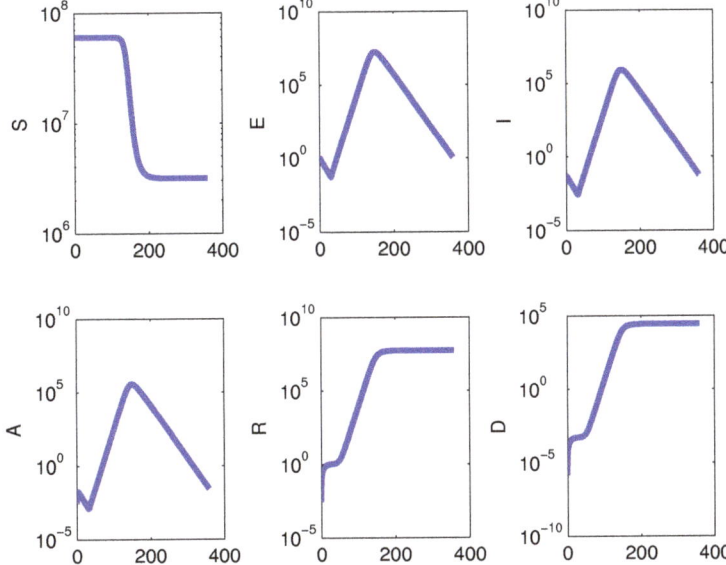

Figure 8. Using a semilogarithmic scale for the vertical axis, we show the results of starting the restrictions at time $t_1 = 1$, setting $\beta_I = 10^{-10}$ and lifting them at time $t_2 = 30$, returning to $\beta_I = 10^{-7}$ one month later, over a one year timespan for the simplified model with no demographics (1) where we take $\Lambda = 0$, $d_p = 0$. Left to right and top to bottom, the subpopulations are: S, E, I, A, R and D, the disease-related deceased class.

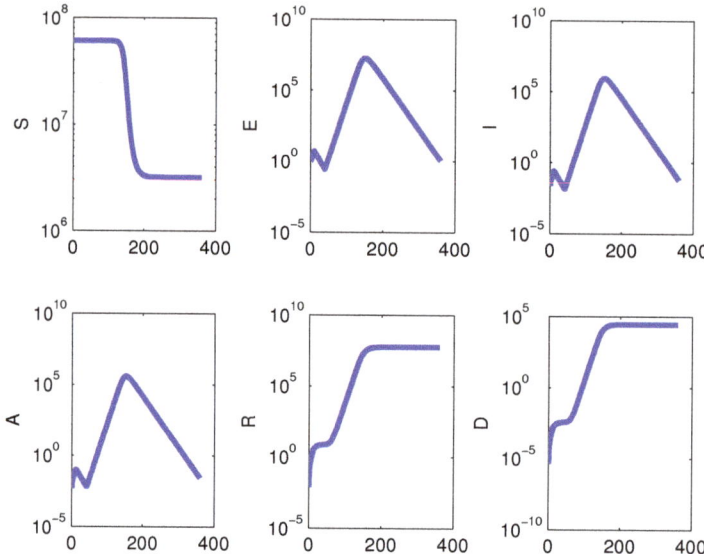

Figure 9. Using a semilogarithmic scale for the vertical axis, we show the results of starting the restrictions at time $t_1 = 10$, setting $\beta_I = 10^{-10}$ and lifting them at time $t_2 = 30$, returning to $\beta_I = 10^{-7}$ one months later, over a one year timespan for the simplified model with no demographics (1) where we take $\Lambda = 0$, $d_p = 0$. Left to right and top to bottom, the subpopulations are: S, E, I, A, R and D, the disease-related deceased class.

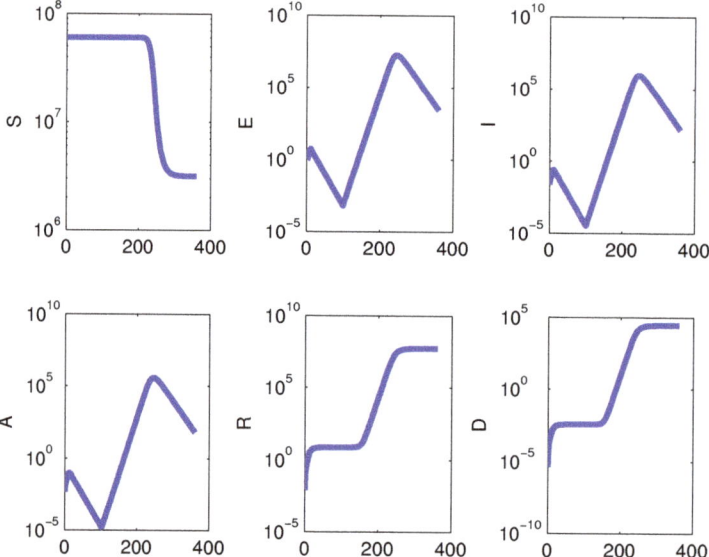

Figure 10. Using a semilogarithmic scale for the vertical axis, we show the results of absolute isolation, $\beta_I = 0$ starting at time $t_1 = 10$, setting $\beta_I = 0$ and lifting it at time $t_2 = 90$, returning to $\beta_I = 10^{-7}$ three months later, over a one year timespan for the model with no demographics (1) where $\Lambda = d_p = 0$. Left to right and top to bottom, the subpopulations are: S, E, I, A, R and D, the disease-related deceased class.

3.4. Investigation of Different Timings for Restrictions' Introduction and Lifting

A further study has been carried out to assess the impact of the time until taking action on the containment measures. All the possible different combinations of simple restriction or total isolation as well as the presence or the absence of demographic effects give essentially the same results. Therefore we present only the results for some selected alternatives, giving the plots in semilogarithmic or total population values, but stressing that for the options not considered, the figures would be the same.

In case the first restriction measure is taken too late, specifically at time $t = 120$, and followed by lifting it either one month or three months later, the epidemic occurs and the measures have no effect whatsoever; see Figure 11, where measures are kept for three months.

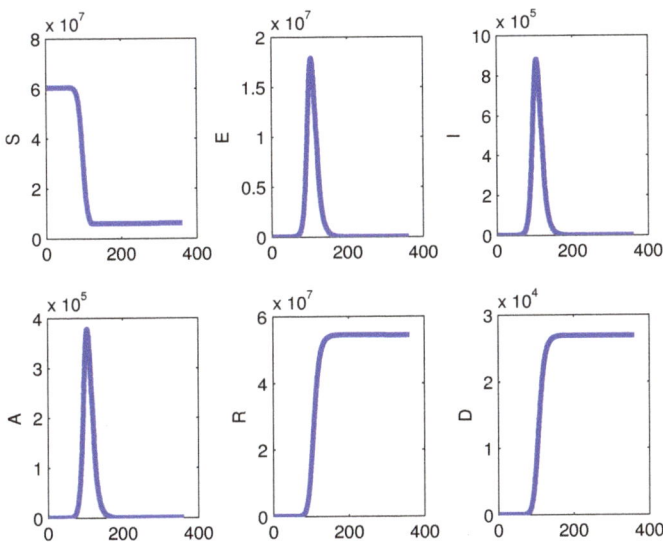

Figure 11. The total populations with the lock-down policy, implemented from time 120 up to time 210, with the reduced contact rate $\beta_I = 10^{-10}$, after which $\beta_I = 10^{-7}$ resumes. The simulation runs over a one year timespan for the model (1). Left to right and top to bottom, the subpopulations are: S, E, I, A, R and D, the disease-related deceased class.

Beginning the restrictions after three months from the start of the epidemic and removing them one month afterwards, causes a second peak about two months later; i.e., six months after the onset of the disease spreading (Figure 12), with a higher number of affected individuals. If instead the lock-down is implemented for three months, the second peak is delayed further, occurring about three months later, Figure 13. Although the pictures are shown on different population scales, absolute values and semilogarithmic, a comparison of the heights of the peaks for the various types of infected subpopulations indicates no difference. Hence, these policies cannot significantly influence the number of people ultimately affected by the disease.

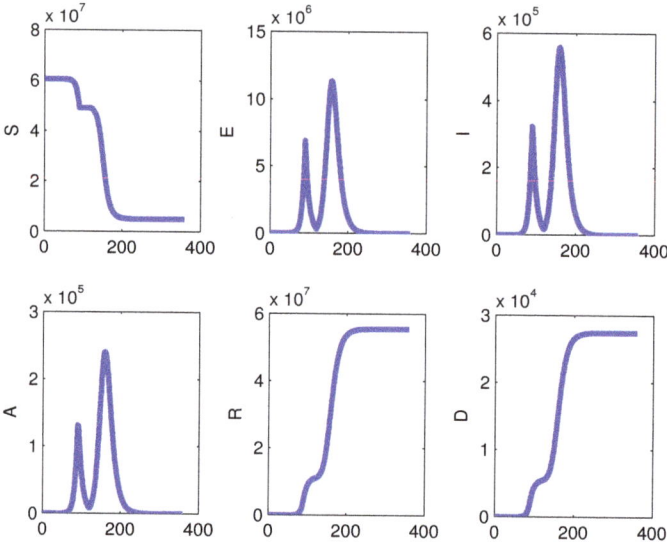

Figure 12. The total populations with the lock-down policy, implemented from time 90 up to time 120, with the total isolation policy $\beta_I = 0$, after which $\beta_I = 10^{-7}$ resumes. The simulation runs over a one year timespan for the simplified model (1) with no demographics, $\Lambda = d_p = 0$. Left to right and top to bottom, the subpopulations are: S, E, I, A, R and D, the disease-related deceased class.

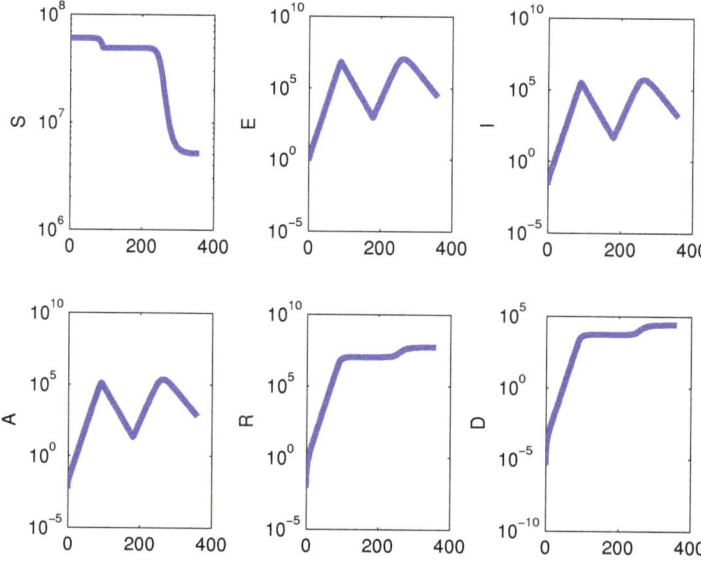

Figure 13. The semilogarithmic plot of the epidemic's spread with the lock-down policy, implemented from time 90 up to time 180, with the reduced contact rate $\beta_I = 10^{-10}$, after which $\beta_I = 10^{-7}$ resumes. The simulation runs over a one year timespan for the model (1) with demographics. Left to right and top to bottom, the subpopulations are: S, E, I, A, R and D, the disease-related deceased class.

3.5. The Intermittent Lock-Down Policy

We finally simulated a policy that attempts to assess the number of infectives at regular times, with a of period one week. If they exceed a threshold, taken to be 10, the lock-down is implemented for a week, and then lifted. Figures 14 and 15 show the results for the case with vital dynamics and in the case of $\Lambda = d_p = 0$. Note that susceptibles in both cases are at a constant value, the vertical scale being extremely small. The infected are kept below the threshold, and the periodic recurrences of the epidemic somewhat change its final impact, as the curves of recovered are reduced by about two orders of magnitude, and above all, the ones of the deceased decrease by about four orders, with respect to the ones found with the one-time lock-down policy. The other relevant change is that here the phenomenon is observed over a longer timespan. Thus the cumulative effects are spread out over a much longer time. This will have some importance to lessening the burden on hospitals. Figure 16 shows the results if the check policy starts immediately at time 1 rather than after a week.

Comparing the population values with the intermittent policy with the one time lock-down, done early enough and implemented for one month, the final outcomes are milder than the latter. Thus the intermittency allows the control of the outbreaks. Susceptibles are almost depleted in the one-time policy; with the intermittent one, however, they are essentially spared from getting the disease; compare Figures 17 and 18.

The intermittency has also been checked with different time intervals. Comparing Figures 19–22, it is seen that the more frequent the checks are implemented, the lower are the peaks in the exposed class, which in turn leads to a smaller cumulative number of recovered and fatalities, at least comparing the policies for the one- and two-weeks alternatives, Figures 19 and 20. For the longer intervals between the checks, again the peaks are higher, the longer the timespan, but it is observed that as time elapses, their heights tend to decrease; see Figures 21 and 22.

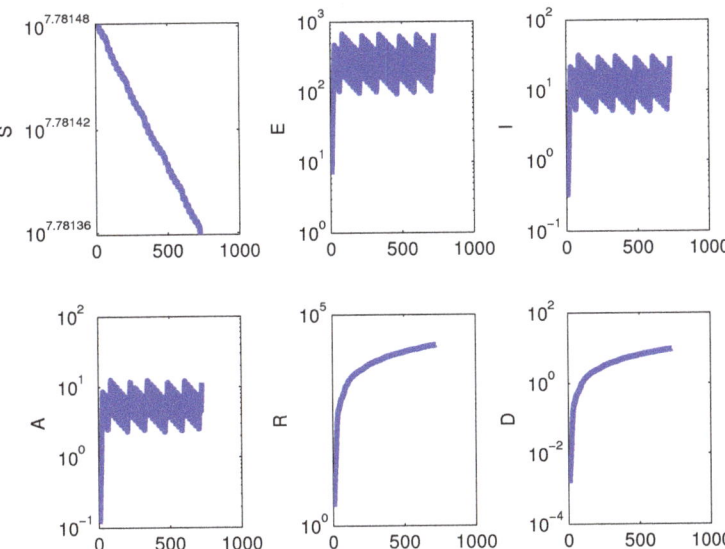

Figure 14. Using a semilogarithmic scale for the vertical axis, we show the results of the intermittent lock-down policy. Here the population is checked every week, starting after a week. If the number of infected is above a small threshold (here taken to be 10) the reduced contact rate $\beta_I = 10^{-10}$ is resumed for a week. The simulation runs over two years timespan for the model with demographics (1). Left to right and top to bottom, the subpopulations are: S, E, I, A, R and D, the disease-related deceased class.

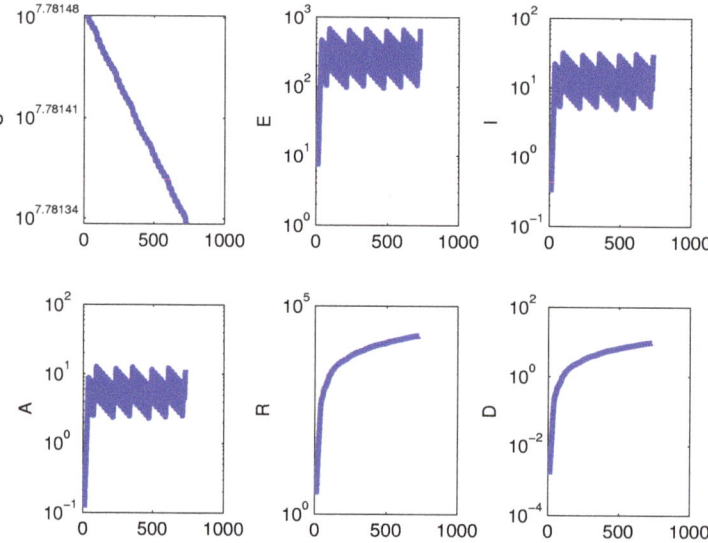

Figure 15. Using a semilogarithmic scale for the vertical axis, we show the results of the intermittent lock-down policy. Here the population is checked every week, starting after a week. If the number of infected is above a small threshold (here taken to be 10) the reduced contact rate $\beta_I = 10^{-10}$ is resumed for a week. The simulation runs over two years timespan for the simplified model with no demographics (1) where we take $\Lambda = 0$, $d_p = 0$. Left to right and top to bottom, the subpopulations are: S, E, I, A, R and D, the disease-related deceased class.

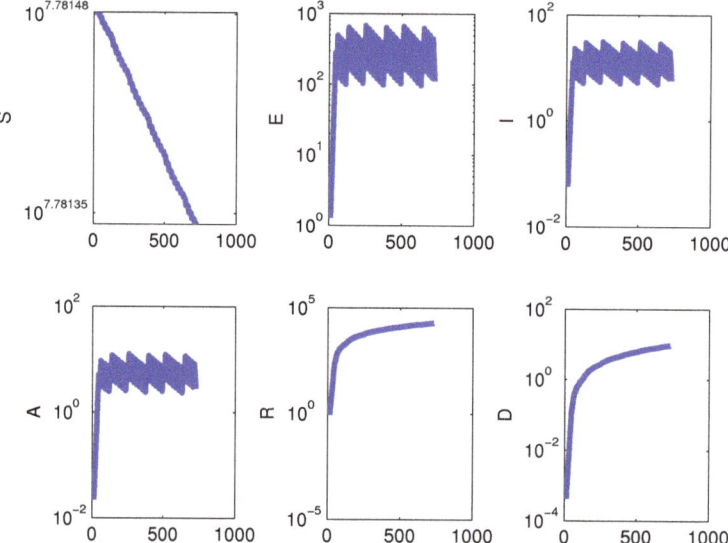

Figure 16. Using a semilogarithmic scale for the vertical axis, we show the results of the intermittent lock-down policy, implemented from time 1. Here the population is checked every week. If the number of infected is above a small threshold (here taken to be 10) the reduced contact rate $\beta_I = 10^{-10}$ is resumed for a week. The simulation runs over two years timespan for the simplified model with no demographics (1) where we take $\Lambda = 0$, $d_p = 0$. Left to right and top to bottom, the subpopulations are: S, E, I, A, R and D, the disease-related deceased class.

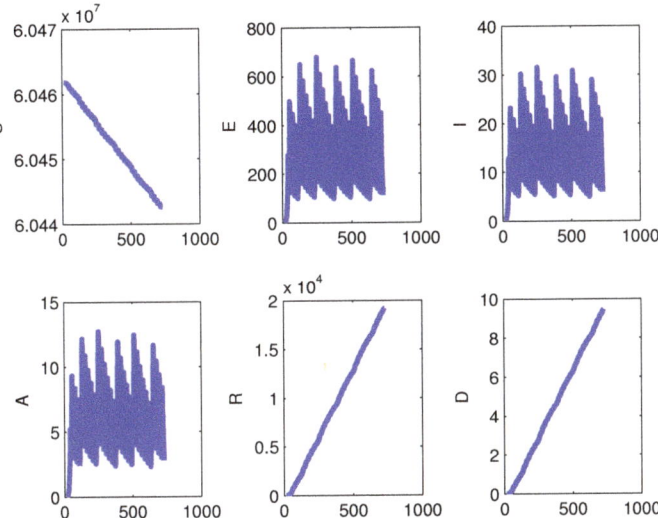

Figure 17. The total populations with the intermittent lock-down policy, implemented from time 1. Here the population is checked every week. If the number of infected is above a small threshold (here taken to be 10) the reduced contact rate $\beta_I = 10^{-10}$ is resumed for a week. The simulation runs over two years timespan for the simplified model with no demographics (1) where we take $\Lambda = 0$, $d_p = 0$. Left to right and top to bottom, the subpopulations are: S, E, I, A, R and D, the disease-related deceased class.

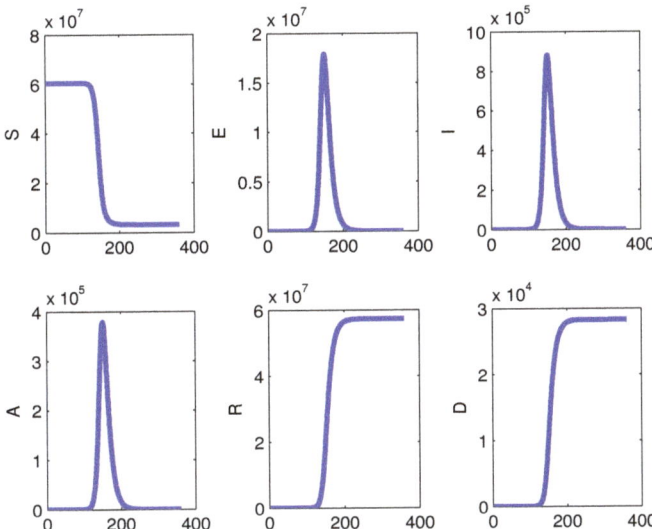

Figure 18. The total populations with the lock-down policy, implemented from time 1 up to time 30, with the reduced contact rate $\beta_I = 10^{-10}$, after which $\beta_I = 10^{-7}$ resumes. The simulation runs over a one year timespan for the simplified model with no demographics (1) where we take $\Lambda = 0$, $d_p = 0$. Left to right and top to bottom, the subpopulations are: S, E, I, A, R and D, the disease-related deceased class.

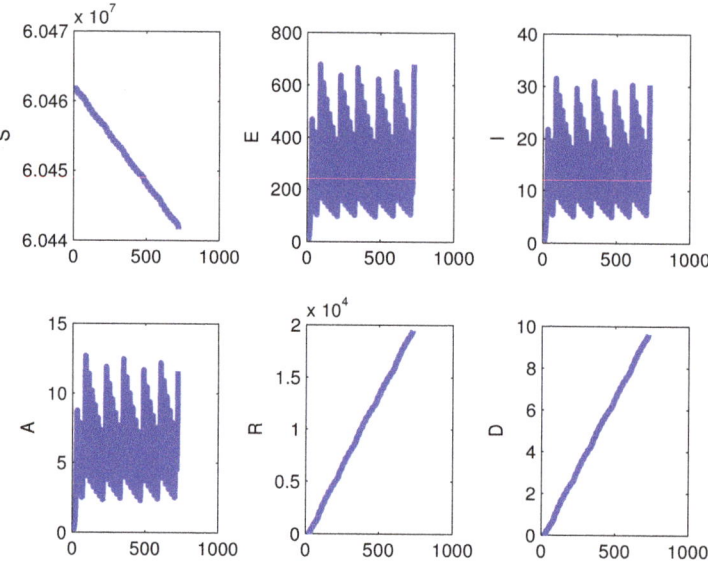

Figure 19. The population values with the lock-down policy, implemented after the first week with the reduced contact rate $\beta_I = 10^{-10}$, after which $\beta_I = 10^{-7}$ resumes. The check for possible repeated implementation is implemented every week afterwards. The simulation runs over a two year timespan for the model (1) with no demographics, i.e., $\Lambda = d_p = 0$. Left to right and top to bottom, the subpopulations are: S, E, I, A, R and D, the disease-related deceased class.

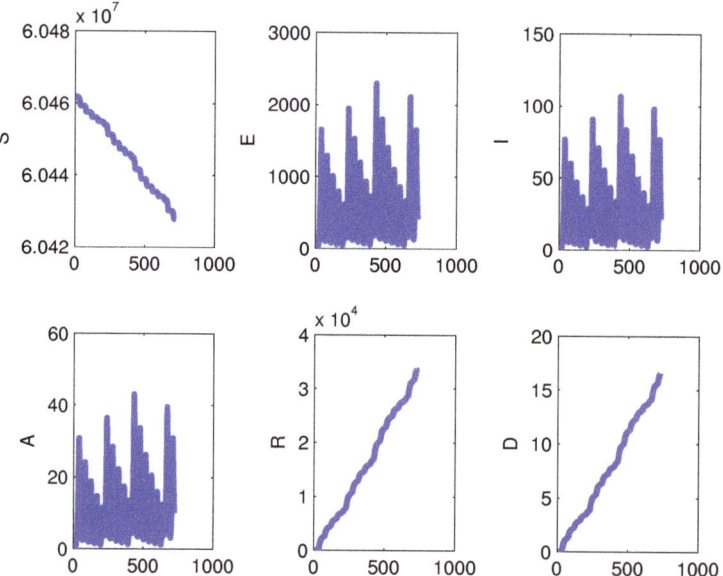

Figure 20. The population values with the lock-down policy, implemented after the first two weeks with the reduced contact rate $\beta_I = 10^{-10}$, after which $\beta_I = 10^{-7}$ resumes. The check for possible repeated implementation is implemented every two weeks afterwards. The simulation runs over a two year timespan for the model (1) with no demographics, i.e., $\Lambda = d_p = 0$. Left to right and top to bottom, the subpopulations are: S, E, I, A, R and D, the disease-related deceased class.

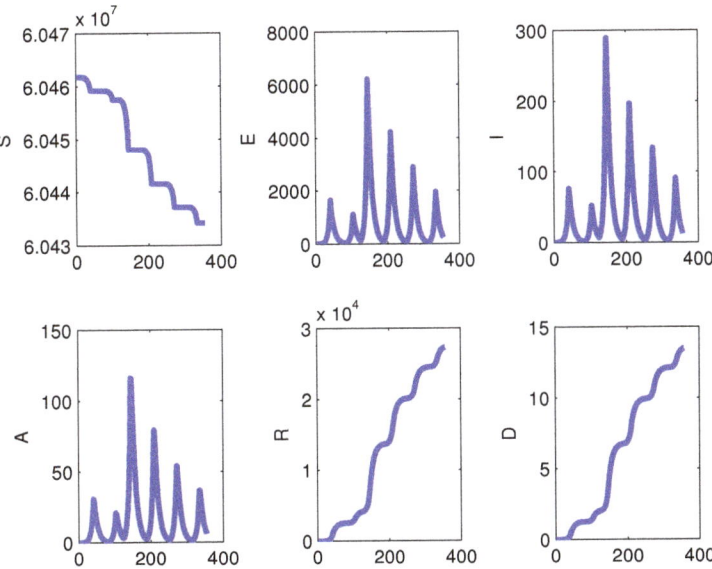

Figure 21. The population values with the lock-down policy, implemented after the first thee weeks with the reduced contact rate $\beta_I = 10^{-10}$, after which $\beta_I = 10^{-7}$ resumes. The check for possible repeated implementation is implemented every three weeks afterwards. The simulation runs over a two year timespan for the model (1) with no demographics, i.e., $\Lambda = d_p = 0$. Left to right and top to bottom, the subpopulations are: S, E, I, A, R and D, the disease-related deceased class.

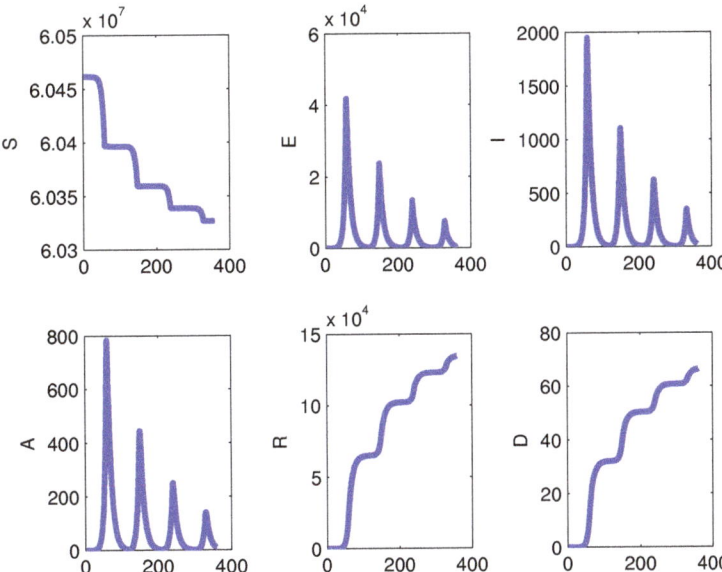

Figure 22. The population values with the lock-down policy, implemented after the first month with the reduced contact rate $\beta_I = 10^{-10}$, after which $\beta_I = 10^{-7}$ resumes. The check for possible repeated implementation is implemented every month afterwards. The simulation runs over a two years timespan for the model (1) with no demographics, i.e., $\Lambda = d_p = 0$. Left to right and top to bottom, the subpopulations are: S, E, I, A, R and D, the disease-related deceased class.

4. Materials and Methods

Here we develop a mathematical model of coronavirus, which is a zoonotic disease. Its biological characteristics indicate that the virus transmission occurred first from infected animals to humans [5], and then spread among populations worldwide by contact with infected individuals, to make it a pandemic.

Let $N(t)$ denote the total population. It is partitioned into the following five disjoint classes of individuals:

$S(t)$: The susceptible class, the individuals who have not yet been exposed to the virus.

$E(t)$: The exposed class, describing people who have become in contact with the virus, but are in incubation period and not yet able to spread the disease; possible presymptomatic individuals that can transmit the infection [13–15] are assumed to have already moved to the asymptomatic class defined below.

$I(t)$: The symptomatic infectious class, individuals that manifest symptoms and can spread the disease.

$A(t)$: The asymptomatic infectious class; those persons that can spread the disease even without having explicit symptoms.

$R(t)$: The removed class, that includes the people that recovered from the disease.

Thus, $N(t) = S(t) + E(t) + I(t) + A(t) + R(t)$.

The basic mechanisms underlying the model are shown in Figure 23. The model is formulated taking into account all the possible interactions among the compartments that were described above.

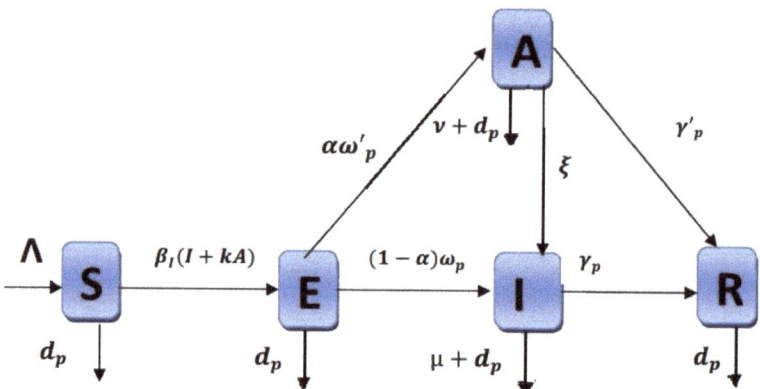

Figure 23. The basic interactions among the compartments.

Under the quasi-steady-state assumption of the total human population, we impose that susceptible individuals are recruited at the constant rate Λ, become infected by direct contact with an infectious individual at rate β_I, which is scaled by a factor k to account for the possibility that the latter is asymptomatic. Finally, all human individuals are subject to natural mortality d_p. These considerations are incorporated in the first equation of the system (1).

Individuals that contract the disease are accounted for in the second equation of (1). They become exposed, i.e., they cannot yet spread the virus, which needs an incubation period within the body of its hosts. In this class enter the susceptibles that were contaminated in the two ways described earlier. People leave it by becoming infectious, and either showing symptoms, thereby migrating into class I, or not, therefore, finding themselves in class A. The progression rates into these two classes are ω_p and ω'_p. Furthermore, we assume that a fraction α becomes asymptomatic and $1 - \alpha$ instead will manifest symptoms.

The third equation models the symptomatic infectious, recruited from the exposed class at rate $(1 - \alpha)\omega$ as described above. Furthermore, there could be asymptomatic individuals that become symptomatic at rate ξ. They leave this class by either progressing to the recovered class at rate γ_p, or dying, naturally or by causes related to the disease at rate μ.

The asymptomatic individuals modeled in the fourth equation appear from the exposed ones, and leave the class by overcoming the disease at rate γ'_p, dying naturally or by disease-related causes at rate ν, or eventually showing the symptoms, for which they migrate into class I.

Recovered individuals are those that have healed from the disease. They are subject only to natural mortality. We assume that they have also become immune so that they are unaffected if become in contact with the infectious.

Note that in the simulations also the cumulative class of disease-related deceased people is shown, although the dead are not explicitly accounted for in the model. They indeed represent a sink, and thus do not contribute to the disease propagation. Incidentally, instead, in cultures where the deceased are kept for a while before burial and become in contact with the relatives, it may be necessary to introduce this class in the model, as another potential source of infection.

Taking into account the above considerations, the model dynamics is regulated by the following system of nonlinear ordinary differential equations:

$$\begin{aligned}
\frac{dS}{dt} &= \Lambda - \beta_I S(I + kA) - d_p S, \\
\frac{dE}{dt} &= \beta_I S(I + kA) - (1-\alpha)\omega_p E - \alpha \omega'_p E - d_p E, \\
\frac{dI}{dt} &= (1-\alpha)\omega_p E - (\gamma_p + d_p + \mu)I + \xi A, \\
\frac{dA}{dt} &= \alpha \omega'_p E - (\gamma'_p + d_p + \nu)A - \xi A, \\
\frac{dR}{dt} &= \gamma_p I + \gamma'_p A - d_p R,
\end{aligned} \quad (1)$$

or alternatively, excluding completely the demographic features, by setting $\Lambda = 0$ and $d_p = 0$ in (1). All the parameters are nonnegative and their meaning is summarized in Table 4. Note that in view of the definitions,

$$\frac{1}{\omega_p}, \quad \frac{1}{\omega'_p}, \quad \frac{1}{\gamma_p}, \quad \frac{1}{\gamma'_p},$$

represent respectively the incubation period before manifesting symptoms, the latent period before becoming asymptomatic infectious, the infectious period for symptomatic infection and the infectious period for asymptomatic infection.

Table 4. Model parameters and their meaning.

Λ	susceptibles recruitment rate
d_p	natural mortality
β_I	disease transmission rate
k	transmissibility ratio between asymptomatics and symptomatics
μ	disease-related mortality for infected
ν	disease-related mortality for asymptomatics
ω_p	progression rate from exposed to symptomatic
ω'_p	progression rate from exposed to asymptomatic
α	fraction of exposed that turn asymptomatic
ξ	progression rate from asymptomatic to symptomatic
γ_p	recovery rate from symptomatic infection
γ'_p	recovery rate from asymptomatic infection

Theorem 1. *The system trajectories are bounded. Letting*

$$M = \max\left\{N(0), \frac{\Lambda}{d_p}\right\}$$

the set

$$\Gamma = \{(S, E, I, A, R) : S + E + I + A + R \leq M, \quad S > 0, \; E \geq 0, \; I \geq 0, \; A \geq 0, \; R \geq 0\}. \tag{2}$$

represents their ultimate attractor. In particular, if $N(0) < \Lambda d_p^{-1}$, $M = \Lambda d_p^{-1}$.

Proof. From the system (1) it follows that the total population evolves as follows:

$$\frac{dN}{dt} + d_p N = \Lambda - \nu A - \mu I \leq \Lambda.$$

Solving the differential inequality easily gives

$$N(t) \leq N(0) \exp(-d_p t) + \frac{\Lambda}{d_p}[1 - \exp(-d_p t)] \leq M,$$

so that all subpopulations, being nonnegative, are bounded as well. □

Note that Γ is positively invariant since all solutions of system (1) originating in Γ remain there for all $t > 0$, in view of the existence and uniqueness of its solutions.

4.1. System's Equilibria Assessment

The equilibrium points of the model are obtained by equating the right hand side of system (1) to zero. The solution of the so-obtained algebraic system gives three equilibrium points: the coronavirus-free equilibrium $C_0 = (S_0, 0, 0, 0, 0,)$, the coronavirus-symptomatic-infected-free equilibrium $C_I = (S_I, E_I, 0, A_I, R_I)$ with conditions $\alpha = 1$ and $\xi = 0$, and the fully coronavirus endemic equilibrium $C^* = (S^*, E^*, I^*, A^*, R^*)$ when either $\alpha \neq 1$ or $\xi \neq 0$. Specifically, for the former two we have:

$$S_0 = \frac{\Lambda}{d_p}, \quad E_I = \frac{1}{B_T}\left(\Lambda - \frac{d_p B_T C_T H_T}{\beta_I \omega_p' D_T}\right), \quad S_I = \frac{\Lambda - B_T E^*}{d_p}, \quad A_I = \left(\frac{\omega_p'}{H_T}\right)E_I, \quad R_I = \left(\frac{\gamma_p'}{d_p}\frac{\omega_p'}{H_T}\right)E_I,$$

where

$$B_T = (1-\alpha)\omega_p + \alpha\omega_p' + d_p, \quad C_T = \gamma_p + \mu + d_p, \quad D_T = \xi + k(\gamma_p + \mu + d_p), \quad H_T = \gamma_p' + \nu + \xi + d_p. \tag{3}$$

The feasibility conditions for C_I are

$$\Lambda > \frac{d_p B_T C_T H_T}{\beta_I \omega_p' D_T}, \quad \alpha = 1 \quad \text{and} \quad \xi = 0. \tag{4}$$

For the fully endemic equilibrium we find

$$E^* = \frac{1}{B_T}\left(\Lambda - \frac{d_p B_T C_T H_T}{\beta_I\left[(1-\alpha)\omega_p H_T + \alpha\omega'_p D_T\right]}\right), \qquad S^* = \frac{\Lambda - B_T E^*}{d_p},$$

$$I^* = \left(\frac{(1-\alpha)\omega_p H_T + \alpha\omega'_p \xi}{C_T H_T}\right) E^*, \qquad A^* = \left(\frac{\alpha\omega'_p}{H_T}\right) E^*$$

$$\text{and } R^* = \left(\frac{\gamma_p}{d_p}\left(\frac{(1-\alpha)\omega_p H_T + \alpha\omega'_p \xi}{C_T H_T}\right) + \frac{\gamma'_p}{d_p}\left(\frac{\alpha\omega'_p}{H_T}\right)\right) E^*,$$

with feasibility condition

$$\Lambda > \frac{d_p B_T C_T H_T}{\beta_I\left[(1-\alpha)\omega_p H_T + \alpha\omega'_p D_T\right]}, \quad \text{and either} \quad \alpha \neq 1 \quad \text{or} \quad \xi \neq 0. \tag{5}$$

4.2. The Basic Reproduction Number

The basic reproduction number R_0 for system (1) is found using the next generation matrix method [16]. The reduced system of (1) may be written in compact form as: $X' = F(X) - V(X)$ where $X = (E, I, A)$

$$F(E, I, A) = \begin{pmatrix} \beta_I S(I + kA) \\ 0 \\ 0 \end{pmatrix}, \quad V(E, I, A) = \begin{pmatrix} -(1-\alpha)\omega_p E - \alpha\omega'_p E - d_p E \\ (1-\alpha)\omega_p E - (\gamma_p + d_p + \mu)I + \xi A \\ \alpha\omega'_p E - (\gamma'_p + d_p + \nu)A - \xi A \end{pmatrix}.$$

The Jacobian matrices of $F(X)$ and $V(X)$ at the disease-free equilibrium point C_0 are

$$J_F(C_0) = \begin{pmatrix} 0 & \beta_I S_0 & \beta_I S_0 k \\ 0 & 0 & 0 \\ 0 & 0 & 0 \end{pmatrix}$$

and

$$J_V(C_0) = \begin{pmatrix} -B_T & 0 & 0 \\ (1-\alpha)\omega_p & -C_T & \xi \\ \alpha\omega'_p & 0 & -H_T \end{pmatrix}.$$

We find that

$$J_V^{-1}(C_0) = \begin{pmatrix} \dfrac{-1}{B_T} & 0 & 0 \\ \dfrac{-[(1-\alpha)\omega_p H_T + \alpha\omega'_p \xi]}{C_T B_T H_T} & \dfrac{-1}{C_T} & \dfrac{-\xi}{C_T H_T} \\ \dfrac{-\alpha\omega'_p}{B_T H_T} & 0 & \dfrac{-1}{H_T} \end{pmatrix}.$$

The next generation matrix is

$$-J_F(C_0) J_V^{-1}(C_0) = \begin{pmatrix} \beta_I S_0 \dfrac{(1-\alpha)\omega_p H_T + \alpha\omega'_p D_T}{C_T B_T H_T} & \dfrac{\beta_I S_0}{C_T} & \dfrac{\beta_I S_0 D_T}{C_T H_T} \\ 0 & 0 & 0 \\ 0 & 0 & 0 \end{pmatrix}.$$

Thus

$$R_0 = \rho(-J_F(C_0) J_V^{-1}(C_0)) = \beta_I S_0 \frac{(1-\alpha)\omega_p H_T + \alpha\omega'_p D_T}{C_T B_T H_T}. \tag{6}$$

The conditions (4) (resp. (5)) are equivalent to $R_0 > 1$ for $\alpha = 1$ and $\xi = 0$ (resp. $R_0 > 1$ for either $\alpha \neq 1$ or $\xi \neq 0$).

We have the following theorem

Theorem 2. *System (1) has the following equilibria:*

1. *The coronavirus-free equilibrium $C_0 = (S_0, 0, 0, 0, 0) = \left(\frac{\Lambda}{d_p}, 0, 0, 0, 0\right)$ which exists always.*
2. *In addition, if $R_0 > 0$ then system (1) admits another nontrivial equilibrium, in fact:*
   When $\alpha = 1$ and $\xi = 0$, it is the coronavirus-symptomatic-infected-free equilibrium $C_I = (S_I, E_I, I_I, A_I, R_I)$.
   When either $\alpha \neq 1$ or $\xi \neq 0$, it is the fully coronavirus endemic equilibrium $C^* = (S^*, E^*, I^*, A^*, R^*)$.

4.3. System's Equilibria Stability

4.3.1. Local Stability

In this subsection we investigate the local stability of the system's equilibria.

Theorem 3. *Letting*

$$a_2 = B_T + C_T + H_T, \tag{7}$$
$$a_1 = H_T[B_T + C_T] + B_T C_T - \beta_I S_0((1-\alpha)\omega_p + k\alpha\omega'_p)$$
$$a_0 = B_T C_T H_T - \beta_I S_0 \left((1-\alpha)\omega_p H_T + \alpha\omega'_p D_T\right).$$

1. *The coronavirus-free equilibrium $C_0 = (S_0, 0, 0, 0, 0)$ of the system (1) is locally asymptotically stable if*

$$\Lambda < \frac{d_p}{\beta_I} \frac{B_T C_T H_T}{(1-\alpha)\omega_p H_T + \alpha\omega'_p D_T}, \ (\text{resp. } R_0 < 1). \tag{8}$$

2. *If $\Lambda > \frac{d_p}{\beta_I} \frac{B_T C_T H_T}{(1-\alpha)\omega_p H_T + \alpha\omega'_p D_T}$, (resp. $R_0 > 1$), then the coronavirus-free equilibrium C_0 of the system (1) is unstable.*

Proof. The Jacobian matrix of system (1) at the coronavirus-free equilibrium C_0 is:

$$J(C_0) = \begin{pmatrix} -d_p & 0 & -\frac{\beta_I \Lambda}{d_p} & -\frac{k\beta_I \Lambda}{d_p} & 0 \\ 0 & -B_T & \frac{\beta_I \Lambda}{d_p} & \frac{k\beta_I \Lambda}{d_p} & 0 \\ 0 & (1-\alpha)\omega_p & -C_T & \xi & 0 \\ 0 & \alpha\omega'_p & 0 & -H_T & 0 \\ 0 & 0 & \gamma_p & \gamma'_p & -d_p \end{pmatrix}.$$

At point C_0, the eigenvalues of J are $-d_p$ of multiplicity order two and the roots of the following characteristic polynomial of a three by three submatrix of J whose coefficients a_i, $i = 0, \ldots, 2$ are given in (7):

$$\lambda^3 + a_2 \lambda^2 + a_1 \lambda + a_0 = 0. \tag{9}$$

It is evident that $a_2 > 0$. From condition (8) the following inequalities are also satisfied

$$a_0 = B_T C_T H_T - \beta_I S_0 \left[(1-\alpha)\omega_p H_T + \alpha\omega'_p D_T\right]$$
$$= \frac{\beta_I}{d_p} \left[(1-\alpha)\omega_p H_T + \alpha\omega'_p D_T\right] \left(\frac{d_p}{\beta_I} \frac{B_T C_T H_T}{\left[(1-\alpha)\omega_p H_T + \alpha\omega'_p D_T\right]} - \Lambda\right) > 0,$$

$$\begin{aligned}
\left[(1-\alpha)\omega_p H_T + \alpha\omega'_p D_T\right] a_1 &= [H_T(B_T+C_T)+B_T C_T]\left[(1-\alpha)\omega_p H_T+\alpha\omega'_p D_T\right]\\
&\quad -\beta_I S_0\left[(1-\alpha)\omega_p H_T+\alpha\omega'_p D_T\right][(1-\alpha)\omega_p+k\alpha\omega'_p]\\
&= [H_T(B_T+C_T)+B_T C_T]\left[(1-\alpha)\omega_p H_T+\alpha\omega'_p D_T\right]\\
&\quad + [a_0 - B_T C_T H_T][(1-\alpha)\omega_p+k\alpha\omega'_p]\\
&= H_T(B_T+C_T)(1-\alpha)\omega_p H_T + (H_T C_T + B_T C_T)\alpha\omega'_p\xi\\
&\quad + a_0[(1-\alpha)\omega_p+k\alpha\omega'_p] > 0
\end{aligned}$$

and

$$\begin{aligned}
\left[(1-\alpha)\omega_p H_T+\alpha\omega'_p D_T\right] a_1 a_2 &= H_T(B_T+C_T)(1-\alpha)\omega_p H_T a_2 + (H_T C_T+B_T C_T)\alpha\omega'_p\xi\, a_2\\
&\quad + a_0[(1-\alpha)\omega_p+k\alpha\omega'_p]a_2\\
&= H_T(B_T+C_T)(1-\alpha)\omega_p H_T a_2 + (H_T C_T+B_T C_T)\alpha\omega'_p\xi\, a_2\\
&\quad + a_0(1-\alpha)\omega_p(B_T+C_T+H_T) + a_0 k\alpha\omega'_p(B_T+C_T+H_T)\\
&> B_T C_T\alpha\omega'_p\xi a_2 + a_0(1-\alpha)\omega_p H_T + a_0\alpha\omega'_p(kC_T+\xi) - a_0\alpha\omega'_p\xi\\
&> B_T C_T H_T\alpha\omega'_p\xi + a_0[(1-\alpha)\omega_p H_T + \alpha\omega'_p D_T] - B_T C_T H_T\alpha\omega'_p\xi\\
&= a_0[(1-\alpha)\omega_p H_T + \alpha\omega'_p D_T].
\end{aligned}$$

Thus, $a_i > 0$, $i = 0,\ldots,2$ and $a_2 a_1 > a_0$.

Then, according to the Routh–Hurwitz criterion, all the roots of the characteristic Equation (9) have negative real parts. Therefore, the coronavirus-free equilibrium point C_0 is locally asymptotically stable under condition (8). □

Since we can deduce the stability of the coronavirus symptomatic infected-free equilibrium C_I from the stability of the coronavirus endemic equilibrium C^* simply by taking $\alpha = 1$ and $\xi = 0$ in the latter, we now just analyze the coronavirus endemic equilibrium C^*.

Theorem 4. *Let*

$$\begin{cases}
c_3 = \beta_I(I^*+kA^*)+d_p+H_T+B_T+C_T > 0,\\
c_2 = [\beta_I(I^*+kA^*)+d_p](H_T+B_T+C_T)+B_T C_T+H_T(B_T+C_T)\\
\quad -[\alpha\omega'_p k+(1-\alpha)\omega_p]\beta_I S^*,\\
c_1 = [\beta_I(I^*+kA^*)+d_p][B_T C_T+H_T(B_T+C_T)]+H_T B_T C_T\\
\quad -[\alpha\omega'_p(kd_p+D_T)+(1-\alpha)\omega_p(d_p+H_T)]\beta_I S^*,\\
c_0 = [\beta_I(I^*+kA^*)+d_p]H_T B_T C_T - d_p[\alpha\omega'_p D_T+(1-\alpha)\omega_p H_T]\beta_I S^*.
\end{cases} \quad (10)$$

The coronavirus endemic equilibrium C^* is locally asymptotically stable if

$$\Lambda > \frac{d_p}{\beta_I}\frac{B_T C_T H_T}{(1-\alpha)\omega_p H_T + \alpha\omega'_p D_T}, \quad (\text{resp. } R_0 > 1). \quad (11)$$

Proof. The Jacobian matrix of system (1) at the coronavirus endemic equilibrium C^* is:

$$J(C^*) = \begin{pmatrix}
-\beta_I(I^*+kA^*)-d_p & 0 & -\beta_I S^* & -\beta_I S^* k & 0\\
\beta_I(I^*+kA^*) & -B_T & \beta_I S^* & \beta_I S^* k & 0\\
0 & (1-\alpha)\omega_p & -C_T & \xi & 0\\
0 & \alpha\omega'_p & 0 & -H_T & 0\\
0 & 0 & \gamma_p & \gamma'_p & -d_p
\end{pmatrix}.$$

At point C^*, the eigenvalues of J are $-d_p$ and the roots of the characteristic polynomial of a three by three submatrix of J. The characteristic equation, in which the coefficients c_i, $i = 0, \ldots, 3$ are given in (10), is:

$$\lambda^4 + c_3\lambda^3 + c_2\lambda^2 + c_1\lambda + c_0 = 0. \tag{12}$$

It is evident that $c_3 > 0$. From condition (11) the following inequalities are also satisfied.

$$\begin{aligned}
c_0 &= [\beta_I(I^* + kA^*) + d_p]H_T B_T C_T - \beta_I[(1-\alpha)\omega_p H_T + \alpha\omega'_p D_T]d_p S^* \\
&= \beta_I\left[(1-\alpha)\omega_p H_T + \alpha\omega'_p(\xi + kC_T)\right] B_T E^* + d_p H_T C_T B_T \\
&\quad - \beta_I[(1-\alpha)\omega_p H_T + \alpha\omega'_p D_T](\Lambda - B_T E^*) \\
&= \beta_I\left[(1-\alpha)\omega_p H_T + \alpha\omega'_p D_T\right]\left(2B_T E^* + \frac{d_p H_T C_T B_T}{\beta_I[(1-\alpha)\omega_p H_T + \alpha\omega'_p D_T]} - \Lambda\right) \\
&= \beta_I\left[(1-\alpha)\omega_p H_T + \alpha\omega'_p D_T\right] B_T E^* > 0,
\end{aligned}$$

$$\begin{aligned}
c_1 &= [\beta_I(I^* + kA^*) + d_p][B_T C_T + H_T(B_T + C_T)] + H_T B_T C_T \\
&\quad - [(1-\alpha)\omega_p(d_p + H_T) + \alpha\omega'_p(kd_p + D_T)]\beta_I S^* \\
&= \left[\beta_I\left(\frac{(1-\alpha)\omega_p H_T + \alpha\omega'_p D_T}{C_T H_T}\right) E^* + d_p\right][B_T C_T + H_T(B_T + C_T)] \\
&\quad - \beta_I[(1-\alpha)\omega_p + \alpha\omega'_p k](\Lambda - B_T E^*) \\
&= \beta_I\left(\frac{(1-\alpha)\omega_p H_T(B_T + C_T)}{C_T} + \frac{\alpha\omega'_p \xi B_T}{C_T} + \frac{\alpha\omega'_p D_T(B_T + H_T)}{H_T}\right) E^* \\
&\quad + d_p[B_T C_T + H_T(B_T + C_T)] - \beta_I[(1-\alpha)\omega_p + \alpha\omega'_p k](\Lambda - 2B_T E^*) \\
&= \beta_I\left(\frac{(1-\alpha)\omega_p H_T(B_T + C_T)}{C_T} + \frac{\alpha\omega'_p \xi B_T}{C_T} + \frac{\alpha\omega'_p D_T(B_T + H_T)}{H_T}\right) E^* \\
&\quad + d_p[B_T C_T + H_T(B_T + C_T)] - \beta_I[(1-\alpha)\omega_p + \alpha\omega'_p k]\left(\frac{2d_p B_T C_T H_T}{\beta_I\left[(1-\alpha)\omega_p H_T + \alpha\omega'_p D_T\right]} - \Lambda\right) \\
&= \beta_I[(1-\alpha)\omega_p H_T + \alpha\omega'_p D_T]\left(\frac{C_T H_T + B_T(C_T + H_T)}{C_T H_T}\right) E^* \\
&\quad + d_p\left(\frac{(B_T + C_T)H_T^2(1-\alpha)\omega_p + H_T B_T \alpha\omega'_p \xi + C_T(B_T + H_T)\alpha\omega'_p D_T}{[(1-\alpha)\omega_p H_T + \alpha\omega'_p D_T]}\right) > 0,
\end{aligned}$$

$$
\begin{aligned}
c_2 &= [\beta_I(I^* + kA^*) + d_p](H_T + B_T + C_T) + B_T C_T + H_T(B_T + C_T) \\
&\quad - [\alpha\omega'_p k + (1-\alpha)\omega_p]\beta_I S^* \\
&= \left[\beta_I\left(\frac{(1-\alpha)\omega_p H_T + \alpha\omega'_p D_T}{C_T H_T}\right)E^* + d_p\right](H_T + B_T + C_T) + B_T C_T + H_T(B_T + C_T) \\
&\quad - [\alpha\omega'_p k + (1-\alpha)\omega_p]\beta_I \frac{\Lambda - B_T E^*}{d_p} \\
&= \beta_I\left(\frac{(1-\alpha)\omega_p(H_T + B_T)}{C_T} + \frac{\alpha\omega'_p \xi(H_T + B_T + C_T)}{C_T H_T} + \frac{\alpha\omega'_p k(B_T + C_T)}{H_T}\right)E^* \\
&\quad + d_p(H_T + B_T + C_T) + B_T C_T + H_T(B_T + C_T) \\
&\quad + \beta_I[(1-\alpha)\omega_p + \alpha\omega'_p k]\left(\frac{(d_p + B_T)E^* - \Lambda}{d_p}\right) \\
&= \beta_I\left(\frac{(1-\alpha)\omega_p(H_T + B_T)}{C_T} + \frac{\alpha\omega'_p \xi(H_T + B_T + C_T)}{C_T H_T} + \frac{\alpha\omega'_p k(B_T + C_T)}{H_T}\right)E^* \\
&\quad + d_p(H_T + B_T + C_T) + B_T C_T + H_T(B_T + C_T) \\
&\quad + \frac{\beta_I[(1-\alpha)\omega_p + \alpha\omega'_p k]}{B_T}\left(\Lambda - \frac{(d_p + B_T)B_T C_T H_T}{\beta_I\left[(1-\alpha)\omega_p H_T + \alpha\omega'_p D_T\right]}\right) \\
&= \beta_I(B_T + C_T + H_T)\left(\frac{(1-\alpha)\omega_p H_T + \alpha\omega'_p D_T}{C_T H_T}\right)E^* \\
&\quad + d_p(H_T + B_T + C_T) + C_T H_T + B_T\left(\frac{(1-\alpha)\omega_p H_T^2 + H_T \alpha\omega'_p \xi + \alpha\omega'_p C_T D_T}{\left[(1-\alpha)\omega_p H_T + \alpha\omega'_p D_T\right]}\right) > 0
\end{aligned}
$$

and

$$
\begin{aligned}
c_1(c_3 c_2 - c_1) &= \beta_I\left(\frac{(1-\alpha)\omega_p H_T + \alpha\omega'_p D_T}{C_T H_T}\right)E^* c_1 c_2 \\
&\quad + \beta_I\left[(H_T + C_T + d_p)H_T + C_T(C_T + d_p)\right]\left(\frac{\left[(1-\alpha)\omega_p H_T + \alpha\omega'_p D_T\right]}{C_T H_T}\right)E^* c_1 \\
&\quad + \beta_I(H_T + B_T + C_T + d_p)B_T\left(\frac{(1-\alpha)\omega_p H_T + \alpha\omega'_p D_T}{C_T H_T}\right)E^* c_1 \\
&\quad + (H_T + B_T + C_T)(H_T + B_T + C_T + d_p)d_p c_1 + C_T H_T(C_T + H_T + B_T)c_1 \\
&\quad + B_T(B_T + C_T + H_T)\left(\frac{\left[(1-\alpha)\omega_p H_T^2 + \alpha\omega'_p C_T D_T\right] + \alpha\omega'_p \xi H_T}{\left[(1-\alpha)\omega_p H_T + \alpha\omega'_p D_T\right]}\right)c_1 \\
&> \beta_I^2 B_T E^{*2}\left\{\beta_I\left[(1-\alpha)\omega_p H_T + \alpha\omega'_p D_T\right]\left(\frac{(1-\alpha)\omega_p H_T + \alpha\omega'_p D_T}{C_T H_T}\right)^2 E^* \right. \\
&\quad \left. + 2(d_p + H_T + B_T + C_T)\left[(1-\alpha)\omega_p H_T + \alpha\omega'_p D_T\right]\left(\frac{(1-\alpha)\omega_p H_T + \alpha\omega'_p D_T}{C_T H_T}\right)\right\} \\
&\quad + \beta_I B_T(d_p + H_T + B_T + C_T)^2\left[(1-\alpha)\omega_p H_T + \alpha\omega'_p D_T\right]E^* \\
&= c_0 c_3^2.
\end{aligned}
$$

Thus, $c_i > 0, i = 0, \ldots, 3$ and $c_1(c_3 c_2 - c_1) > c_0 c_3^2$. Then, according to the Routh–Hurwitz criterion, all the roots of the characteristic Equation (12) have negative real parts. Therefore, the coronavirus endemic equilibrium point C^* is locally asymptotically stable under condition (11). □

From Theorem 4 the following result is reached.

Theorem 5. *Let*

$$\begin{cases} b_3 &= \beta_I k A_I + d_p + H_T + B_T + C_T > 0, \\ b_2 &= [\beta_I k A_I + d_p](H_T + B_T + C_T) + B_T C_T + H_T(B_T + C_T) \\ &\quad - \omega'_p k \beta_I S_I, \\ b_1 &= [\beta_I k A_I + d_p][B_T C_T + H_T(B_T + C_T)] + H_T B_T C_T \\ &\quad - \omega'_p (k d_p + D_T) \beta_I S_I, \\ b_0 &= [\beta_I k A_I + d_p] H_T B_T C_T - d_p \omega'_p D_T \beta_I S^*. \end{cases} \quad (13)$$

The coronavirus symptomatic-infected-free equilibrium C_I of the system (1) is locally asymptotically stable if

$$\Lambda > \frac{d_p}{\beta_I} \frac{B_T C_T H_T}{\omega'_p D_T}, \ (\text{resp. } R_0 > 1). \quad (14)$$

Proof. The result can easily obtained from Theorem 4 by taking $\alpha = 1$ and $\xi = 0$. □

Additionally, from the previous discussion, we can claim the following result:

Theorem 6. *There is a transcritical bifurcation between C_0 and C^*.*

4.3.2. Global Stability

Next, we address the issue of global stability of the coronavirus–free equilibrium, employing as a tool a suitably constructed Lyapunov function and La Salle's Invariance Principle.

Theorem 7. *The coronavirus-free equilibrium C_0 of model (1) is globally asymptotically stable if*

$$\Lambda < \frac{d_p B_T C_T H_T}{\beta_I[(1-\alpha)\omega_p H_T + D_T \alpha \omega'_p]}, \ (\text{resp. } R_0 < 1). \quad (15)$$

Proof. First, the four equations of (1) are independent of R, therefore, the last equation of (1) can be omitted without loss of generality. Hence, let us consider the following function:

$$P = \frac{1}{2S_0}(S - S_0)^2 + E + \frac{B_T}{[(1-\alpha)\omega_p H_T + D_T \alpha \omega'_p]}(H_T I + D_T A) \quad (16)$$

It is easily seen that the above function is nonnegative and also $P = 0$ if and only if $S = S_0$, $E = 0$, $I = 0$ and $A = 0$. Further, calculating the time derivative of P along the positive solutions of (1), we find:

$$\begin{aligned}
\frac{dP}{dt} &= \frac{1}{S_0}(S-S_0)(-\beta_I S(I+kA) - d_p(S-S_0)) + \beta_I S(I+kA) - B_T E \\
&\quad + \frac{B_T H_T((1-\alpha)\omega_p E - C_T I + \xi A)}{[(1-\alpha)\omega_p H_T + D_T \alpha \omega'_p]} + \frac{B_T D_T(\alpha \omega'_p E - H_T A)}{[(1-\alpha)\omega_p H_T + D_T \alpha \omega'_p]} \\
&= -\frac{d_p}{S_0}(S-S_0)^2 + \beta_I [2S - \frac{S^2}{S_0} - S_0](I+kA) + \beta_I S_0(I+kA) \\
&\quad - \frac{B_T C_T H_T I}{[(1-\alpha)\omega_p H_T + D_T \alpha \omega'_p]} - \frac{B_T(-\xi + D_T) H_T A}{[(1-\alpha)\omega_p H_T + D_T \alpha \omega'_p]} \\
&= -\frac{d_p}{S_0}(S-S_0)^2 + \beta_I [2S - \frac{S^2}{S_0} - S_0](I+kA) \\
&\quad + \left(\beta_I S_0 - \frac{B_T C_T H_T}{[(1-\alpha)\omega_p H_T + D_T \alpha \omega'_p]} \right)(I+kA) \\
&= -\frac{d_p}{S_0}(S-S_0)^2 + \beta_I [2S - \frac{S^2}{S_0} - S_0](I+kA) \\
&\quad + \frac{\beta_I}{d_p}\left(\Lambda - \frac{d_p B_T C_T H_T}{\beta_I[(1-\alpha)\omega_p H_T + D_T \alpha \omega'_p]} \right)(I+kA).
\end{aligned}$$

From condition (15) we can show that the coefficients of the term $I + kA$ in the last equality are negative. Further, we have $2S - \frac{S^2}{S_0} - S_0 = -\frac{S^2 - 2SS_0 + S_0^2}{S_0} = -\frac{(S-S_0)^2}{S_0} \leq 0$ for all $S \geq 0$. Thus, we have $\frac{dP}{dt} \leq 0$ for all $(S, E, I, A) \in \mathbb{R}^4_+$ and $\frac{dP}{dt} = 0$ if and only if $(S, E, I, A) = (S_0, 0, 0, 0)$. Thus, the only invariant set contained in $\mathbb{R}^4_+$ is $\{(S_0, 0, 0, 0)\}$. Hence, La Salle's theorem implies convergence of the solutions (S, E, I, A) to $(S_0, 0, 0, 0)$. From the last equation if (1) we can show obviously that R converge also to 0. Therefore C_0 is globally asymptotically stable if $R_0 < 1$. □

4.4. Numerical Simulations

The calculation of the value of R_0 according to (6) with the parameter values used in the numerical simulations gives $R_0 = 3.1402$, in line with the current estimates [11,12].

4.4.1. Simulations Methodology

We use a simple own-developed driver code calling the Matlab intrinsic routine ode45, implementing the classical Runge–Kutta 45 integration method for ordinary differential equations.

At first, we consider only the demographic simulation and show that the population is essentially at the same level during a year. This fact is substantiated also by the simulation results, for which there is scant difference between those of the model (1) and the ones obtained by using its no-demographic counterpart, where Λ and d_p are both set to zero.

We then perform three sets of simulations describing different possible scenarios. The first one considers lock-down, i.e., decreasing the contact rate significantly, but not to zero, as some essential activities are still open. Then the total isolation policy, for which the contact rate is set to zero. Finally an intermittent closure policy, for which when infectives reappear in a significant way, temporary lock-down measures resume again.

4.4.2. Data Acquisition

We use data published on official websites about the epidemic's spread in Italy collected between 29 January and 28 March 2020, a period that spans 61 days, incremented by more recent information [17].

Using the day as the base time unit, we assume that the average incubation period lies in the interval between two and 14 days, with a mean of 8 days. Based on the percentage of the reported symptomatic infected patients, the proportion of symptomatic in the infected class α is estimated to be in the interval $[0.01, 0.3]$. The correction k for asymptomatics to diffuse the disease is set in the range $k \in [0:005; 0:2]$. There have been 27,359 deaths between 15 February and 29 April [17], with changes in the number of fatalities every day. Dividing the fatal cases by the timespan, one gets 370 daily fatalities, which gives a rate 0.0027. Using this value in the simulation, puts the total losses to about 10^5. But we observed that apparently children hardly get the disease, the younger and adult people have it generally in a mild form and fatalities occur mainly for the elderly people, compare with Figure 3 of [18]. In view of the fact that there is no age structure in this model, we corrected this value by taking a third of the above result to set the disease mortality rate at the final value $\mu = 0.001$, which gives a reasonable estimate for the losses in the timespan, in rough agreement with the actual tallies. We neglect altogether mortality for the asymptomatics, setting $\nu = 0$. Based on the officially published data we estimate $\gamma_p = 0.1764$, $\gamma'_p = 0.6024$. For the initial values, the total population is obtained from the report published by the official cite of worldometers [19], $S = 60461826$. To avoid demographic effects, we set the susceptible recruitment rate Λ in order that on the timespan of the simulation the total population N does not change much.

4.4.3. The Pure Demographic Case

We simulate first the population model without disease. In so doing, we varied the parameter Λ until a satisfactory behavior of N, the total population was found. With $\Lambda = 500$ there is little variation of N during a whole year, the population remains roughly stable around the level $60,400,000$, see Figure 24. In this way the demographic effects are sort of removed, and we can concentrate mainly on the epidemics. Actually, the number of newborns per day in Italy would be about four times higher, but as mentioned, we just would like here to hide the demographics from the simulations and not have a picture more adherent to reality.

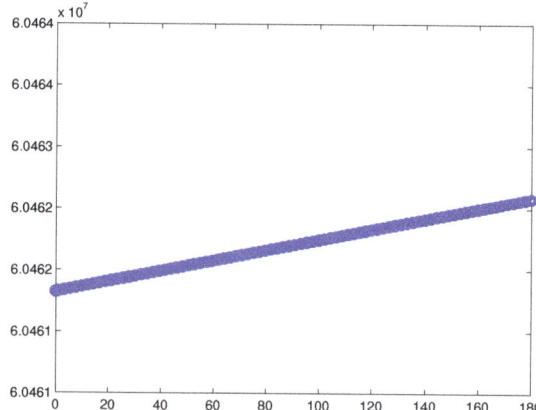

Figure 24. The susceptible population behavior over a year, without disease. It does not vary much as the vertical scale is rather small, the range of variation being around 3000, over a population of 60×10^6.

4.4.4. Epidemics Spread in the Absence of Measures

Here we introduced the disease, with incidence $\beta_I = 10^{-7}$. The result is shown in Figure 25 for absolute numbers, and in Figure 6 in semilogarithmic scale. In this case no measures are assumed to be taken to counteract the epidemics. These results are reported in order be able to compare the simulations with restrictions to what would happen if the containment measures were not taken.

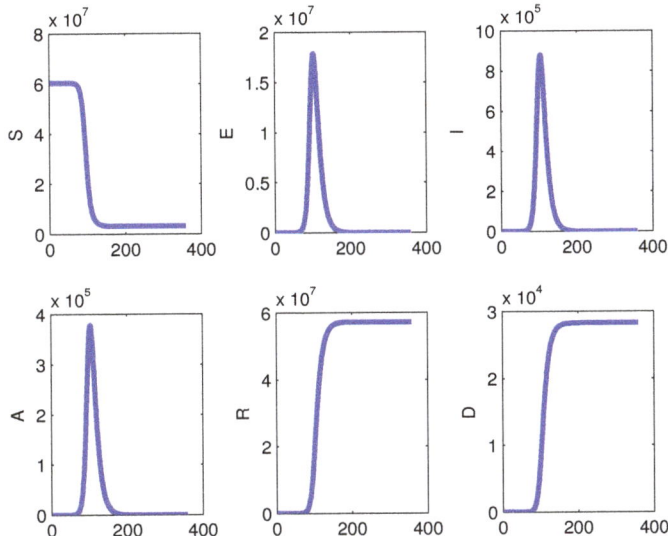

Figure 25. The epidemic's effect on the population in the absence of measures for $\beta_I = 10^{-7}$, over a period of one year. Left to right and top to bottom, the total population sizes: S, E, I, A, R and D, the disease-related deceased class.

4.4.5. Containment Measures for the Epidemics

Finally, we considered the introduction of the distancing policy. It is assumed to start at time t_1 and end at time $t_1 + t_2$. Two forms of containment measures are considered, substantially reducing the contact rate, or even setting it equal to zero, meaning the extreme measure of total individuals isolation.

In particular, we present the experience of using the reference value of the contact rate $\beta_I = 10^{-7}$, then reducing it to $\beta_I = 10^{-10}$ during the interval $[t_1, t_1 + t_2]$. We then reset it to its previous reference value after time $t_1 + t_2$. We monitored the epidemics evolution over six months.

The alternative, milder choice $\beta_I = 10^{-8}$ is also used, for comparison.

The simulations are then repeated with total isolation, setting $\beta_I = 0$ during the implementation of the restrictions.

A comparison of the results with the model obtained by disregarding the demographic parameters, i.e., setting $\Lambda = d_p = 0$ is also performed in the same way as done for the model (1).

Different timings for taking both the first restriction measure and for lifting it are then investigated, using all the above alternatives.

Finally an intermittent restrictive policy is examined, for which when the infected are observed to trespass a threshold, distancing measures are taken. Here again lock-down or total isolation produce essentially the same results. The use of different timings for the introduction of the restrictions is also scrutinized.

Author Contributions: Model formulation, Y.B. and E.V.; methodology, Y.B., M.H. and E.V.; formal analysis, Y.B., M.H. and E.V.; writing—original draft preparation, Y.B.; simulations, writing—review and editing, E.V. All authors have read and agreed to the published version of the manuscript.

Funding: E.V. was partially funded by the local research project "Questioni attuali di approssimazione numerica e loro applicazioni" of the Dipartimento di Matematica, Universitá di Torino. This research has been undertaken within the framework of the COST Action: CA 16227—Investigation and Mathematical Analysis of Avant-garde Disease Control via Mosquito Nano-Tech-Repellents. This work was partially supported by the Ministry of Higher Education and Scientific Research of Algeria (MESRS) and General Direction of Scientific Research and Technological Development (DGRSDT) through Research Project-University Formation (PRFU: C00L03UN220120190001 and PRFU: C00L03UN220120180004).

Acknowledgments: The authors would like to thank the reviewers for their valuable comments and suggestions that greatly improved the presentation of this work.

Conflicts of Interest: The authors declare no conflict of interest.

References

1. Cecconi, M.; Forni, G.; Mantovani, A. *COVID-19: An Executive Report, "Commissione Salute, March 25th"*; Accademia Nazionale dei Lincei: Roma, Italy, 2020.
2. Herbert, W. Hethcote, The Mathematics of Infectious Diseases. *SIAM Rev.* **2000**, *42*, 599–653. [CrossRef]
3. Magal, P.; Webb, G. Predicting the number of reported and unreported cases for the COVID-19 epidemic in South Korea, Italy, France and Germany. *medRxiv* **2020**. [CrossRef]
4. Buonomo, B. Effects of information-dependent vaccination behavior on coronavirus outbreak: Insights from a SIRI model. *Ricerche di Matematica* **2020**, doi:10.1007/s11587-020-00506-8 [CrossRef]
5. Chen, T.M.; Rui, J.; Wang, Q.P.; Zhao, Z.Y.; Cui, J.A.; Yin, L. A mathematical model for simulating the transmission of Wuhan novel Coronavirus. *bioRxiv* **2020**. [CrossRef]
6. Jia, J.; Ding, J.; Liu, S.; Liao, G.; Li, J.; Duan, B.; Wang, G.; Zhang, R. Modeling the control of Covid-19: Impact of policy interventions and meteorological factors. *Electron. J. Differ. Equ.* **2020**, *2020*, 1–24. Available online: https://www.researchgate.net/publication/339786734 (accessed on 27 February 2020).
7. Murray, J.D. *Mathematical Biology*; Springer: Berlin/Heidelberg, Germany, 2002.
8. Perko, L. *Differential Equations and Dynamical Systems*; Springer: Berlin/Heidelberg, Germany, 2001.
9. Strogatz, S. *Nonlinear Dynamics and Chaos*; Perseus Books: Reading, MA, USA, 1994.
10. Adamik, B.; Bawiec, M.; Bezborodov, V.; Bock, W.; Bodych, M.; Burgard, J.; Götz, T.; Krueger, T.; Migalska, A.; Pabjan, B.; et al. Mitigation and herd immunity strategy for COVID-19 is likely to fail. *medRxiv* **2020**, doi:10.1101/2020.03.25.20043109 [CrossRef]
11. Liu, G.; Wilder-Smith, A.; Rocklöv, J. The reproductive number of COVID-19 is higher compared to SARS coronavirus. *J. Travel Med.* **2020**, *27*, taaa021. doi:10.1093/jtm/taaa021. [CrossRef] [PubMed]
12. COVID-19 Pandemic in Italy. Available online: https://en.wikipedia.org/wiki/COVID-19_pandemic_in_Italy (accessed on 29 April 2020).
13. Banerjee, M.; Tokarev, A.; Volpert, V. Immuno-epidemiological model of two-stage epidemic growth. *Math. Model. Nat. Phenom.* **2020**, *15*, 27. [CrossRef]
14. Kochańczyk, M.; Grabowski, F.; Lipniacki, T. Dynamics of COVID-19 pandemic at constant and time-dependent contact rates. *Math. Model. Nat. Phenom.* **2020**, *15*, 28 [CrossRef]
15. Volpert, V.; Banerjee, M.; d'Onofrio, A.; Lipniacki, T.; Petrovskii, S.; Tran, V.C. Coronavirus—Scientific insights and societal aspects. *Math. Model. Nat. Phenom.* **2020**, *15*, E2. [CrossRef]
16. Van den Driessche, P.; Watmough, J. A simple SIS epidemic model with a backward bifurcation. *J. Math. Biol.* **2000**, *40*, 525–540. [CrossRef] [PubMed]
17. Italy Coronavirus: Cases and Deaths–Worldometer. Available online: https://www.worldometers.info/coronavirus/country/italy/ (accessed on 27 February 2020).
18. Verity, R.; Okell, L.C.; Dorigatti, I.; Winskill, P.; Whittaker, C.; Imai, N.; Cuomo-Dannenburg, G.; Thompson, H.; Walker, P.G.; Fu, H.; et al. Estimates of the severity of coronavirus disease 2019: A model-based analysis. *Lancet* **2020**, doi:10.1016/S1473-3099(20)30243-71. [CrossRef]
19. Reported Cases and Deaths by Country, Territory, or Conveyance. Available online: https://www.worldometers.info/coronavirus/#countries (accessed on 27 February 2020).

© 2020 by the authors. Licensee MDPI, Basel, Switzerland. This article is an open access article distributed under the terms and conditions of the Creative Commons Attribution (CC BY) license (http://creativecommons.org/licenses/by/4.0/).

Article

A Delayed Epidemic Model for Propagation of Malicious Codes in Wireless Sensor Network

Zizhen Zhang [1,*], Soumen Kundu [2] and Ruibin Wei [1]

1. School of Management Science and Engineering, Anhui University of Finance and Economics, Bengbu 233030, China; rbwxy@126.com
2. Department of Mathematics, National Institute of Technology Durgapur, Durgapur 713209, India; soumenkundu75@gmail.com
* Correspondence: zzzhaida@163.com

Received: 30 March 2019; Accepted: 26 April 2019; Published: 1 May 2019

Abstract: In this paper, we investigate a delayed SEIQRS-V epidemic model for propagation of malicious codes in a wireless sensor network. The communication radius and distributed density of nodes is considered in the proposed model. With this model, first we find a feasible region which is invariant and where the solutions of our model are positive. To show that the system is locally asymptotically stable, a Lyapunov function is constructed. After that, sufficient conditions for local stability and existence of Hopf bifurcation are derived by analyzing the distribution of the roots of the corresponding characteristic equation. Finally, numerical simulations are presented to verify the obtained theoretical results and to analyze the effects of some parameters on the dynamical behavior of the proposed model in the paper.

Keywords: boundedness; delay; Hopf bifurcation; Lyapunov functional; stability; SEIQRS-V model

1. Introduction

Malicious codes are harmful programs which reproduce themselves from one computer to others without any user interaction [1–3]. Specially, they have the ability to transmit directly from device to device through wireless technology such as Bluetooth or Wi-Fi. With the increasing rapid advent of wireless technology and the Internet of Things, the threat from malicious codes have become increasingly serious. According to 2017 Cybercrime Report [4], hundreds of thousands—and possibly millions—of people can be hacked via their wirelessly connected and 'The Big Data Bang' is an IoT (Internet of Things) world that will explode from 2 billion objects (smart devices which communicate wirelessly) in 2006 to a projected 200 billion by 2020 according to Intel. Thus, there has been an urgent need to investigate the malicious propagation dynamics in wireless sensor networks especially in the aftermath of the Yahoo hack and Equifax breach. In the past decades, some mathematical models describing malicious codes propagation are proposed to study viruses' behavior. For example, the classic epidemic models [5–9], the models with graded infection rate [10–18], the stochastic models [19–23] and some other models [2,24–26].

The common problem of the above models is that the characteristics of networks like communication radius, and distributed density of nodes are not considered in models. Thus, computer virus models considering the characteristics of networks have drawn the attention of scholars both at home and abroad. In [27], Feng et al. formulated an improved SIRS epidemic model considering communication radius and distributed density of nodes in wireless sensor network. In [28], Srivastava et al. proposed an SIDR model for worm propagation in wireless sensor network and they considered the dead nodes, the communication radius and node density in the proposed model. Nwokoye et al. [29,30] investigated an SEIRS-V worm model with different forms. Ojha et al. [31] proposed a modified SIQRS worm propagation model by introducing quarantined compartment into the model proposed by Feng et al. in [27]. Very recently, based on the model proposed in [29,30,32], Nwokoye and Umeh [33] formulated the following modified SEIQRS-V epidemic model for propagation of malicious codes in wireless sensor network:

$$\begin{cases} \frac{dS(t)}{dt} = A - \frac{\beta\sigma\pi r^2}{L^2}S(t)I(t) - (d_1+\rho)S(t) \\ \qquad\quad +\varphi R(t) + \varepsilon V(t), \\ \frac{dE(t)}{dt} = \frac{\beta\sigma\pi r^2}{L^2}S(t)I(t) - d_1 E(t) - \theta E(t), \\ \frac{dI(t)}{dt} = \theta E(t) - (d_1+d_2+\eta_1+\alpha)I(t), \\ \frac{dQ(t)}{dt} = \alpha I(t) - (d_1+d_2+\eta_2)Q(t), \\ \frac{dR(t)}{dt} = \eta_1 I(t) + \eta_2 Q(t) - (d_1+\varphi)R(t), \\ \frac{dV(t)}{dt} = \rho S(t) - (d_1+\varepsilon)V(t), \end{cases} \qquad (1)$$

where $S(t)$, $E(t)$, $I(t)$, $Q(t)$, $R(t)$ and $V(t)$ denote the numbers of the susceptible, exposed, infectious, quarantined, recovered, and vaccinated nodes at time t, respectively. A is the entering rate of nodes into the sensor network; d_1 is the death rate of the nodes due to hardware or software failure; d_2 is the death rate due to attack of of malicious codes; r is the communication radius of the nodes; $L \times L$ is the area in which the nodes distributed; β is the contact rate of the infectious nodes; σ is the distribution density of nodes; ρ, φ, ε, θ, η_1 and η_2 are the state transition rates.

When malicious codes spread in networks, there are different forms of delay, including immunity period delay, latent period delay, cleaning-virus period delay etc. In [34], Keshri and Mishra considered a dynamic model on the transmission of malicious signals in wireless sensor network with latent period delay and the temporary immunity period delay. They showed that the two delays play a positive role in controlling a malicious attack. In [35], Zhang and Bi investigated the Hopf bifurcation of a delayed computer virus model with the effect of external computers by using the latent period delay as the bifurcation parameter. Zhao and Bi studied a delayed SEIR computer virus spreading model with limited anti-virus ability and analyzed the effect of the cleaning-virus period delay on the model [36]. In [37], Chai and Wang analyzed the Hopf bifurcation of a delayed SEIRS epidemic model with vertical transmission in network by taking the different combinations of the latent period delay and the temporary immunity period delay as the bifurcation parameter. In [38], Dai et al. proposed a delayed computer virus propagation model with saturation incidence rate and temporary immunity period delay and studied stability and Hopf bifurcation.

Motivated by the work about delayed computer virus models in [34–38], we incorporate the latent period delay into system (1) and obtain the following delayed SEIQRS-V epidemic model for propagation of malicious codes in wireless sensor network:

$$\begin{cases} \frac{dS(t)}{dt} &= A - \frac{\beta\sigma\pi r^2}{L^2}S(t)I(t) - (d_1+\rho)S(t), \\ & \quad +\varphi R(t) + \varepsilon V(t), \\ \frac{dE(t)}{dt} &= \frac{\beta\sigma\pi r^2}{L^2}S(t)I(t) - d_1 E(t) - \theta E(t-\tau), \\ \frac{dI(t)}{dt} &= \theta E(t-\tau) - (d_1+d_2+\eta_1+\alpha)I(t), \\ \frac{dQ(t)}{dt} &= \alpha I(t) - (d_1+d_2+\eta_2)Q(t), \\ \frac{dR(t)}{dt} &= \eta_1 I(t) + \eta_2 Q(t) - (d_1+\varphi)R(t), \\ \frac{dV(t)}{dt} &= \rho S(t) - (d_1+\varepsilon)V(t), \end{cases} \qquad (2)$$

subject to the initial conditions $S(\theta) = \phi_1(\theta) > 0$, $E(\theta) = \phi_2(\theta) > 0$, $I(\theta) = \phi_3(\theta) > 0$, $Q(\theta) = \phi_4(\theta) > 0$, $R(\theta) = \phi_5(\theta) > 0$, $V(\theta) = \phi_6(\theta) > 0$, $\theta \in [-\tau, 0)$, $\phi_i(0) > 0$, $i = 1,2,3,4,5,6$, and τ is the latent period delay of malicious codes.

The structure of this paper is as follows. In the next section, it is shown that the solution of system (2) is positive and bounded in a feasible region $\bar{R}$, which is invariant. In Section 3, the condition for local asymptotical stability is examined by constructing a suitable Lyapunov functional. Section 4 deals with local stability and existence of Hopf bifurcation. Some numerical simulations are carried out to illustrate the obtained theoretical results and effect of some parameters on behaviors of the model in Section 5. The paper finally ends with conclusion in Section 6.

2. Positivity and Boundedness

In this section we shall discuss about the positivity and boundedness of solution of the system (2). For this we assume the function $\bar{V}$ as:

$$\bar{V}(t) = S(t) + E(t) + I(t) + Q(t) + R(t) + V(t). \qquad (3)$$

Taking the derivative of (3) and using (2) we get,

$$\dot{\bar{V}}(t) = A - d_1 S(t) - d_1 E(t) - (d_1+d_2)(I(t)+Q(t)) - d_1 R(t) - d_1 V(t), \qquad (4)$$

where $S(t), E(t), I(t), Q(t), R(t), V(t) \geq 0$.
If $E(t) = 0, I(t) = 0, Q(t) = 0, R(t) = 0$ and $V(t) = 0$ from (4) we get

$$\limsup_{t \to \infty} \bar{V}(t) \leq \frac{A}{d_1}. \qquad (5)$$

Also, if $\bar{V}(t) > \frac{A}{d_1}$ then $\dot{\bar{V}}(t) < 0$. Therefore, we get $0 < \bar{V} \leq \frac{A}{d_1}$, i.e., we get a feasible region $\bar{R}$ as $\bar{R} = \{(S(t), E(t), I(t), Q(t), R(t), V(t)) \in R^6 : 0 < S(t)+E(t)+I(t)+Q(t)+R(t)+V(t) \leq \frac{A}{d_1}\}$.

Thus we see that the solution of system (2) is bounded and independent of the initial condition. So the feasible region $\bar{R}$ is an invariant set. Also, as $A > 0, d_1 > 0, \frac{A}{d_1} > 0$, i.e., the feasible region $\bar{R}$ is positive. Hence all solutions of system (2) will come to the field $\bar{R}$ or will remain in $\bar{R}$.

3. Lyapunov Stability Analysis

In this section the linear stability of the system (2) has been discussed by constructing a suitable Lyapunov functional given in Equation (7). By direct computation, it can be concluded that if the basic reproduction number

$$R_0 = \frac{A\pi r^2 \beta \theta \sigma (d_1 + \varepsilon)}{L^2 d_1 (d_1 + \theta)(d_1 + \varepsilon + \rho)(d_1 + d_2 + \alpha + \eta_1)} > 1,$$

then, system (2) has a unique endemic equilibrium $P_*(S_*, E_*, I_*, Q_*, R_*, V_*)$, where

$$S_* = \frac{L^2(d_1 + \theta)(d_1 + d_2 + \alpha + \eta_1)}{\beta \theta \sigma \pi r^2},$$

$$E_* = \frac{(d_1 + \varphi)(d_1 + d_2 + \eta_2)(d_1 + d_2 + \alpha + \eta_1)[A - S_* d_1(d_1 + \varepsilon + \rho)/(d_1 + \varepsilon)]}{\alpha \eta_2 \varphi \theta + (d_1 + d_2 + \eta_2)[(d_1 + \theta)(d_1 + \varphi)(d_1 + d_2 + \alpha + \eta_1) - \theta \varphi \eta_1]},$$

$$I_* = \frac{\theta E_*}{d_1 + d_2 + \alpha + \eta_1}, Q_* = \frac{\alpha I_*}{d_1 + d_2 + \eta_2}, R_* = \frac{\eta_1 I_* + \eta_2 Q_*}{d_1 + \varphi}, V_* = \frac{\rho S_*}{d_1 + \varepsilon}.$$

For this let $u_1(t) = S(t) - S^*$, $u_2(t) = E(t) - E^*$, $u_3(t) = I(t) - I^*$, $u_4(t) = Q(t) - Q^*$, $u_5(t) = R(t) - R^*$ and $u_6(t) = V(t) - V^*$, then the system (2) transform into

$$\begin{cases} \frac{du_1(t)}{dt} = -\alpha_{11} u_1 - \alpha_{12} u_3 + \phi u_5 + \varepsilon u_6, \\ \frac{dp_1(t)}{dt} = \alpha_{21} u_1 - (d_1 + \theta) u_2 + \alpha_{12} u_3, \\ \frac{dp_2(t)}{dt} = \theta u_2 - (d_1 + d_2 + \eta_1 + \alpha) u_3, \\ \frac{du_4(t)}{dt} = \alpha u_3 - (d_1 + d_2 + \eta_2) u_4, \\ \frac{du_5(t)}{dt} = \eta_1 u_3 + \eta_2 u_4 - (d_1 + \phi) u_5, \\ \frac{du_6(t)}{dt} = \rho u_1 - (d_1 + \varepsilon) u_6, \end{cases} \quad (6)$$

where $p_1(t) = u_2 - \theta \int_{t-\tau}^{t} u_2(s) ds$, $p_2(t) = u_3 + \theta \int_{t-\tau}^{t} u_2(s) ds$, $\alpha_{11} = d_1 + \rho + \alpha_{21}$, $\alpha_{12} = \frac{\beta \sigma \pi r^2 S^*}{L^2}$, $\alpha_{21} = \frac{\beta \sigma \pi r^2 I^*}{L^2}$.

Now, following the steps as in [39], we shall check the stability of the system by assuming a suitable Lyapunov function $w(u)(t)$ as follows:

$$w(u)(t) = \sum (k_i w_j(u)(t)), \quad (7)$$

where $k_i, i = 1, \cdots, 21$ are given in Appendix A and $w_j(u)(t), j = 1, \cdots, 21$ are given in Appendix B.

As all the parameters are assumed positive and chosen in such a way that $k_i > 0, i = 1, \cdots, 21$ and $w(u)(t) > 0$. Taking the derivative of Equation (7), and using Equation (6) we get

$$\frac{d}{dt} w(u)(t) \leq (\sum \Lambda_i u_j^2), i = 1, \cdots, 6; j = 1, \cdots, 6. \quad (8)$$

where expression for $\Lambda_i, i = 1, \cdots, 6$ are given in Appendix C.

Theorem 1. *If the value of the delays τ satisfy the conditions $\Lambda_i < 0, i = 1, \cdots, 6$ then the endemic equilibrium point $P^*(S^*, E^*, I^*, Q^*, R^*, V^*)$ of system (2) is locally asymptotically stable. Otherwise if any one of $\Lambda'_i s (i = 1, 2, \cdots, 6.)$ becomes positive then the system will be unstable.*

Proof of Theorem 1. Let $\Lambda = \max\{\Lambda_i, i = 1, \cdots, 6\}$. Then for $t > T$, from Equation (8) we get $w(u)(t) + \Lambda \int_T^t (\sum u_i^2(s)) ds \leq w(u)(T)$, $i = 1, \cdots, 6$ for $t \geq T$, implies $\sum u_i^2 \in L_1[T, \infty]$, $i = 1, \cdots, 6$. It is easy to conclude from (6) and the boundedness of $u(t)$ that $\sum u_i^2(t)(i = 1, \cdots, 6)$ is uniformly continuous. Using Barbalat's lemma in [38], we can say that

$$\lim_{t \to \infty} \{\sum u_i^2, i = 1, \cdots, 6\} = 0. \tag{9}$$

So the internal solution of Equation (6) as well as solutions of system (2) is asymptotically stable, i.e., the endemic equilibrium P^* of system (2) is locally asymptotically stable. Hence, this completes the proof.

We remark that as $\Lambda_i, i = 1, \cdots, 6$ depends on the delay τ and the local stability condition for P^* of the system (2) is preserved for small τ satisfying $\Lambda_i < 0, i = 1, \cdots, 6$. □

4. Existence of Hopf Bifurcation

The characteristic equation at the endemic equilibrium P_* can be obtained as follows

$$\lambda^6 + U_5 \lambda^5 + U_4 \lambda^4 + U_3 \lambda^3 + U_2 \lambda^2 + U_1 \lambda + U_0 \\ + (V_5 \lambda^5 + V_4 \lambda^4 + V_3 \lambda^3 + V_2 \lambda^2 + V_1 \lambda + V_0) e^{-\lambda \tau} = 0, \tag{10}$$

with

$$\begin{aligned}
U_0 &= \alpha_{22}\alpha_{33}\alpha_{44}\alpha_{55}(\alpha_{11}\alpha_{66} + \alpha_{16}\alpha_{61}), \\
U_1 &= -[\alpha_{16}\alpha_{61}(\alpha_{22}\alpha_{33}(\alpha_{44} + \alpha_{55}) + \alpha_{44}\alpha_{55}(\alpha_{22} + \alpha_{33})) \\
&\quad + \alpha_{11}\alpha_{22}\alpha_{33}(\alpha_{44}\alpha_{55} + \alpha_{44}\alpha_{66} + \alpha_{55}\alpha_{66}) \\
&\quad + \alpha_{44}\alpha_{55}\alpha_{66}(\alpha_{11}\alpha_{22} + \alpha_{11}\alpha_{33} + \alpha_{22}\alpha_{33})], \\
U_2 &= (\alpha_{11}\alpha_{22} + \alpha_{11}\alpha_{33} + \alpha_{22}\alpha_{33})(\alpha_{44}\alpha_{55} + \alpha_{44}\alpha_{66} + \alpha_{55}\alpha_{66}) \\
&\quad + \alpha_{11}\alpha_{22}\alpha_{33}(\alpha_{44} + \alpha_{55} + \alpha_{66}) + \alpha_{44}\alpha_{55}\alpha_{66}(\alpha_{11} + \alpha_{22} + \alpha_{33}) \\
&\quad + \alpha_{16}\alpha_{61}(\alpha_{22}\alpha_{33} + \alpha_{44}\alpha_{55} + (\alpha_{22} + \alpha_{33})(\alpha_{44} + \alpha_{55})), \\
U_3 &= -[\alpha_{16}\alpha_{61}(\alpha_{22} + \alpha_{33} + \alpha_{44} + \alpha_{55}) + \alpha_{11}\alpha_{22}\alpha_{33} + \alpha_{44}\alpha_{55}\alpha_{66} \\
&\quad + (\alpha_{11} + \alpha_{22} + \alpha_{33})(\alpha_{44}\alpha_{55} + \alpha_{44}\alpha_{66} + \alpha_{55}\alpha_{66}) \\
&\quad + (\alpha_{44} + \alpha_{55} + \alpha_{66})(\alpha_{11}\alpha_{22} + \alpha_{11}\alpha_{33} + \alpha_{22}\alpha_{33})], \\
U_4 &= \alpha_{16}\alpha_{61} + \alpha_{11}\alpha_{22} + \alpha_{11}\alpha_{33} + \alpha_{22}\alpha_{33} \\
&\quad + \alpha_{44}\alpha_{55} + \alpha_{44}\alpha_{66} + \alpha_{55}\alpha_{66} \\
&\quad + (\alpha_{11} + \alpha_{22} + \alpha_{33})(\alpha_{44} + \alpha_{55} + \alpha_{66}), \\
U_5 &= -(\alpha_{11} + \alpha_{22} + \alpha_{33} + \alpha_{44} + \alpha_{55} + \alpha_{66}), \\
V_0 &= \alpha_{21}\alpha_{66}\beta_{32}(\alpha_{16} - \alpha_{15})(\alpha_{43}\alpha_{54} + \alpha_{44}\alpha_{53}) \\
&\quad + \alpha_{44}\alpha_{55}(\alpha_{16}\alpha_{33}\alpha_{61}\beta_{22} + \alpha_{13}\alpha_{21}\alpha_{66}\beta_{32} + \alpha_{11}\alpha_{33}\alpha_{66}\beta_{22}),
\end{aligned}$$

$$\begin{aligned}
V_1 &= \alpha_{21}\beta_{32}(\alpha_{16}-\alpha_{15})(\alpha_{43}\alpha_{54}+\alpha_{44}\alpha_{53})\\
&\quad -\alpha_{21}\alpha_{53}\beta_{32}(\alpha_{44}+\alpha_{66})(\alpha_{15}-\alpha_{16})\\
&\quad -\alpha_{16}\alpha_{61}\beta_{22}(\alpha_{33}\alpha_{44}+\alpha_{33}\alpha_{55}+\alpha_{44}\alpha_{55})\\
&\quad -\alpha_{13}\alpha_{21}\beta_{32}(\alpha_{44}\alpha_{55}+\alpha_{44}\alpha_{66}+\alpha_{55}\alpha_{66}),\\
&\quad -\beta_{22}[\alpha_{11}\alpha_{33}\alpha_{44}(\alpha_{55}+\alpha_{66})+\alpha_{11}\alpha_{55}\alpha_{66}(\alpha_{33}+\alpha_{44})+\alpha_{33}\alpha_{44}\alpha_{55}\alpha_{66}],\\
V_2 &= (\alpha_{16}\alpha_{61}\beta_{22}(\alpha_{33}+\alpha_{44}+\alpha_{55})+\alpha_{13}\alpha_{21}\beta_{32}(\alpha_{44}+\alpha_{55}+\alpha_{66}))\\
&\quad +\alpha_{21}\alpha_{53}\beta_{32}(\alpha_{16}-\alpha_{15})\\
&\quad +\beta_{22}[\alpha_{33}\alpha_{44}(\alpha_{55}+\alpha_{66})+\alpha_{55}\alpha_{66}(\alpha_{33}+\alpha_{44})]\\
&\quad +\alpha_{11}\beta_{22}[\alpha_{33}\alpha_{44}+\alpha_{55}\alpha_{66}+(\alpha_{33}+\alpha_{44})(\alpha_{55}+\alpha_{66})],\\
V_3 &= -[\alpha_{11}\beta_{22}(\alpha_{33}+\alpha_{44}+\alpha_{55}+\alpha_{66})+\alpha_{13}\alpha_{21}\beta_{32}+\alpha_{16}\alpha_{61}\beta_{22}\\
&\quad +\beta_{22}(\alpha_{33}\alpha_{44}+\alpha_{55}\alpha_{66}+(\alpha_{33}+\alpha_{44})(\alpha_{55}+\alpha_{66}))],\\
V_4 &= \beta_{22}(\alpha_{11}+\alpha_{22}+\alpha_{33}+\alpha_{44}+\alpha_{55}+\alpha_{66}),\; V_5 = -\beta_{22},\\
\alpha_{11} &= -(\frac{\beta\sigma\pi r^2}{L^2}I_*+d_1+\rho),\\
\alpha_{13} &= -\frac{\beta\sigma\pi r^2}{L^2}S_*,\alpha_{15}=\varphi,\alpha_{16}=\varepsilon,\\
\alpha_{21} &= \frac{\beta\sigma\pi r^2}{L^2}I_*,\alpha_{22}=-d_1,\\
\alpha_{23} &= \frac{\beta\sigma\pi r^2}{L^2}S_*,\beta_{22}=-\theta,\\
\alpha_{33} &= -(d_1+d_2+\eta_1+\alpha),\beta_{32}=\theta,\\
\alpha_{43} &= \alpha,\alpha_{44}=-(d_1+d_2+\eta_2),\\
\alpha_{53} &= \eta_1,\alpha_{54}=\eta_2,\\
\alpha_{55} &= -(d_1+\varphi),\alpha_{61}=\rho,\\
\alpha_{66} &= -(d_1+\varepsilon).
\end{aligned}$$

To guarantee the existence of Hopf bifurcation of system (2), we need some assumptions and they are listed in the following for clarity.

Assumption (H_1):

$$D_1 = U_{00} > 0, \tag{11}$$

$$D_2 = \begin{vmatrix} U_{05} & 1 \\ U_0 & U_{04} \end{vmatrix} > 0 \tag{12}$$

$$D_3 = \begin{vmatrix} U_{05} & 1 & 0 \\ U_{03} & U_{04} & U_{05} \\ U_{01} & U_{02} & U_{03} \end{vmatrix} > 0, \tag{13}$$

$$D_4 = \begin{vmatrix} U_{05} & 1 & 0 & 0 \\ U_{03} & U_{04} & U_{05} & 1 \\ U_{01} & U_{02} & U_{03} & U_{04} \\ 0 & U_{00} & U_{01} & U_{02} \end{vmatrix} > 0, \tag{14}$$

$$D_5 = \begin{vmatrix} U_{05} & 1 & 0 & 0 & 0 \\ U_{03} & U_{04} & U_{05} & 1 & 0 \\ U_{01} & U_{02} & U_{03} & U_{04} & U_{05} \\ 0 & U_{00} & U_{01} & U_{02} & U_{03} \\ 0 & 0 & 0 & U_{00} & U_{01} \end{vmatrix} > 0, \tag{15}$$

where

$$\begin{aligned} U_{00} &= U_0 + V_0, U_{01} = U_1 + V_1, \\ U_{02} &= U_2 + V_2, U_{03} = U_3 + V_3, \\ U_{04} &= U_4 + V_4, U_{05} = U_5 + V_5. \end{aligned}$$

Assumption (H_2):
Equation (16) has at least one positive root v_0,

$$v^6 + U_{15}v^5 + U_{14}v^4 + U_{13}v^3 + U_{12}v^2 + U_{11}v + U_{10} = 0, \tag{16}$$

where

$$\begin{aligned} U_{10} &= U_0^2 - V_0^2, \\ U_{11} &= U_1^2 - 2U_0U_2 + 2V_0V_2 - V_1^2, \\ U_{12} &= U_2^2 + 2U_0U_4 + 2U_1U_3 + 2V_1V_3 - V_2^2 - 2V_0V_4, \\ U_{13} &= U_3^2 + 2U_1U_5 - 2U_0 - 2U_2U_4 + 2V_1V_5 + 2V_2V_4 - V_3^2, \\ U_{14} &= U_4^2 + 2U_2 - 2U_3U_5 + 2V_3V_5 - V_4^2, \\ U_{15} &= U_5^2 - 2U_4 - V_5^2. \end{aligned}$$

Assumption (H_3):
$g'(\nu_0) \neq 0$, where $g(\nu) = \nu^6 + U_{15}\nu^5 + U_{14}\nu^4 + U_{13}\nu^3 + U_{12}\nu^2 + U_{11}\nu + U_{10}$.

Theorem 2. *For system (2), if the conditions (H_1)-(H_3) hold, then $P_*(S_*, E_*, I_*, Q_*, R_*, V_*)$ is locally asymptotically stable when $\tau \in [0, \tau_0)$; system (2) undergoes a Hopf bifurcation at $P_*(S_*, E_*, I_*, Q_*, R_*, V_*)$ when $\tau = \tau_0$ and τ_0 is defined as in Equation (21).*

Proof of Theorem 2. When $\tau = 0$, Equation (10) becomes

$$\lambda^6 + U_{05}\lambda^4 + U_{04}\lambda^4 + U_{03}\lambda^3 + U_{02}\lambda^2 + U_{01}\lambda + U_{00} = 0, \tag{17}$$

Obviously, $U_{05} = U_5 + V_5 = \frac{\beta\sigma\pi r^2}{L^2}I_* + \alpha + \varphi + \varepsilon + \theta + \rho + \eta_1 + \eta_2 > 0$. Thus, according to the Hurwitz criterion, it can be concluded that system (2) is locally asymptotically stable when $\tau = 0$, if the following the condition (H_1) holds.

For $\tau > 0$, let $\lambda = i\omega(\omega > 0)$ be a root of Equation (10). Then,

$$\begin{cases} (V_5\omega^5 - V_3\omega^3 + V_1\omega)\sin\tau\omega + (V_4\omega^4 - V_2\omega^2 + V_0)\cos\tau\omega = \omega^6 - U_4\omega^4 + U_2\omega^2 - U_0, \\ (V_5\omega^5 - V_3\omega^3 + V_1\omega)\cos\tau\omega - (V_4\omega^4 - V_2\omega^2 + V_0)\sin\tau\omega = U_3\omega^3 - U_5\omega^5 - U_1\omega. \end{cases} \tag{18}$$

Thus, one can obtain

$$\omega^{12} + U_{15}\omega^{10} + U_{14}\omega^8 + U_{13}\omega^6 + U_{12}\omega^4 + U_{11}\omega^2 + U_{10} = 0, \quad (19)$$

Let $\omega^2 = \nu$, then, Equation (19) becomes

$$\nu^6 + U_{15}\nu^5 + U_{14}\nu^4 + U_{13}\nu^3 + U_{12}\nu^2 + U_{11}\nu + U_{10} = 0. \quad (20)$$

If the condition (H_2) holds, then, Equation (19) has one positive root $\omega_0 = \sqrt{\nu_0}$ such that Equation (10) has a pair of purely imaginary roots $\pm i\omega_0$. From Equation (21), we obtain

$$\tau_0 = \frac{1}{\omega_0} \times \arccos\left\{\frac{G_1(\omega_0)}{G_2(\omega_0)}\right\}, \quad (21)$$

with

$$\begin{aligned}
G_1(\omega_0) &= (V_4 - U_5V_5)\omega_0^{10} + (U_5V_3 - U_3V_5 - U_4V_4 - V_2)\omega_0^8 \\
&\quad + (U_2V_4 + U_4V_2 - U_1V_5 - U_3V_3 - U_5V_1 + V_0)\omega_0^6 \\
&\quad + (U_1V_3 + U_3V_1 - U_0V_4 - U_2V_2 - U_4V_0)\omega_0^4 \\
&\quad + (U_0V_2 + U_2V_0 - U_1V_1)\omega_0^2 - U_0V_0, \\
G_2(\omega_0) &= V_5\omega_0^{10} + (V_4^2 - 2V_3V_5)\omega_0^8 + (V_3^2 + 2V_1V_5 - 2V_2V_4)\omega_0^6 \\
&\quad + (V_2^2 + 2V_0V_4 + 2V_1V_3)\omega_0^4 + (V_1^2 - 2V_0V_2)\omega_0^2 + V_0^2.
\end{aligned}$$

Differentiating both sides of Equation (10) with respect to τ yields

$$\left[\frac{d\lambda}{d\tau}\right]^{-1} = -\frac{(6\lambda^5 + 5U_5\lambda^4 + 4U_4\lambda^3 + 3U_3\lambda^2 + 2U_2\lambda + U_1)}{\lambda(\lambda^6 + U_5\lambda^5 + U_4\lambda^4 + U_3\lambda^3 + U_2\lambda^2 + U_1\lambda + U_0)} + \frac{5V_5\lambda^4 + 4V_4\lambda^3 + 3V_3\lambda^2 + 2V_2\lambda + V_1}{\lambda(V_5\lambda^5 + V_4\lambda^4 + V_3\lambda^3 + V_2\lambda^2 + V_1\lambda + V_0)} - \frac{\tau}{\lambda}.$$

Further,

$$\text{Re}\left[\frac{d\lambda}{d\tau}\right]^{-1}_{\tau=\tau_0} = \frac{g'(\nu_0)}{G_2(\omega_0)}.$$

Obviously, if the condition (H_3) is satisfied, then $\text{Re}[\frac{d\lambda}{d\tau}]^{-1}_{\tau=\tau_0} \neq 0$. Based on the discussion above and the Hopf bifurcation theorem in [40], Theorem 2 can be proved. □

5. Numerical Simulations

In this section, we present some numerical simulations to support our obtained theoretical results. Choosing $A = 1000$, $\beta = 0.009$, $\sigma = 0.5$, $r = 1$, $L = 10$, $d_1 = 0.05$, $\rho = 0.65$, $\varphi = 0.05$, $\varepsilon = 0.55$, $\theta = 0.45$, $d_2 = 0.035$, $\eta_1 = 0.35$, $\alpha = 0.1$ and $\eta_2 = 0.07$, then Equation (2) becomes

$$\begin{cases} \frac{dS(t)}{dt} = 1000 - 1.4130e - 004S(t)I(t) \\ \quad\quad\quad -0.7S(t) + 0.05R(t) + 0.55V(t), \\ \frac{dE(t)}{dt} = 1.4130e - 004S(t)I(t) - 0.05E(t) - 0.45E(t-\tau), \\ \frac{dI(t)}{dt} = 0.45E(t-\tau) - 0.535I(t), \\ \frac{dQ(t)}{dt} = 0.1I(t) - 0.155Q(t), \\ \frac{dR(t)}{dt} = 0.35I(t) + 0.07Q(t) - 0.1R(t), \\ \frac{dV(t)}{dt} = 0.65S(t) - 0.6V(t), \end{cases} \quad (22)$$

from which one can obtain $R_0 = 2.2819 > 1$ and the unique endemic equilibrium $P_*(4207, 1663.8, 1399.5, 902.9032, 5530.3, 4557.6)$. It can be verified that system (22) is locally asymptotically stable when $\tau = 0$.

For $\tau = 0$, by some computations with the aid of Matlab software package, we obtain $\omega_0 = 0.0558$, $\tau_0 = 13.1047$ and $g'(v_0) = 0.0029 > 0$. Thus, the conditions for existence of Hopf bifurcation are satisfied. Based on Theorem 1, we can see that $P_*(4207, 1663.8, 1399.5, 902.9032, 5530.3, 4557.6)$ is locally asymptotically stable when $\tau \in [0, \tau_0 = 13.1047)$. This can be shown as in Figure 1. However, $P_*(4207, 1663.8, 1399.5, 902.9032, 5530.3, 4557.6)$ will lose its stability when the value of τ passes through the critical threshold τ_0, a Hopf bifurcation occurs, which can be seen from Figure 2. The bifurcation phenomenon can be also illustrated by the bifurcation diagrams in Figure 3. In what follows, we are interested to study the effect of some other parameters on the dynamics of system (22).

(i) Effect of η_1 and η_2: In Figure 4, we can see that the number of infectious nodes decreases when the values of η_1 and η_2 increase. And the system changes its behavior from limit cycle to stable focus as we increase the value of η_1 and η_2, which can be shown as in Figure 5.

(ii) Effect of φ and ε: In the same manner, we can see from Figures 6 and 7 that the number of infectious nodes increases when the values of φ and ε increase. Also, we observe that system changes its behavior from stale focus to limit cycle as we increase the value of φ and ε.

(iii) Effect of r and L: As is shown in Figures 8 and 9, the number of infectious nodes increases when the value of r increases and the value of L decreases. In other words, as the density of sensor node increases, the number of infectious nodes increases. In addition, r and L effect the dynamic behavior of system (22) when their value changes. That is, system changes its behavior from stable focus to limit cycle as we increase the value of r and decrease the value of L.

In addition, in the presence of delay, the Lyapunov exponents (LE) have been derived numerically. For a non zero value of τ, LE for different species have been plotted in Figure 10. As all LEs are negative, then the system is stable.

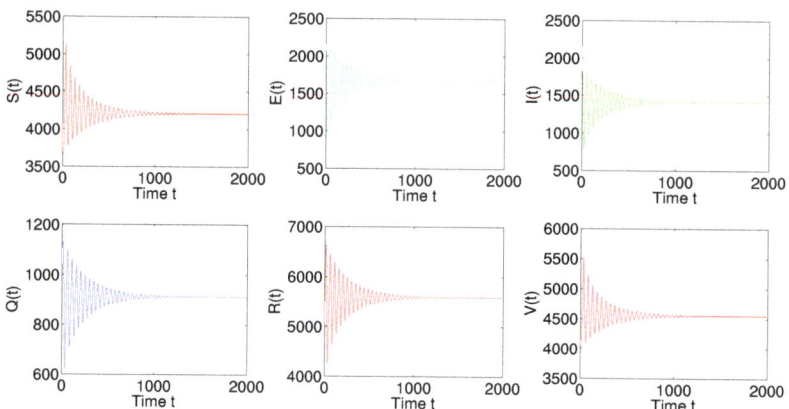

Figure 1. Time plots of S, E, I, Q, R and V with $\tau = 12.85 < \tau_0 = 13.1047$.

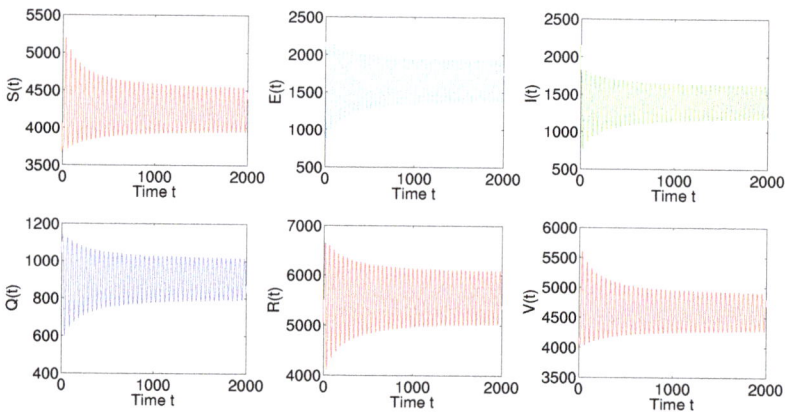

Figure 2. Time plots of S, E, I, Q, R and V with $\tau = 13.75 > \tau_0 = 13.1047$.

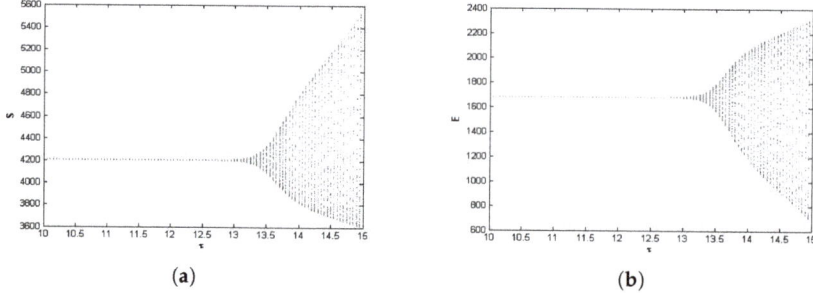

Figure 3. Cont.

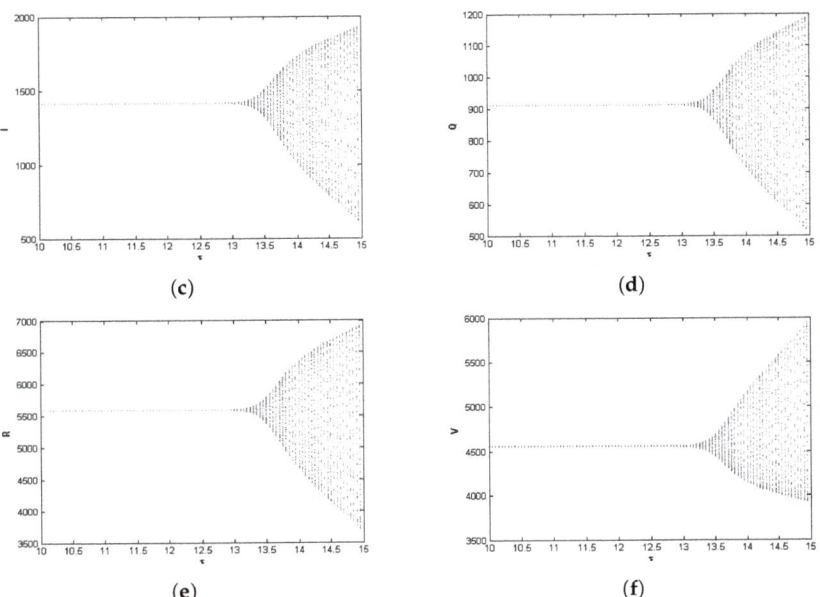

Figure 3. Bifurcation diagram with respect to time delay of system (22): (**a**) $S - \tau$, (**b**) $E - \tau$, (**c**) $I - \tau$, (**d**) $Q - \tau$, (**e**) $R - \tau$, (**f**) $V - \tau$.

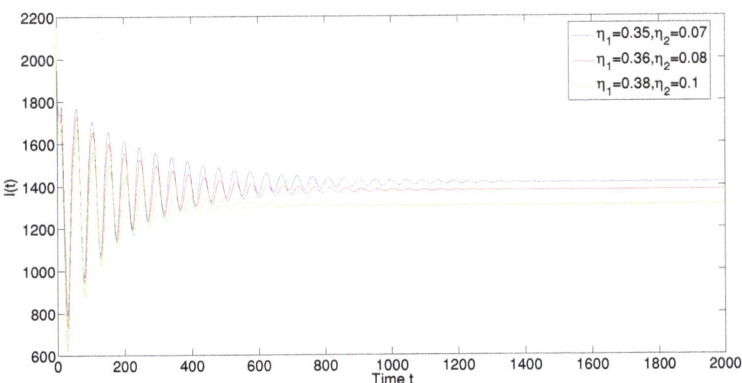

Figure 4. Time plots of I for different η_1 and η_2 at $\tau = 12.85$.

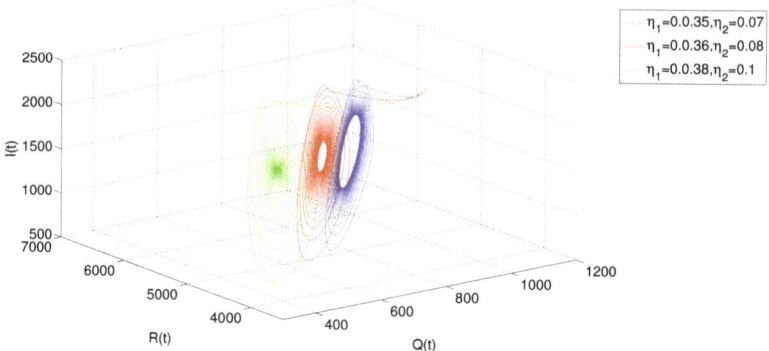

Figure 5. Dynamic behavior of system (22): projection on I-Q-R for different η_1 and η_2 at $\tau = 13.75$.

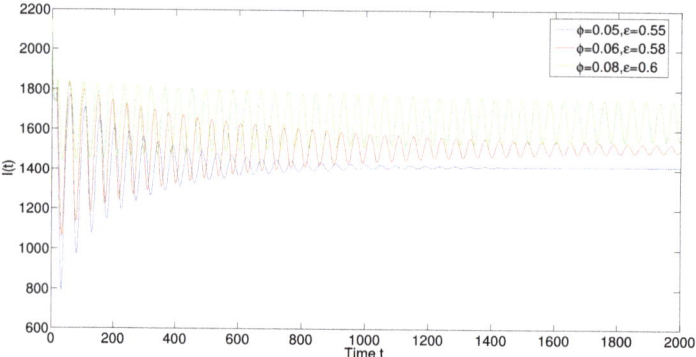

Figure 6. Time plots of I for different φ and ε at $\tau = 12.85$.

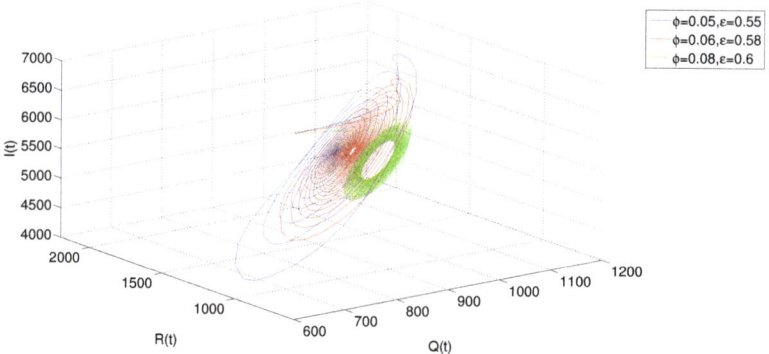

Figure 7. Dynamic behavior of system (22): projection on I-Q-R for different φ and ε at $\tau = 12.85$.

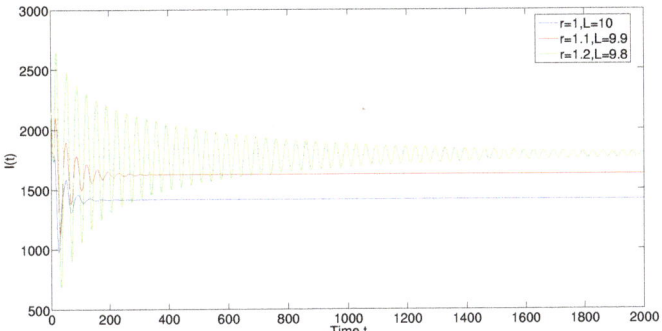

Figure 8. Time plots of I for different φ and ε at $\tau = 8.85$.

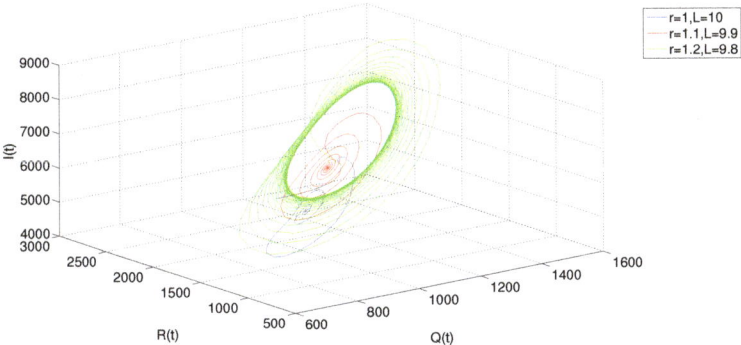

Figure 9. Dynamic behavior of system (22): projection on I-Q-R for different φ and ε at $\tau = 9.25$.

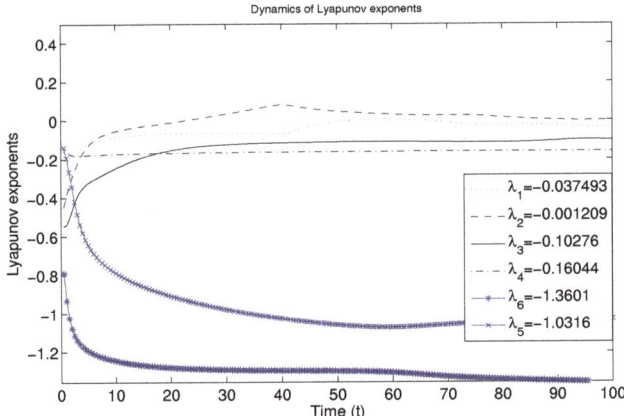

Figure 10. Other parameters are as in the text.

6. Discussion and Conclusions

In this paper, we present a delayed SEIQRS-V epidemic model for propagation of malicious codes in wireless sensor network based on the work in [32] by incorporating the latent period delay of malicious codes. As stated in [41], one of the significant features of malicious codes is their latent characteristics, which implies that the nodes are infected at time $t - \tau$ and they are surviving in the latent period τ and then become infective at time t. In addition, too large time delay may lead to large number of infected nodes, because of which malicious codes propagation persists in the system. Therefore, compared with the model proposed in [33], the delayed model in our paper is more general. It should be also pointed out that there are some proposed epidemic models for propagation of malicious code in a wireless sensor network such as the models in [5,6,9,42,43], but the authors did not consider the characteristics of networks like communication radius and distributed density of nodes in wireless sensor network.

We first find a feasible region which is invariant and where the solutions of our model are positive and the Lyapunov exponent stability is analyzed by constructing a Lyapunov functional. Then, the critical value of time delay τ_0 at which a Hopf bifurcation occurs is obtained by choosing the delay as the bifurcating parameter. It is found that when the time delay is suitably small ($\tau \in [0, \tau_0)$), system (2) is locally asymptotically stable. In this case, the propagation of malicious codes can be controlled easily. However, once the value of the time delay passes through the critical value τ_0, system (2) loses its stability and a family of periodic solutions bifurcate from the endemic equilibrium of system (2). In this case, the propagation of malicious codes will be out of control.

Also, the effects of some crucial parameters on dynamics of system (2) are studied by numerical simulations. As the values of η_1 and η_2 increase, the number of infectious nodes decreases and system (22) changes its behavior from limit cycle to stable focus as we increase the value of η_1 and η_2, it is strongly recommended that users of the wireless sensor network should periodically run antivirus software of the newest version, so that the propagation of malicious codes can be controlled. This phenomenon can also be illustrated by the effects of φ and ε on dynamics of the system. In addition, the number of infectious nodes increases when the density of the sensor node increases. Thus, it can be concluded that the manager of the wireless sensor network should control the number of nodes connected to the network properly.

Author Contributions: All authors have equally contributed to this paper. They have read and approved the final version of the manuscript.

Funding: This research was supported by National Natural Science Foundation of China(Nos.11461024, 61773181), Natural Science Foundation of Inner Mongolia Autonomous Region (No.2018MS01023), Project of Support Program for Excellent Youth Talent in Colleges and Universities of Anhui Province (Nos.gxyqZD2018044, gxbjZD49), Bengbu University National Research Fund Cultivation Project (2017GJPY03).

Conflicts of Interest: The authors declare no conflict of interest.

Appendix A

$$k_1 = k_4 = 2k_5 = 3k_6 = \frac{\tau}{2}\left(-\alpha_{12} + 4\alpha_{21} - 2\alpha_{11} - 3d_1 - 2d_2 + \alpha_{12} + 2\rho\right),$$

$$k_2 = k_3 = 2k_8 = k_7 = k_{10} = \frac{3\tau}{2}\left(\phi + \varepsilon + \alpha_{12} + 3\rho - 3\alpha_{11} - 6d_1 + \alpha - d_2 - \alpha_{21} + \eta_1 - \theta\right),$$

$$k_9 = k_{11} = 2k_{12} = \tau\left(2\alpha + 2\varepsilon - \alpha_{12} - 4d_1 - d_2 - 2\eta_2\right),$$

$$k_{13} = 3k_{14} = 2k_{15} = k_{19} = 2k_{20} = 2\eta_1 + \phi + \varepsilon - 2\alpha_{12} - d_1$$

$$k_{16} = k_{17} = 2k_{18} = k_{21} = 2\eta_2 + 2\varepsilon + 3\phi.$$

Appendix B

$$w_1(u)(t) = u_1^2(t),$$
$$w_2(u)(t) = p_1^2(t) + \theta(d_1 + \theta - 2\alpha_{12}) \int_{t-\tau}^{t} \int_{s}^{t} u_2^2(l) dl ds,$$
$$w_3(u)(t) = p_2^2(t) + \theta(\theta - d_1 - d_2 - \eta_1 - \alpha) \int_{t-\tau}^{t} \int_{s}^{t} u_2^2(l) dl ds,$$
$$w_4(u)(t) = u_4^2(t), w_5(u)(t) = u_5^2(t), w_6(u)(t) = u_6^2(t),$$
$$w_7(u)(t) = u_1(t)p_1(t) + \frac{\theta(\alpha_{11} + \alpha_{12} - \phi - \varepsilon)}{2} \int_{t-\tau}^{t} \int_{s}^{t} u_2^2(l) dl ds,$$
$$w_8(u)(t) = u_1(t)p_2(t) + \frac{\theta(\varepsilon + \phi - \alpha_{12} - \alpha_{11})}{2} \int_{t-\tau}^{t} \int_{s}^{t} u_2^2(l) dl ds,$$
$$w_9(u)(t) = u_1(t)u_4(t), w_{10}(u)(t) = u_1(t)u_5(t), w_{11}(u)(t) = u_1(t)u_6(t),$$
$$w_{12}(u)(t) = p_1(t)p_2(t) + \frac{\theta(\alpha_{12} + \alpha_{21} + \alpha - 2\theta + d_2 + \eta_1)}{2} \int_{t-\tau}^{t} \int_{s}^{t} u_2^2(l) dl ds,$$
$$w_{13}(u)(t) = p_1(t)u_4(t) + \frac{\theta(d_1 + d_2 + \eta_2 - \alpha)}{2} \int_{t-\tau}^{t} \int_{s}^{t} u_2^2(l) dl ds,$$
$$w_{14}(u)(t) = p_1(t)u_5(t) + \frac{\theta(d_1 + \phi - \eta_2 - \eta_1)}{2} \int_{t-\tau}^{t} \int_{s}^{t} u_2^2(l) dl ds,$$
$$w_{15}(u)(t) = p_1(t)u_6(t) + \frac{\theta(d_1 + \varepsilon - \rho)}{2} \int_{t-\tau}^{t} \int_{s}^{t} u_2^2(l) dl ds,$$
$$w_{16}(u)(t) = p_2(t)u_4 + \frac{\theta(\alpha - d_1 - d_2 - \eta_2)}{2} \int_{t-\tau}^{t} \int_{s}^{t} u_2^2(l) dl ds,$$
$$w_{17}(u)(t) = p_2(t)u_5 + \frac{\theta(\eta_1 + \eta_2 - d_1 - \phi)}{2} \int_{t-\tau}^{t} \int_{s}^{t} u_2^2(l) dl ds,$$
$$w_{18}(u)(t) = p_2(t)u_6 + \frac{\theta(\rho - d_1 - \varepsilon)}{2} \int_{t-\tau}^{t} \int_{s}^{t} u_2^2(l) dl ds,$$
$$w_{19}(u)(t) = u_4(t)u_5(t), w_{20}(u)(t) = u_4(t)u_6(t), w_{21}(u)(t) = u_5(t)u_6(t).$$

Appendix C

$$\Lambda_1 = -2\alpha_{11}k_1 - \theta\alpha_{21}\tau k_2 + k_7\left(\alpha_{21} + \frac{\theta\alpha_{11}\tau}{2}\right) - \frac{\alpha_{11}\theta\tau}{2}k_8 + \rho k_{11} + \frac{\alpha_{21}\theta\tau}{2}k_{12} - \frac{\rho\theta\tau}{2}k_{15} + \frac{\rho\theta\tau}{2}k_{18},$$

$$\Lambda_2 = k_2(2\theta(d_1 + \theta)\tau - 2d_1 - 2\theta - 2\theta\alpha_{12}\tau) + k_3(2\theta^2\tau - \theta\tau(d_1 + d_2 + \eta_1 + \alpha))$$
$$+ k_7\frac{\tau\theta(\alpha_{11} + \alpha_{12} - \phi - \varepsilon)}{2} + k_8\frac{\tau\theta(-\alpha_{11} - \alpha_{12} + \phi + \varepsilon)}{2}$$
$$+ k_{12}\{\theta + \frac{\tau\theta(\alpha_{12} + \alpha_{21} + \alpha - 4\theta + d_2 + \eta_1 - d_1)}{2}\} + k_{13}\frac{\tau\theta(d_1 + d_2 + \eta_2 - \alpha)}{2}$$
$$+ k_{14}\frac{\tau\theta(d_1 + \phi - \eta_1 - \eta_2)}{2} + k_{15}\frac{\tau\theta(d_1 + \varepsilon - \rho)}{2} + k_{16}\frac{\tau\theta(\alpha - d_1 - d_2 - \eta_2)}{2}$$
$$+ k_{17}\frac{\tau\theta(\eta_1 + \eta_2 - d_1 - \phi)}{2} + k_{18}\frac{\tau\theta(\rho - d_1 - \varepsilon)}{2},$$

$$\begin{aligned}
\Lambda_3 &= -\theta\tau\alpha_{12}k_2 - (2+\tau\theta)(d_1+d_2+\alpha+\eta_1)k_3 + k_7\frac{\tau\theta\alpha_{12}}{2} - k_8\left(\alpha_{12}+\frac{\alpha_{12}\tau\theta}{2}\right)\\
&\quad + k_{12}\left(\alpha_{12}+\frac{\tau\theta(d_1+d_2+\eta_1+\alpha+\alpha_{12})}{2}\right) - \alpha\theta\tau k_{13} - \frac{\eta_1\theta\tau}{2}k_{14}\\
&\quad + k_{16}\left(\alpha+\frac{\alpha\theta\tau}{2}\right) + k_{17}\left(\eta_1+\frac{\eta_1\theta\tau}{2}\right),\\
\Lambda_4 &= -2(d_1+d_2+\eta_2)k_4 - \frac{\tau\theta\eta_2}{2}k_{14} - \frac{(d_1+d_2+\eta_2)\tau\theta}{2}k_{16} + \frac{\tau\theta\eta_2}{2}k_{17} + \eta_2 k_{19},\\
\Lambda_5 &= -2(d_1+\phi)k_5 - \frac{\tau\theta\phi}{2}k_7 + \frac{\tau\theta\phi}{2}k_8 + \frac{\tau\theta(d_1+\phi)}{2}k_{14} - \frac{\tau\theta(d_1+\phi)}{2}k_{17},\\
\Lambda_6 &= -2(d_1+\varepsilon)k_6 - \frac{\tau\theta\phi}{2}k_7 + \frac{\tau\theta\varepsilon}{2}k_8 + \varepsilon k_{11} + k_{15}\left(\varepsilon+\frac{\tau\theta(d_1+\varepsilon)}{2}\right) - \frac{\tau\theta(d_1+\varepsilon)}{2}k_{18}.
\end{aligned}$$

References

1. Keshri, A.K.; Mishra, B.K.; Mallick, D.K. Library formation of known malicious attacks and their future variants. *Int. J. Adv. Sci. Technol.* **2016**, *94*, 1–12. [CrossRef]
2. Singh, J.; Kumar, D.; Hammouch, Z.; Atangana, A. A fractional epidemiological model for computer viruses pertaining to a new fractional derivative. *Appl. Math. Comput.* **2018**, *316*, 504–515. [CrossRef]
3. Keshri, A.K.; Mishra, B.K.; Mallick, D.K. A predator-prey model on the attacking behavior of malicious objects in wireless nanosensor networks. *Nano Commun. Netw.* **2018**, *15*, 1–16. [CrossRef]
4. Cybercrime-Report. Available online: http://cyberseurityventures.com/2015-wp/wp-content/uploads/2017/10/2017-Cybercrime-Report.pdf (accessed on 23 April 2019).
5. Khanh, N.H. Dynamics of a worm propagation model with quarantine in wireless sensor networks. *Appl. Math. Inf. Sci.* **2016**, *10*, 1739–1746. [CrossRef]
6. Mishra, B.K.; Kershi, N. Mathematical model on the transmission of worms in wireless sensor network. *Appl. Math. Model.* **2013**, *3*, 4103–4111. [CrossRef]
7. Mishra, B.K.; Pandey, S.K. Dynamic model of worms with vertical transmission in computer network. *Appl. Math. Comput.* **2011**, *217*, 8438–8446. [CrossRef]
8. Xiao, X.; Fu, P.; Dou, C.S.; Li, Q.; Hu, G.W.; Xia, S.T. Design and analysis of SEIQR worm propagation model in mobile internet. *Commun. Nonlinear Sci. Numer. Simul.* **2017**, *43*, 341–350. [CrossRef]
9. Keshri, N.; Mishra, B.K. Stability analysis of a predator-prey model in wireless sensor network. *Int. J. Comput. Math.* **2014**, *91*, 928–943.
10. Yang, L.X.; Yang, X.F. The spread of computer viruses under the influence of removable storage devices. *Appl. Math. Comput.* **2012**, *219*, 3914–3922. [CrossRef]
11. Muroya, Y.; Li, H.X.; Kuziya, T. On global stabiity of a nonresident computer virus model. *Acta Math. Sci.* **2014**, *34B*, 1427–1445. [CrossRef]
12. Wang, F.W.; Zhang, Y.K.; Wang, C.G.; Ma, J.F. Stability analysis of an e-SEIAR model with point-to-group worm propagation. *Commun. Nonlinear Sci. Numer. Simul.* **2015**, *20*, 897–904. [CrossRef]
13. Tang, C.Q.; Wu, Y.H. Global exponential stability of nonresident computer virus models. *Nonlinear Anal. Real World Appl.* **2017**, *34*, 149–158. [CrossRef]
14. Fatima, U.; Ali, M.; Ahmed, N.; Rafiq, M. Numerical modeling of susceptible latent breaking-out quarantine computer virus epidemic dynamics. *Heliyon* **2018**, *4*, e00631. [CrossRef]
15. Zhang, X.X.; Li, C.D. A novel computer virus model with generic nonlinear burst rate. In Proceedings of the International Workshop on Complex Systems and Networks, Doha, Qatar, 8–10 December 2017; pp. 325–329.
16. Yang, X.F.; Liu, B.; Gan, C.Q. Global stability of an epidemic model of computer virus. *Abstr. Appl. Anal.* **2014**, *2014*, 456320. [CrossRef]
17. Chen, L.J.; Hattaf, K.; Sun, J.T. Optimal control of a delayed SLBS computer virus model. *Phys. Stat. Mech. Its Appl.* **2015**, *427*, 244–250. [CrossRef]

18. Muroya, Y.; Kuniya, T. Global stability of nonresident computer virus models. *Math. Methods Appl. Sci.* **2015**, *38*, 281–295. [CrossRef]
19. Zhou, H.X.; Guo, W. A stochastic worm model. *Telecommun. Syst.* **2017**, *64*, 135–145. [CrossRef]
20. Amador, J. The stochastic SIRA model for computer viruses. *Appl. Math. Comput.* **2014**, *232*, 1112–1124. [CrossRef]
21. Tafazzoli, T.; Sadeghiyan, B.A. Stochastic model for the size of worm origin. *Secur. Commun. Netw.* **2016**, *9*, 1103–1118. [CrossRef]
22. Jafarabadi, A.; Azgomi, M.A. A stochastic epidemiological model for the propagation of active worms considering the dynamicity of network topology. *Peer-to-Peer Netw. Appl.* **2015**, *8*, 1008–1022. [CrossRef]
23. Zhang, C.M.; Zhao, Y.; Wu, Y.J.; Deng, S.W. A stochastic dynamic model of computer viruses. *Discret. Dyn. Nat. Soc.* **2012**, *2012*, 264874. [CrossRef]
24. Keshri, N.; Gupta, A.; Mishra, B.K. Impact of reduced scale free network on wireless sensor network. *Phys. Stat. Mech. Its Appl.* **2016**, *463*, 236–245. [CrossRef]
25. Hosseini, S.; Azgomi, M.A.; Rahmani, A.T. Malware propagation modeling considering software diversity andimmunization. *J. Comput. Sci.* **2016**, *13*, 49–67. [CrossRef]
26. Zhang, C.M.; Huang, H.T. Optimal control strategy for a novel computer virus propagation model on scale-free networks. *Phys. Stat. Mech. Its Appl.* **2016**, *451*, 251–265. [CrossRef]
27. Feng, L.P.; Song, L.P.; Zhao, Q.S.; Wang, H.B. Modeling and stability analysis of worm propagation in wireless sensor network. *Math. Probl. Eng.* **2015**, *2015*, 129598. [CrossRef]
28. Srivastava, A.P.; Awasthi, S.; Ojha, R.P.; Srivastava, P.K.; Katiyar, S. Stability analysis of SIDR model for worm propagation in wireless sensor network. *Indian J. Sci. Technol.* **2016**, *9*, 1–5. [CrossRef]
29. Nwokoye, C.H.; Ejiofor, W.E.; Orji, R. Investigating the effect of uniform random distribution of nodes in wireless sensor networks usingan epidemic worm model. In Proceedings of the CORI'16, Ibadan, Nigeria, 7–9 September 2016; pp. 58–63.
30. Singh, A.; Awasthi, A.K.; Singh, K.; Srivastava, P.K. Modeling and analysis of worm propagation in wireless sensor networks. *Wirel. Pers. Commun.* **2018**, *98*, 2535–2551. [CrossRef]
31. Ojha, R.P.; Sanyal, G.; Srivastava, P.K.; Sharma, K. Design and analysis of modified SIQRS model for performance study of wireless sensor network. *Scalable Comput.* **2017**, *18*, 229–241. [CrossRef]
32. Mishra, B.K.; Tyagi, I. Defending against malicious threats in wireless sensor network: A mathematical model. *Int. J. Inf. Technol. Comput. Sci.* **2014**, *6*, 12–19. [CrossRef]
33. Nwokoye, C.H.; Umeh, I.I. The SEIQR-V Model: On a More Accurate Analytical Characterization of Malicious Threat Defense. *Int. J. Inf. Technol. Comput. Sci.* **2017**, *12*, 28–37. [CrossRef]
34. Keshri, N.; Mishra, B.K. Two time-delay dynamic model on the transmission of malicious signals in wireless sensor network. *Chaos Solitons Fractals* **2014**, *68*, 151–158. [CrossRef]
35. Zhang, Z.Z.; Bi, D.J. Bifurcation analysis in a delayed computer virus model with the effect of external computers. *Adv. Differ. Equat.* **2015**, *317*, 13. [CrossRef]
36. Zhao, T.; Bi, D.J. Hopf bifurcation of a computer virus spreading model in the network with limited anti-virus ability. *Adv. Differ. Equat.* **2017**, *183*, 16. [CrossRef]
37. Wang, C.L.; Chai, S.X. Hopf bifurcation of an SEIRS epidemic model with delays and vertical transmission in the network. *Adv. Differ. Equat.* **2016**, *10*, 19. [CrossRef]
38. Dai, Y.X.; Lin, Y.P.; Zhao, H.T.; Khalique, C.M. Global stability and Hopf bifurcation of a delayed computer virus propagation model with saturation incidence rate and temporary immunity. *Int. J. Mod. Phys.* **2016**, *30*, 1640009. [CrossRef]
39. Xia, W.J.; Kundu, S.; Maitra, S. Dynamics of a delayed SEIQ epidemic model. *Adv. Differ. Equat.* **2018**, *336*, 21. [CrossRef]
40. Hassard, B.D.; Kazarinoff, N.D.; Wan, Y.H. *Theory and Applications of Hopf Bifurcation*; Cambridge University Press: Cambridge, UK, 1981.
41. Ren, J.G.; Yang, X.F.; Yang, L.X.; Xu, Y.H.; Yang, F.Z. A delayed computer virus propagation model and its dynamics. *Chaos Solitons Fractals* **2012**, *45*, 74–79. [CrossRef]

42. Zhang, Z.Z.; Song, L.M. Dynamics of a delayed worm propagation model with quarantine. *Adv. Differ. Equat.* **2017**, *155*, 13. [CrossRef]
43. Upadhyay, R.K.; Kumari, S. Bifurcation analysis of an e-epidemic model in wireless sensor network. *Int. J. Comput. Math.* **2018**, *95*, 1775–1805. [CrossRef]

© 2019 by the authors. Licensee MDPI, Basel, Switzerland. This article is an open access article distributed under the terms and conditions of the Creative Commons Attribution (CC BY) license (http://creativecommons.org/licenses/by/4.0/).

MDPI
St. Alban-Anlage 66
4052 Basel
Switzerland
Tel. +41 61 683 77 34
Fax +41 61 302 89 18
www.mdpi.com

Mathematics Editorial Office
E-mail: mathematics@mdpi.com
www.mdpi.com/journal/mathematics